UNIVERSITY OF BATH • SCIENCE 16-19

Project Director: J. J. Thompson, CBE

MEDICAL PHYSICS

MARTIN HOLLINS

Nelson

Thomas Nelson and Sons Ltd
Nelson House Mayfield Road
Walton-on-Thames Surrey
KT12 5PL UK

First published by Macmillan Education Ltd 1990
ISBN 0-333-46657-8

This edition published by Thomas Nelson and Sons Ltd 1992

I(T)P Thomas Nelson is an International
Thomson Publishing Company

I(T)P is used under licence

ISBN 0-17-448188-8
NPN 15 14 13 12 11 10

Printed in China.

Contents

The Project: an introduction

The **University of Bath · Science 16–19 Project,** grew out of a reappraisal of how far sixth form science had travelled during a period of unprecedented curriculum reform and an attempt to evaluate future development. Changes were occurring both within the constitution of 16–19 syllabuses themselves and as a result of external pressures from 16+ and below: syllabus redefinition (starting with the common cores), the introduction of AS-level and its academic recognition, the originally optimistic outcome to the Higginson enquiry; new emphasis on skills and processes, and the balance of continuous and final assessment at GCSE level.

This activity offered fertile ground for the School of Education at the University of Bath to join forces with a team of science teachers, drawn from a wide spectrum of educational experience, to create a flexible curriculum model and then develop resources to fit it. This group addressed the task of satisfying these requirements:

- the new syllabus and examination demands of A- and AS-level courses;
- the provision of materials suitable for both the core and options parts of syllabuses;
- the striking of an appropriate balance of opportunities for students to acquire knowledge and understanding, develop skills and concepts, and to appreciate the applications and implications of science;
- the encouragement of a degree of independent learning through highly interactive texts;
- the satisfaction of the needs of a wide ability range of students at this level.

Some of these objectives were easier to achieve than others. Relationships to still-evolving syllabuses demand the most rigorous analysis and a sense of vision – and optimism – regarding their eventual destination. Original assumptions about AS-level, for example, as a distinct though complementary sibling to A-level, needed to be revised.

The Project, though, always regarded itself as more than a provider of materials, important as this is, and concerned itself equally with the process of provision – how material can best be written and shaped to meet the requirements of the educational market-place. This aim found expression in two principal forms: the idea of secondment at the University and the extensive trialling of early material in schools and colleges.

Most authors enjoyed a period of secondment from teaching, which allowed them not only to reflect and write more strategically (and, particularly so, in a supportive academic environment) but, equally, to engage with each other in wrestling with the issues in question.

The Project saw in the trialling a crucial test for the acceptance of its ideas and their execution. Over one hundred institutions and one thousand students participated, and responses were invited from teachers and pupils alike. The reactions generally confirmed the soundness of the model and allowed for more scrupulous textual housekeeping, as details of confusion, ambiguity or plain misunderstanding were revised and reordered.

The test of all teaching must be in the quality of the learning, and the proof of these resources will be in the understanding and ease of accessibility which they generate. The Project, ultimately, is both a collection of materials and a message of faith in the science curriculum of the future.

J.J. Thompson
January 1990

How to use this book

The aim of this book is to provide an introduction to medical physics which I hope you will find interesting and enjoyable. It is particularly geared to the requirements of students who are following an Advanced level or AS–level course in Physics who are taking an option in medical physics.

The book covers the range of topics usually covered in medical physics and is divided into themes. The first three of which link to different parts of a basic study of physics, the fourth concentrating on the practice of medical physics. Each theme is relatively self-contained, so that you can start the book at Chapter 1, 4 or 9. Within a theme the preferred order is as written, but there are cross-references to help if you wish to select your own order. For example, it is possible to study only medical imaging by covering chapters 4 and 8, and themes 3 and 4.

The links to core physics are detailed in the prerequisites for a theme. The detailed content of each chapter is summarised in the learning objectives for each chapter. You should compare the chapter and section headings, together with the learning objectives, with your syllabus, to see which parts of the book are essential to your own course. The learning objectives are intended, also, to help you organise your learning and provide an aid to planning your study of the subject.

The book has been written so that there are a number of activities for you to carry out in addition to reading, which on its own is too passive to promote effective learning. In-text questions are designed to check your knowledge and understanding, as you proceed. Answers to these where appropriate, are given at the back of the book. Summary assignments cover larger topics, with the aim of helping you to consolidate and review the material. Some of the assignments will be examination questions which help you to see if your attainment is at the required level. Some questions do not have a simple, or even single, answer, and are intended for you to discuss with others.

Medical physics is not a fixed body of knowledge to be learnt, but a changing collection of theories and practices with vital implications for the health of human beings. A number of features of the book are designed to convey something of this variety and complexity. Extracts of articles on current developments are included to develop the skills of comprehension and evaluation of both technical and popular sources. Further reading suggestions are given in most chapters and can be supplemented by topical publications, such as journals and newspapers.

Investigations in the book are fairly open-ended; more like practical explorations of the ideas in topics, than standard laboratory procedures. Visits to appropriate hospital departments are invaluable for seeing actual diagnostic equipment in use, as this is rarely inexpensive enough for schools or colleges to purchase.

The case study in Theme 4 is a major group activity in familiarisation with current equipment and procedures in medical imaging, in planning and making decisions. Like any worthwhile learning, it will need a reasonable input of your time and energy to be successful. This simulation has been designed from existing practice in UK hospitals. I would like to express my thanks particularly to the helpful staff involved in medical imaging at Queen Mary's, Roehampton and St. George's, Tooting, while accepting full responsibility for any errors.

Prerequisites

These are presented at the beginning of each theme, to remind you of the scientific content which underlies the material in the theme. You should usually have covered this in your previous GCSE or A-level studies. If you have not you will need to consult a suitable text, or your tutor.

Learning objectives

These are statements of attainment which should apply to you, when you have finished the chapter. They often link closely to statements in a course syllabus, and can be used to help you to make notes for revision, as well as for checking your progress.

Questions

In–text questions may be answered from material just presented in the previous section, or may require additional thought and information.

Summary assignments

These can be used to build up your general understanding of a topic and provide notes for reference when revising for examinations. Answering the questions and summary assignments successfully is the best way of knowing that you have achieved the required understanding – so don't miss them out!

Extracts

Most of the time, we gain information not from text books but from a range of published material – can you comprehend a scientific paper, or evaluate the reliability of a newspaper article? These extracts can help.

Investigations

What is the best way to measure our body temperature? How would we find out our exposure to radiation? What factors affect the clarity of an image? These are some of the questions which are used to further your skills and understanding in this subject.

Case study

The practice of medical imaging is going through great technical changes, at a time when the economics of health care is also under severe scrutiny. In this simulation you have the opportunity to participate in the debates and decision–making.

Further reading

You should find the books and papers selected at the end of each chapter interesting to read, rather that more essential 'stuff' to learn for an examination. Some more technical books and papers are listed at the end if you are really keen!

Index and glossary of technical words

This lists important topics, concepts and terms, including those technical terms which it is important for you to know. These are explained when they are first encountered in the text, indicated in bold. There are a great many medical terms which you do *not* need to learn, if they are not explained you should be able to understand their meaning from the context.

Acknowledgements

The author and publishers wish to thank the following who have kindly given permission for the use of copyright material:

Academic Press, Inc. for material from *Foundations of Biophysics* by A.L. Stanford. American Association for the Advancement of Science for an extract from 'The Global Impact of the Chernobyl Reactor Accident' by L.R. Anspaugh, *Science,* Vol. 242, p.1513, 16th Dec. 1988. John Fairfax & Sons Ltd. for 'How a fun run meant meltdown for Mark Dorrity's body' by Sue-Ellen O'Grady, *Sydney Morning Herald. The Guardian* for material from their 20.8.88 issue. The Controller of Her Majesty's Stationery Office and Macmillan Magazines Ltd. for figure, 'Taken from wet and dry deposition of Chernobyl' by Clark and Smith, *Nature,* Vol. 332, No. 6161, 17.3.88. *The Independent* for 'Radiation more hazardous than previously thought' by Nicholas Schoon, *The Independent,* 19.11.87. Joint Matriculation Board, University of Cambridge Local Examinations Syndicate, and University of London School Examinations Board for questions from specimen and past examination papers. Kingsmoor Publications Ltd, for extracts from 'Survey of the current status of SPECT' by A. Todd–Pokropek, *RAD Magazine.* National Radiological Protection Board for material from their publications. New Scientists Syndication for the extracts from 'New X-ray equipment for safer treatment', *New Scientist,* 31.1.85 and drawings from 'Lasers: the coherent substitute for a scalpel', New Scientist, 20.2.86. *The Observer* for extracts from material from one of their issues, 1988. Robert J. Ott for extracts from 'MUPPET: a new low-cost PET', *RAD Magazine.* St. George's Hospital and Medical School for material by John Griffiths from their Scanner Appeal Brochure.

The author and publishers wish to acknowledge, with thanks, the following photographic sources:

Allergan Optical *p 35;* Allsport *pp 7, 8 left and right, 17;* Amersham International Plc *pp 137 left and right, 169;* Associated Press *p 128;* The Brompton Hospital, Cardiac Department *p 103;* CEL Instruments *p 46;* Colorsport *p 15;* Eastman Dental Hospital *p 112;* Dr P.J. Evennett, Department of Pure and Applied Biology, Leeds University; *p 32 left and right;* Vivian Fifield *p 111 right;* Sally and Richard Greenhill *p 20;* Philip Harris Education *p 59;* William F Hinkes *pp 70, 123, 160, 172;* Hulton Picture Company *p 111;* Illustrated London News *p 3 lower;* The Image Bank *pp 1, 189;* National Medical Slide Bank *pp 51, 88;* Peckham's of Stroud *pp 13 right, 16, 31;* Philips Medical Systems *p 183;* RAD *pp 180 left and right, 188 left;* Royal Marsden Hospital *p 184;* Saint Bartholamew's Hospital, London *p 188 upper right;* Science Photo Library *pp 3 upper, 4 left and right, 13 left, 61 left and right, 80, 90, 91, 95, 109, 125, 136, 141, 157, 168, 171, 175, 185, 188 lower right;* UKAEA *pp 73, 125.*

Every effort has been made to trace all the copyright holders, but if any have been inadvertently overlooked the publishers will be pleased to make the necessary arrangement at the first opportunity.

Introduction

WHAT IS MEDICAL PHYSICS?

A simple definition might be, physics applied to medicine, which then leads to the question what is medicine? A description of medical practice would include the three stages of examination, diagnosis and treatment. Those matters directly concerned with the practice of medicine are often termed **clinical**, as distinct from the more general or theoretical **medical** aspects. For example a medical student usually undergoes a period of scientific pre-clinical education before getting involved with patients in clinical training. Physics is used in all three of these stages of medicine and in both clinical and more general roles.

Lord Butterfield, former Professor of Physic at the University of Cambridge, has recently advised the House of Lords that Britain needs a national centre for health research. He was speaking as an eminent **physician**. The Chelsea **physic** garden in London is full of medicinal plants, not Newtonian apple trees or Hawking black holes.

Examination of the human body in health and disease is a basic part of the work of the physician and the physicist, which accounts for the similarity of the names, coming from the Greek *physike,* the science of nature.

The human body is composed of tissues with a wide range of physical functions, including mechanical, optical, acoustic, electrical, hydraulic, pneumatic and metabolic.There is scope therefore for measurement in almost any field of physics. This book reflects this; it includes a general review of the techniques of instrumentation in Chapter 4 and covers a wide range of examples from the clinical thermometer to the computed tomograph.

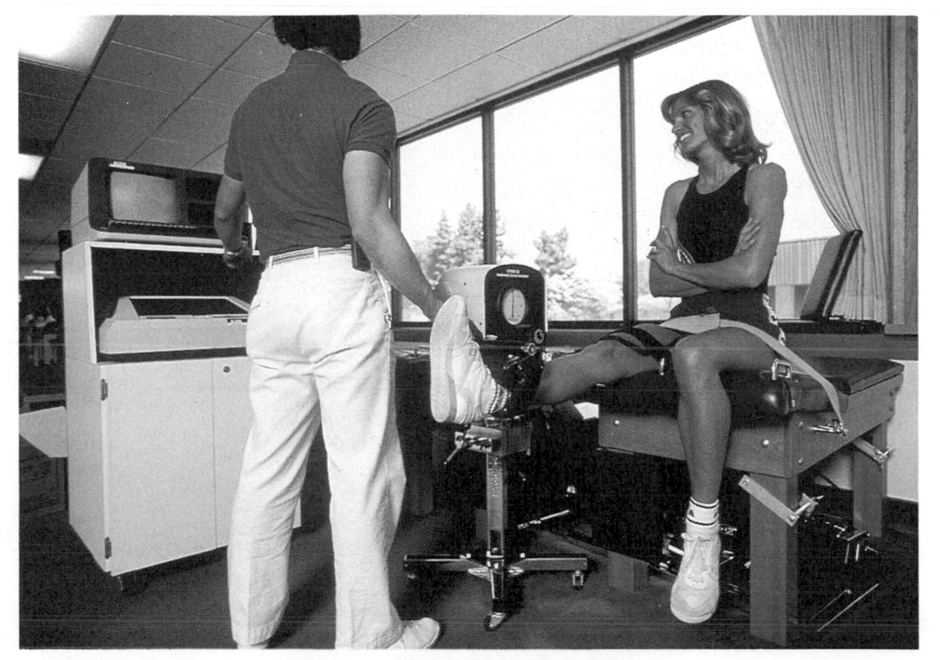

Medical examination of an athlete in training

Diagnosis in medicine is based on the patient 'history' or physical examination, with an emphasis on what is abnormal or malfunctioning. The invention and development of new techniques for diagnosis has to a large extent defined the practice of medical physics. It can be argued that its origin was the discovery of X-rays; now the emphasis is on radiation protection and magnetic resonance, and extensive use of computers and microelectronics. Technology has enabled a better diagnosis to be made by offering a wider range of examination methods, leading to fuller information about the patient.

This is the part of medical physics most emphasised in this book, especially in Themes 3 and 4, which deal with some of the most advanced techniques available in hospitals today.

Treatment or therapy is understandably more under the control of the clinician (specialist doctor) than the physicist or technician. Examples are given of a wide range of physical therapies, throughout this book, but their treatment(!) is brief. It is not the purpose of this book to teach you medicine!

WHAT DO MEDICAL PHYSICISTS DO?

The role of the physicist in medicine is wide-ranging and often crosses scientific and technological disciplines. In the UK the scientific aspects of the work are supported by the Institute of Physical Sciences in Medicine, the title emphasising the broad nature of the interests. Specialist areas within this can be identified; a particular hospital physics department may be involved in most of these at any one time.

Medical or clinical engineering includes the replacement of body parts such as heart valves and joints, and the fitting of powered artificial limbs. There has been a growth in recent years of electronic aids for the disabled, not only to assist in physical actions such as locomotion and eating, but also to produce sophisticated communication aids. The term **bio-engineering**, sometimes used for this work, is really much wider, including engineering in any biological system, overlapping with biotechnology. Chapter 1 deals with mechanical aspects of the body. The manufacture and fitting of hearing aids and lenses for defective vision is not usually included under this heading, being dealt with by specialists such as audiologists, opthalmologists and opticians. The physics of these aids is covered in Chapters 2 and 3.

Aiding the communication skills of a physically disabled girl.

Biomedical measurement and medical electronics is a rapidly growing field as a result of developments in microelectronics. There is a considerable amount of local cooperation between clinician and physicist or electronic engineer to produce new equipment to meet special requirements. For example the pH in the oesophagus can be monitored by a radio link to the measuring device, to enable surgeons to diagnose problems with the acid contents of the stomach refluxing. Such a project can only succeed with the collaboration of several different specialists. Examples will be found in the chapters of Theme 2. A common role for the physicist is the evaluation of the performance and potential hazards of major equipment, for example a new X-ray scanner. This aspect of the job is considered in more detail in Theme 4.

Medical ultrasonics, using echoes of high-frequency sound from the body, has become a very common method for investigating many features of soft tissue, especially in obstetrics and cardiology. The physicist provides technical backup for routine procedures and is involved in the development of new techniques, such as the increased use of computers in the formation of dynamic images. This is covered in Chapter 8.

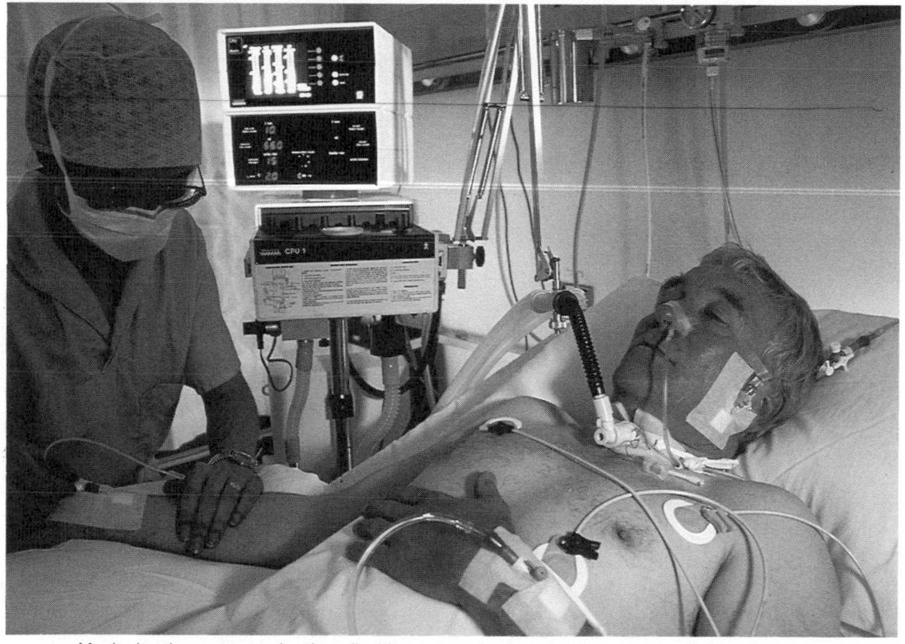

Monitoring the oxygen and carbon dioxide content of exhaled air from a patient in intensive care.

Radiology began with the application of X-rays to 'see' into the body, immediately after their discovery by Roentgen, perhaps the most dramatic event in the history of medical physics. The subsequent realisation of the damaging effects of radiation has reduced the extent of their use, but recent technological advances have re-established an important place for X-rays in

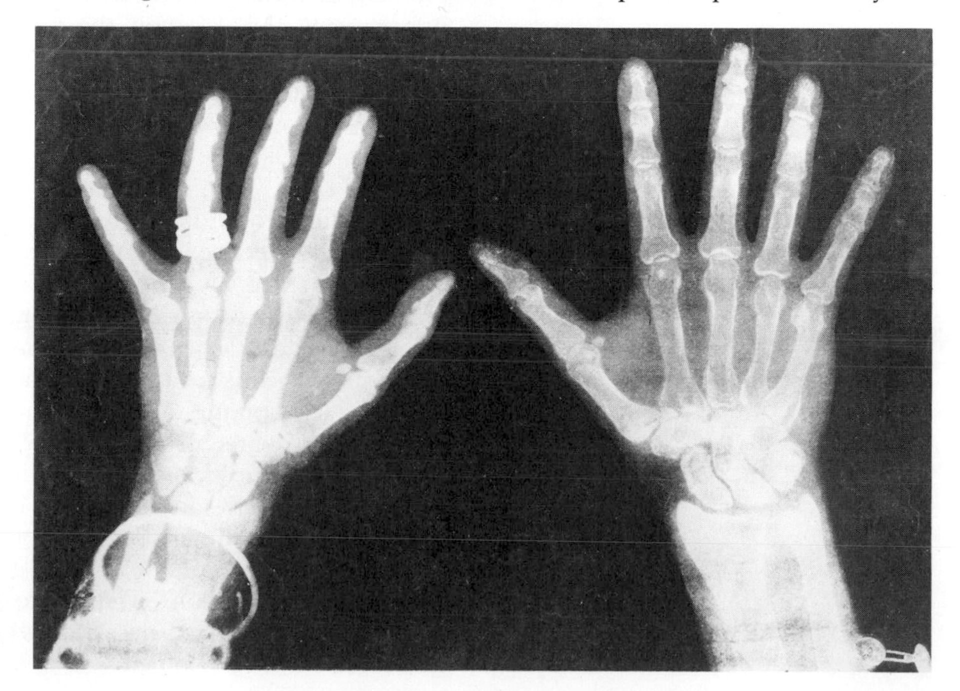

The hands of the Duke and Duchess of York; radiographs published in the *Illustrated London News*, 18 July 1896, just a few months after Roentgen's discovery was announced.

the modern imaging department. These changes give considerable scope for the role of the physicist in research and development. These issues are presented in Themes 3 and 4.

Nuclear medicine is the use of radioisotopes to measure and image body functions and organs. It was born after the development of the atomic bomb in the second world war, as a result of some physicists searching for a beneficial use for artificial radioactivity to 'atone' for the destructive power of the bomb. The use of radioisotopes often complements radiology, and the physicist will be involved in similar tasks such as the production of computer software to provide more clinically useful information. This is described in Chapters 10 and 12. Both types of radiation sources can be used therapeutically at high doses.

Radiation protection for both patients and staff from the unintended irradiation by sources in medical use is the responsibility of the physicist. Specialists in radiation physics have an important role in monitoring and evaluating the effects of other radiation exposures, such as those from nuclear reactor accidents. This is dealt with in Chapter 11.

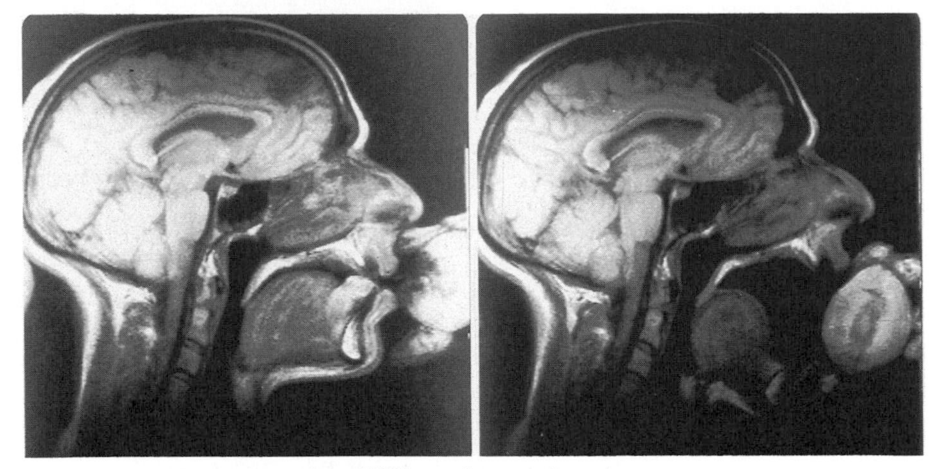

Lateral NMR scan of a man eating a plum.

Computing is now an integral part of much of the routine processing of biomedical measurement, as described in Chapter 4. In smaller hospital departments this is often the responsibility of the physicist. In larger departments considerable development work will be undertaken by computer scientists and electronic engineers. These staff may also have a role in the wider implementation of data processing throughout the hospital. This is reviewed in Theme 4.

WHAT ARE THE CAREER OPPORTUNITIES IN MEDICAL PHYSICS?

The development of physics within hospitals has in the past been largely a response to local need. This is reflected, in the UK, in the range of sizes of departments and the variety of functions that they carry out. In the USA and many other countries there is no separate department; physicists and technicians are attached to appropriate medical departments.

Physical scientists may enter the profession directly after graduating (with a good honours degree or equivalent). An increasing number study for a higher degree or enter medical physics with experience of work in another field. Physical science technicians enter with a training qualification such as BTEC, in the UK.

In-service training leading to professional qualifications, is organised in the UK by the Institute of Physical Sciences in Medicine and the Hospital Physicists Association. There are similar organisations for medical physicists in other countries, from whom further details of the type of work and careers can be obtained.

HIGH TECH MEDICINE – IS THERE NO ALTERNATIVE?

You may gain the impression from reading this book that all the developments in medicine at the present time involve complex expensive equipment; that progress is only to be achieved by the use of 'high' technology. That would certainly be a limited view, for a number of reasons. This book has tried to include current medical developments which involve physics. It does not attempt to describe their relative importance in securing the health of the population. In measurement and diagnosis the stethoscope and sphygmomanometer are undoubtedly of far greater importance than the CT scanner or Gamma camera. There has also been little opportunity in the space available to describe the 'alternative ' treatments such as hydrotherapy (exercise in water), or osteopathy (manipulation of bones), though the merits of these and other 'low tech' methods are now being increasingly recognised.

There is a crisis in health care in many western countries – the financial demands made of the services continue to increase with no indication that these will ever be fully met. (In less prosperous countries the situation is of course even more serious.) In these circumstances high tech medicine becomes something that will always have to be rationed. Who decides who gets what treatment? In Chapter 13 you will have an opportunity to relate these issues to the provision of a medical imaging service.

QUESTION FOR DISCUSSION

In *High Technology Medicine, benefits and burdens,* Bryan Jennett, an experienced neurosurgeon, teacher and administrator reviews these issues and includes the following list of inappropriate uses of high technology:

Unnecessary – because the desired objective can be achieved by simpler means.

Unsuccessful – because the patient has a condition too advanced to respond to treatment.

Unsafe – because the complications outweigh the probable benefits.

Unkind – because the quality of life after rescue is not good enough or its duration for long enough to have justified the intervention.

Unwise – because it diverts resources from activities that would yield greater benefits to other unknown patients.

What examples can you think of from your own experience, that might come into each of these categories? Do all of the group discussing this agree or do different personal experiences affect our views? How would a health service decide what is appropriate?

Further reading

Physical sciences in medicine Institute of Physical Sciences in Medicine

An illustrated booklet of career information on medical physics.

High technology medicine, benefits and burdens, Jennett, B. Oxford (paperback) 1986

A well-written, balanced account with wide-ranging examples from the author's own experience.

Man modified, Fishlock, D. Paladin

Subtitled 'an exploration of the man/machine interface', the book is a readable account of the interaction of the human body with technology, with examples ranging from transplants to computers.

Theme 1

BODY PHYSICS

How well do you know your own body? Well enough to know its capabilities and limitations, perhaps. You may have been helped in this by your previous studies in science, linking personal experience with objective knowledge. You will also know something of the complexity of the body, and the range of sciences which are used in studying its functions. In this theme we will study those parts of the body where physics has a substantial contribution to make in answering vital questions such as:

Why do so many people suffer from bad backs?
What are a person's daily energy needs?
Why can we not see colours when light intensities are low?
How does loud noise damage our hearing?
And will the vaulter's pole absorb his kinetic energy supply?

Prerequisites

Before starting this theme you should have some familiarity with the following:

- The basic structure of the human body.
- The concepts of force, work, energy, power, temperature and waves.
- The principle of moments and the resolution of forces.
- The nature of light and of sound.

(You will need to refer to GCSE or A-level texts at some points if these are unfamliar to you – or forgotten!)

Daley Thompson breaks his pole, vaulting at the 1988 Olympic Games.

Chapter 1

HUMAN MECHANICS

The balance of the gymnast and the strength of the weightlifter (Fig 1.1) are two examples of the body's abilities. You may envy their achievements or be thankful that you don't have their bodies! This chapter may not enable you to gain the former or avoid the latter, but you should learn how the body functions in both these and more everyday situations.

Fig 1.1 The body balanced, and the body strong. Olga Korbut, champion Olympic gymnast and Nain Suleymonogin, champion Olympic weightlifter.

LEARNING OBJECTIVES

After studying this chapter you should be able to:

1. describe how muscles and bones operate to produce body movements;
2. give examples of levers in the muscular-skeletal system of the body;
3. provide a simple analysis of:
   (a) the forces involved in standing, lifting and bending,
   (b) the interactions of the body with the ground in walking and running;
4. recall typical human energy needs and energy values of typical food types;
5. estimate the power typically provided by muscle;
6. describe how the body maintains a constant temperature, and give examples of how this may be assisted.

1.1 INTRODUCTORY ANATOMY

Muscles and skeleton

The human body is supported by a framework of some 200 bones connected together to make the skeleton. Bone is a living material; stiff, light and strong. Individual bones have the optimum properties required for their particular function. For example the femur, which must withstand forces of up to thirty times body weight, is a thick-walled, hollow cylinder with internal

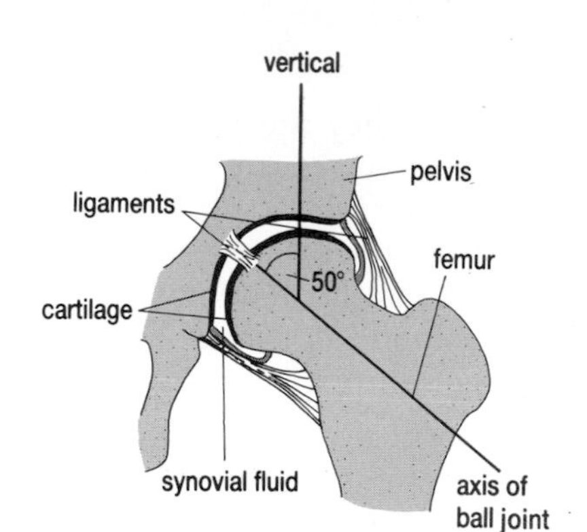

Fig 1.2 Structure of the hip joint.

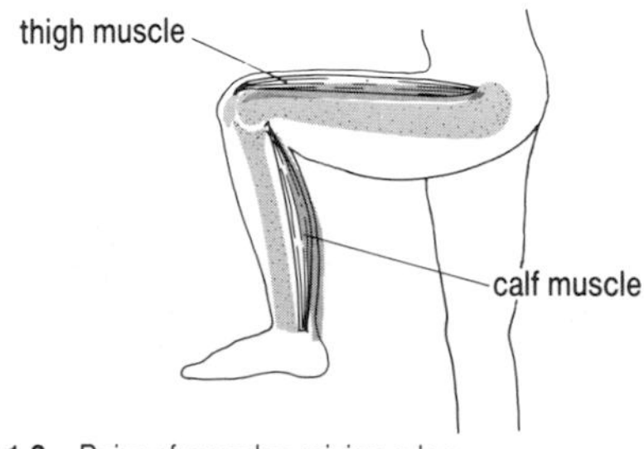

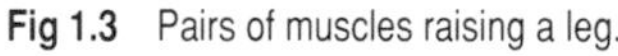

Fig 1.3 Pairs of muscles raising a leg.

cross struts. This gives an excellent strength-to-weight ratio. The skeleton is held together at joints by ligaments. There are several kinds of joints to allow the considerable flexibility of human movement. These range from the plane joints of the toes through pivot (neck), hinge (elbow), ellipsoid (wrist) and saddle (ankle), to the ball and socket of the shoulder and hip which gives maximum freedom of movement (Fig 1.2). The illustration shows features common to all joints, namely the layers of the tough cartilage and the interposed synovial fluid, which prevent bone damage during movement.

Movement of the skeleton is effected by the muscles, which are attached to the bones by tendons. Muscle is a flexible fibrous material which surrounds the skeleton. (You can see the fibres in the 'meat' of a butcher's joint such as a leg of lamb.) Movement is produced by a tensing of the muscle which causes it to become shorter and fatter and pull on the bone. To restore the bone to its original position there is always an opposing muscle acting in antagonism. Hence to raise the leg as shown in Fig 1.3 the rear calf muscle and the front thigh muscles contract whilst the opposing muscles relax.

Some body figures

number of bones in the skeleton	206
number of joints in the skeleton	187
types of joints	6
number of voluntary muscles (the involuntary muscles are those of the heart and the intestines)	620
percentage of body weight which is bone	~17
percentage of body weight which is muscle	~40
strength of bone compared to reinforced concrete	× 4

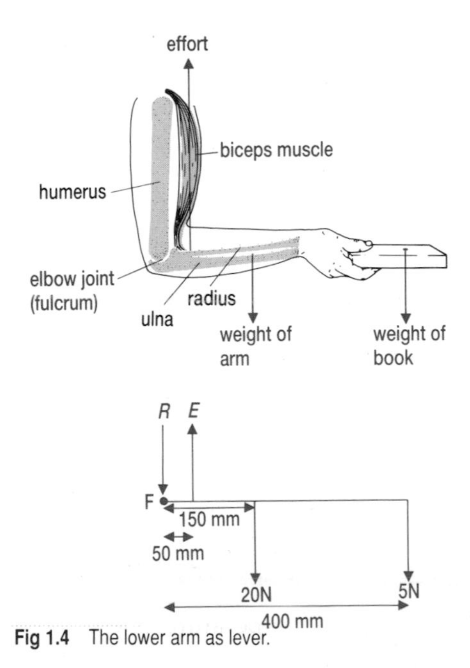

Fig 1.4 The lower arm as lever.

Forces and levers

The tensed muscle exerts a force on the bone at the place where it is attached by the tendon. This produces a turning effect about the joint. The system acts as a lever with the joint as fulcrum. The muscle provides the effort (E) and the load (L) may be the weight of the part of the body being moved, or may include an external load. From the principle of moments for a body in equilibrium:

$$L \times \text{distance of load from fulcrum} = E \times \text{distance of effort from fulcrum}$$

Often the effort is applied closer to the fulcrum than the load, giving the system a mechanical advantage of less than one, since:

$$\text{mechanical advantage} = \frac{\text{load}(L)}{\text{effort}(E)} = \frac{\text{distance of } E \text{ from fulcrum}}{\text{distance of } L \text{ from fulcrum}}$$

This arrangement, illustrated in Fig 1.4 gives a greater control to the body movement, a small movement of the biceps muscle, with a large force, produces a much larger movement of the book.

QUESTION

1.1 (a) The head can be tipped back by the pull of the muscles on the back of the neck, with the top of the spine, the atlas vertebra, as fulcrum. Draw a simple diagram of this lever, showing the forces acting.

(b) Draw a similar diagram for the action of the calf muscle when a person stands on tiptoe, pivoting on the ball of the foot.

Solving lever problems

We can work out the forces in muscles when they are producing movements, by considering only the part of the body involved. Look at Fig 1.4, where we can take moments about the joint as the fulcrum. In this treatment it is assumed that the force of the humerus on the lower arm acts through the fulcrum, at a single point in the joint and that the tendon is acting at a single point on the lower arm. When the forearm is bent at right angles, with the dimensions shown in Fig 1.4(b), the contractile force of the biceps (*E*), and the reaction of the upper arm, can be calculated by taking moments about F.

$$E \times 50\text{ mm} = (20\text{ N} \times 150\text{ mm}) + (5\text{ N} \times 400\text{ mm})$$
$$50E = 3000\text{ N} + 2000\text{ N}$$
$$E = 100\text{ N}$$

Resolving forces vertically:

$$E = R + 20\text{ N} + 5\text{ N}$$
$$R = 75\text{ N}$$

QUESTION

1.2 A champion weightlifter can raise a mass of 250 kg. His forearm has a length from elbow joint to palm of hand of 0.5 m, and a weight of 30 N. Assuming that the force of the biceps muscle acts at a point 0.07 m from the fulcrum, calculate the maximum values of the force in each biceps and the reaction in each upper arm.
(Take g, the gravitational field strength as 10 N kg^{-1})

The vertebral column

The backbone provides the main support to the body, and is required to make a great variety of movements. It consists of 33 vertebra, of which nine are fused together to make up the sacrum and the coccyx, and the remaining 24 are separate, covered with cartilage and interspersed with tough fibrous pads called discs. It is the discs which permit the free bending of the spine, whilst cushioning the bones from the compression of body weight and protecting the spinal cord which runs down the centre of the vertebrae. The discs are subjected to considerable wear throughout life and they also lose some of their toughness with age. If the stresses become too large a disc can slip out of position. It is not therefore surprising that about 80 per cent of people suffer from back pain at some time in their lives. The lowest part, the lumbosacral region is the most susceptible. Because, as Fig 1.5 shows, it is here that the normal backbone deviates most from the vertical. The sacrum is attached firmly to the pelvis so has much more limited movement, whereas the lumbar vertebrae are connected to each other only through the discs.

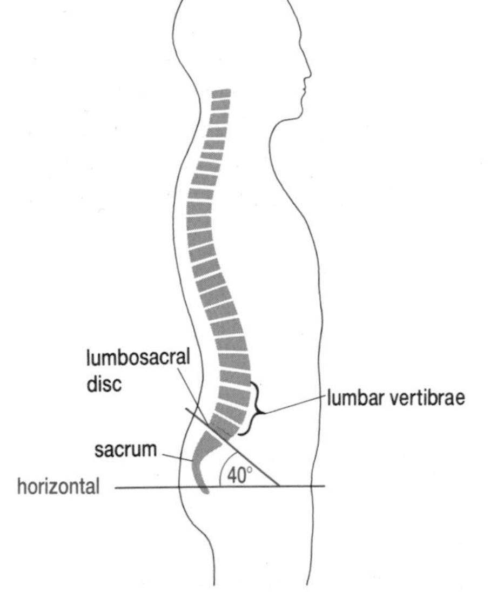

Fig 1.5 The vertebral column.

1.2 STANDING, BENDING AND LIFTING

Standing

In normal posture the lumbosacral disc lies at an angle of about 40° to the horizontal; this is called the lumbar lordosis. The weight of the body on this disc is about 60 per cent of the total body weight, *W*. It is called the superincumbent weight, (the weight lying on). This will be balanced by the support force of the sacrum, *S*. These forces can be resolved to give a **compressive stress** (squashing), perpendicular to the disc, and a **shear stress** (twisting) parallel to the bone surface (Fig 1.6).

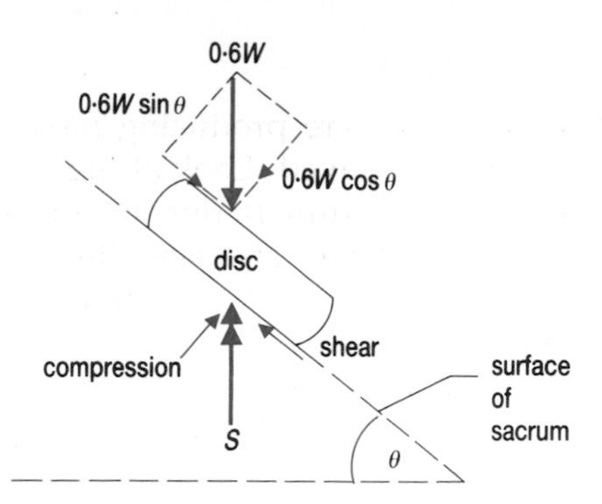

Fig 1.6 Forces on the lumbosacral disc.

$$\text{compressive stress} = S\cos\theta$$
$$\text{shear stress} = S\sin\theta$$

For normal posture

$$\text{compressive stress} = 0.6W \times \cos 40° = 0.46W$$
$$\text{shear stress} = 0.6W \times \sin 40° = 0.39W$$

The discs are designed to withstand compression rather than shear, so this shear stress is potentially damaging. The situation is made worse if a person has poor posture so that the lumbar lordosis is greater than 40°, which may be caused by weakened muscles, for example as a result of pregnancy.

When we stand on two legs our body weight is equally divided between each leg. At the head of the femur, where the leg connects to the body, the superincumbent weight is about 0.7*W*. The axis of the ball joint is usually at an angle of about 50° to the vertical (Fig 1.2).

When the body is supported on one leg the body's centre of gravity has to shift so that it is over the supporting foot. The superincumbent weight then has a moment about the femoral head, which increases the total force on it to about $2\frac{1}{2}$ times body weight (compared with just over $\frac{1}{3}$ body weight when standing on two legs).

QUESTION

1.3 Resolve the force at the femoral joint to find the compressive and shear stresses on this joint when a person of mass 70 kg stands on two legs. How does this compare with the stress on the lumbosacral disc?

Bending

Damage to the back occurs most commonly during bending and lifting loads. A simple analysis of the forces involved shows why this is so. A range of muscles, the erector spinae, link the vertebrae of the spine to the sacrum and ileum of the pelvis. We can simplify the arrangement by considering the spine as a rigid column, pivoting on the sacrum at the lumbosacral joint (Fig 1.7). The superincumbent weight, W_b, of the trunk, head and arms acts at their combined centre of gravity, G, which is about two-thirds the way up the trunk. This is balanced by the lift provided by the muscles to the upper part of the body, which can be represented as a single force through G. X-ray observations indicate that this force is applied at about 10° to the vertebral column.

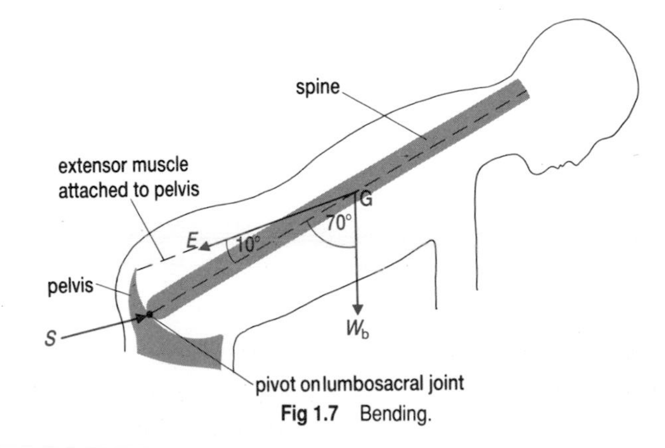

Fig 1.7 Bending.

Let us now calculate the tensile force *E*, of the muscles and the compressive force on the lumbosacral disc, *S*, when the body is bent at an angle of 70° to the vertical (Fig 1.7). The three forces W_b, *E* and *S* are in equilibrium, so the line of action of *S* is along the column, so that it passes through G.

Average body weight for a person is about 700 N, so

$$W_b = 0.6\ W = 420\ \text{N}$$

Resolving forces perpendicular to the column,

$$E \sin 10° = W_b \sin 70°$$
$$E = 2275\ \text{N}$$

a considerably greater force than body weight because of the acute angle of the muscles' action.

Resolving forces parallel to the column,

$$S = E \cos 10° + W_b \cos 70°$$
$$S = 2385\ \text{N}$$

This substantial force reduces the thickness of the disc by about 20 per cent.

Lifting

We can extend this treatment to consider the forces involved when bending and lifting an object, for example a chair of weight 300 N. Fig 1.8 shows the forces involved, with the load acting at the top of the trunk in line with the shoulder joint (the distance L from the pivot). There are now four forces acting and so S acts at a small angle to the axis of the column. There is a compressive component S_c and a shear component S_p.
Taking moments about the fulcrum, F:

$$E \times 0.67L \sin 10° = W_b \times 0.67L \sin 70° + 300\ \text{N} \times L \sin 70°$$
$$E = 4711\ \text{N}$$

Resolving forces parallel to the column,

$$S_c = E \cos 10° + W_b \cos 70° + 300\ \text{N} \cos 70°$$
$$S_c = 4886\ \text{N}$$

Resolving forces perpendicular to the column,

$$S_p + W_b \sin 70° + 300\ \text{N} \sin 70° = E \sin 10°$$
$$S_p = 141\ \text{N}$$

The total reaction is given by:

$$S^2 = S_c^2 + S_p^2$$
$$S = 4888\ \text{N}$$

This acts at an angle θ to the column, where:

$$\tan \theta = S_p/S_c$$
$$\theta = 1.5°$$

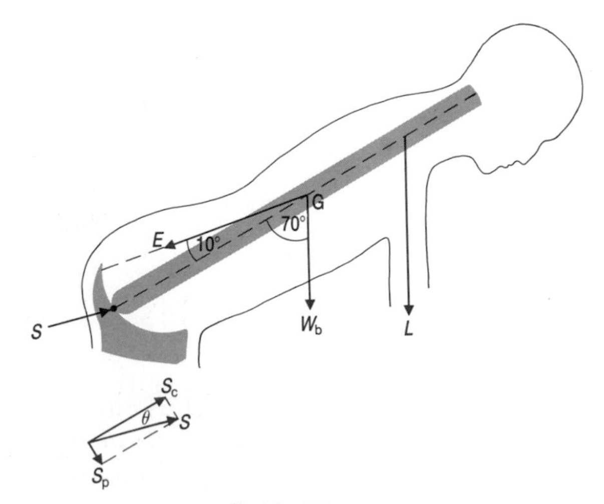

Fig 1.8 Lifting.

The disc under pressure

A normal lumbar vertebral disc will contract elastically under compression up to a force of about 1000 N. Beyond this it deforms plastically and will rupture when it has contracted by about 35 per cent, at a load of about 15 000 N. Rupture causes the fluid in the disc to be extruded and cause pressure on adjacent nerves in the spinal cord (Fig 1.9). This is what causes the intense pain and muscle spasm associated with disc trouble. Treatments include rest, strengthening the muscles by exercise, skilled manipulation and surgery.

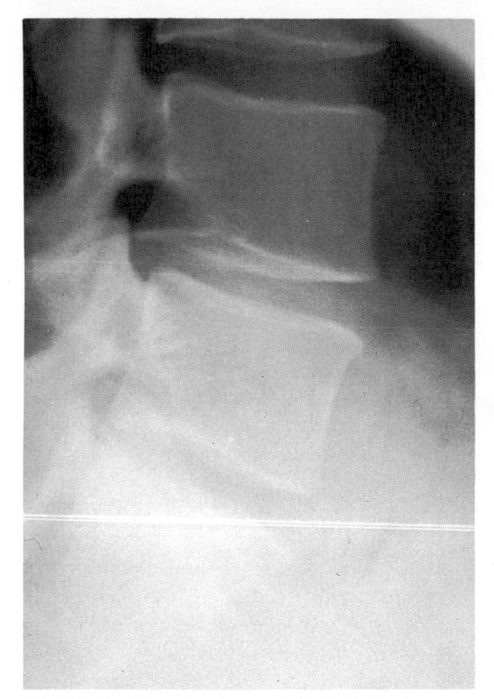

Fig 1.9 An X-ray of a damaged disc in the lumbar spine.

Fig 1.10 A good posture for lifting.

QUESTION

1.4 (a) From the above figures, work out the ratio of the compressive stresses in the lumbosacral disc, for
(i) standing
(ii) bending
(iii) lifting a weight of 300 N.

(b) Explain how a vertical posture when standing and when lifting objects (as shown in Fig 1.10), helps to avoid a slipped or ruptured disc.

1.3 WALKING AND RUNNING

Ground forces

A body standing on the ground is supported by a force equal and opposite to the body's weight. This force V, will therefore have a magnitude of W and act vertically upwards. When the body is moved relative to the ground the force will change. When the foot lands on the ground at the end of a step, a force is needed to decelerate it to rest. When the foot is thrust off the ground for the next step an accelerating force is required. In each case the ground force will be greater than W, and will be the resultant of a vertical force V, and friction F, as illustrated in Fig 1.11. The forces will of course increase as the speed with which the foot meets and leaves the ground increases.

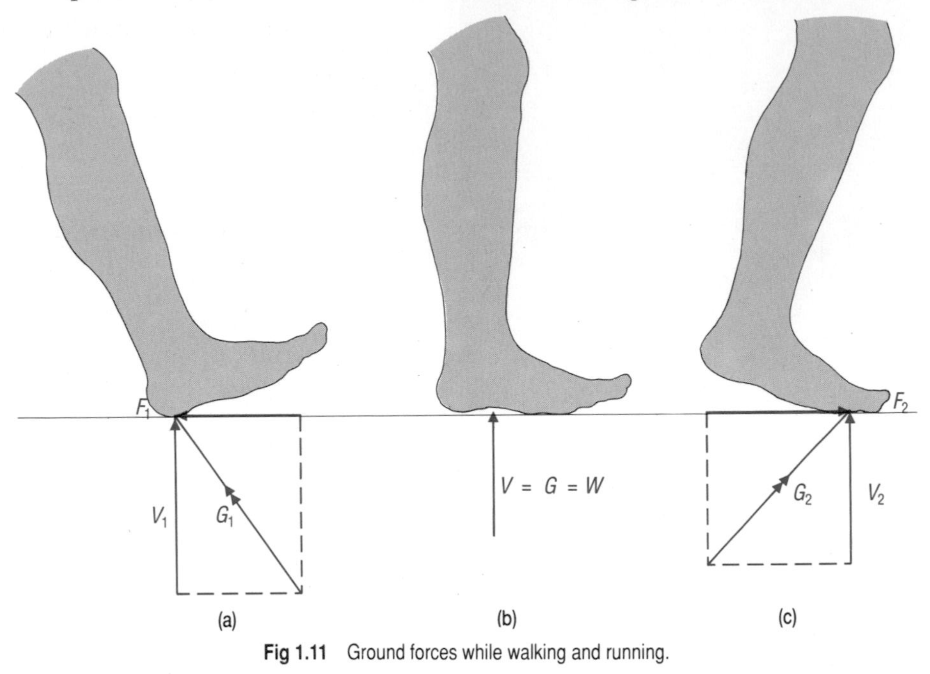

Fig 1.11 Ground forces while walking and running.

The maximum size of the vertical force depends on the material of the ground. If this is soft it may undergo considerable deformation in order to provide the force needed. The maximum size of the frictional force depends on the nature of the surfaces and the size of the vertical force. This is expressed as,

$$F = \mu V$$

where μ is the coefficient of static friction between two surfaces. Thus if the horizontal force needed is greater than μV the foot will slip. Typical values of μ for shoes on ground are 0.6 to 0.75.

The resultant force of the ground G, has magnitude given by:

$$G^2 = V^2 = F^2$$

and direction given by:

$$\tan \theta = F/V$$

and the horizontal forces required for walking are about 15 per cent to 20 per cent of body weight.

QUESTION

1.5 (a) With reference to the previous section explain:
- (i) why slipping does not normally occur while walking.
- (ii) why slipping is more likely to occur when the stride is long and when θ is small.

(b) Imagine you have to cross a wet and muddy ploughed field on a cross-country run. Use the above account to explain why your options are to slip across the surface slowly or plough through, ankle deep, with a great effort.

Walking

The forward acceleration of each leg requires a horizontal force given by:

Force = mass × acceleration.

There is also a rotation of the leg as it moves forward, in which the turning force or torque produces an angular acceleration. The relationship here is given by:

$$\text{torque} = \text{moment of inertia} \times \text{angular acceleration}$$

where the moment of inertia depends both on the total mass of the leg and its distribution. Thus any change such as the wearing of heavy boots, will result in extra energy expenditure and possibly muscle fatigue.

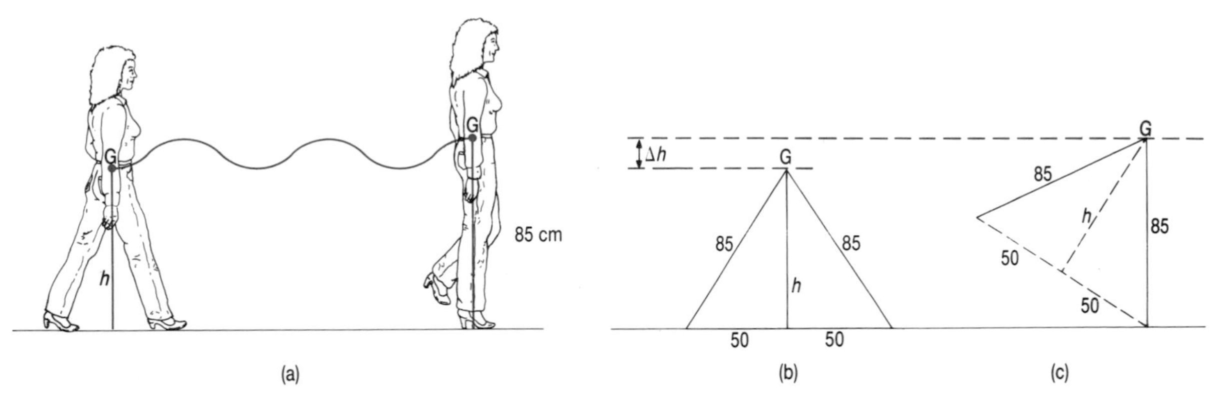

Fig 1.12 Changing the height of centre of gravity while walking.

Fig 1.13 Conserving angular momentum while running.

Deceleration requires a similar application of force and torque, though this is assisted by gravity. A similar forward swinging movement occurs at the knee. Work is also done in raising and lowering the centre of gravity G, during each step. This is shown in Fig 1.12 which represents a walker of leg length 85 cm taking a stride of 1 m.

The rise in G is

$$\Delta h = 85 - h = 85 - \sqrt{85^2 - 50^2}$$
$$= 16.25$$

This is an over estimate as we actually bend our knees at the position where G is highest, which reduces the rise to about half of this value.

Running

Accelerations and torques in running are much greater, in order to produce the higher speed. A fast sprint (about 9 m s^{-1}) is six times faster than a typical walking speed. This is achieved by a combination of an increase in stride frequency and in stride length. To offset the considerable angular acceleration of the swinging leg, the opposite arm is swung forward. This causes the upper and lower parts of the body to twist in opposite directions to conserve angular momentum (Fig 1.13).

1.4 ENERGY EXPENDITURE

Energy is rather like wealth – a description of a situation or condition, rather than a thing in itself. We are usually interested in changes of energy and of wealth, not absolute values, and the currency we use is the joule. The human body expends energy doing mechanical work and maintaining a constant body temperature. The income is from the food we eat. Excess of income over expenditure is stored in the body, mainly as fat. Excess of expenditure over income leads to fatigue, ill health and eventually the possibility of death.

Doing work

Work is done by the body in moving against gravity, external forces of friction and air or water resistance, and internal resistance of the body. In walking for example, the raising of the centre of gravity with each step, as described in the previous section, results in an energy expenditure of:

$$\begin{aligned}\text{work done per stride} &= \text{weight} \times \text{rise} \\ E &= mg\Delta h \\ &= 70 \text{ kg} \times 10 \text{ N kg}^{-1} \times 0.08 \text{ m} \\ &= 56 \text{ J (joules)} \\ \text{work done per km} &= 56 \text{ kJ (for strides of 1 m)}\end{aligned}$$

Assuming a walking speed of 5 km h^{-1}

$$\begin{aligned}\text{power} &= 5 \text{ km h}^{-1} \times 56\,000 \text{ J} / 3600 \text{ s h}^{-1} \\ P &= 78 \text{ W (watts)}\end{aligned}$$

In jumping the muscles do work to produce kinetic energy ($E_k = \frac{1}{2}mv^2$) at take off. This is converted to potential energy ($E_p = mgh$) as the body is propelled upwards. In high jumping and pole-vaulting the vertical height of the centre of gravity is all-important. In long jump and triple jump speed and height are both important in carrying the jumper forward.

INVESTIGATION

The mechanics of jumping

This is an opportunity to measure performance and study how it may be improved (in yourself or a suitable volunteer!).

Fig 1.14 Measuring jump height with chalk marks.

Standing jump

1. Measure take-off speed u, when jumping vertically. The normal laboratory ticker-timer will probably not be suitable for this. You could devise a photographic method, but a simpler way is to measure the height Δh jumped (Fig 1.14) and use the equation

$$v^2 = u^2 + 2gh$$

For example, a good jump would be 1.0 m, giving a take-off speed of 4.43 m s^{-1}.

Long jump

A simple model of long jumping is that the athlete runs as fast as possible and jumps vertically at the take-off point.

2. Measure maximum sprint speed v_{max} and length of long jump s.

3. Check whether this simple model is appropriate for your jump, using the equations for a projectile:

$$h = ut + \tfrac{1}{2}gt^2$$
$$s = v_{max}t$$

For example, Carl Lewis can run at approximately 10 m s^{-1}.
Assuming his vertical take-off speed is that of a 'good' jumper as in **1.** above, then:

$$0 = 4.43t - 4.9t^2$$
$$t = 0.904 \text{ s}$$
$$s = 9.04 \text{ m.}$$

Lewis won the 1988 Olympic gold medal with a jump of 8.72 m, so this simple model seems to give a reasonable prediction for a top athlete. How do your results compare?

High jump

4. By considering the conversion of E_k to E_p, derive an expression to show that the height jumped Δh depends on the square of the take-off velocity.

The height jumped is the distance that the centre of gravity is moved upwards. For a standing adult the distance of the centre of gravity from the ground h_c is about 1 m. Assuming it does not rise much above the bar, we can calculate the expected jump height from this expression. For example, with a take-off velocity for a standing start of 4.43 m s^{-1}, Δh is 1 m, giving a bar clearance H of 2m. The 1988 women's Olympic record (by Louise Ritter) was 2.03 m, so you can see that horizontal running speed has little contribution to make.

5. Calculate the height of your centre of gravity above the ground (about 55 per cent of full height for women, 57 per cent for men). Measure the bar clearance height at various take-off speeds. Compare the values of Δh calculated from $H - h_c$ with the predictions from $H = u^2/2g$. For the successful high jumper technique and body position are much more important than sprint speed (Fig. 1.15).

Fig 1.15 Sylvia Costa, champion high jumper, keeping her centre of gravity as low as possible.

QUESTION

1.6 With reference to the previous investigation explain:
(a) why good long jumpers are fast sprinters,
(b) why good high jumpers are tall,
(c) the function of aerial leg kicks in long jumping,
(d) the function of the diving or flopping technique in high jumping.

Basal metabolic rate

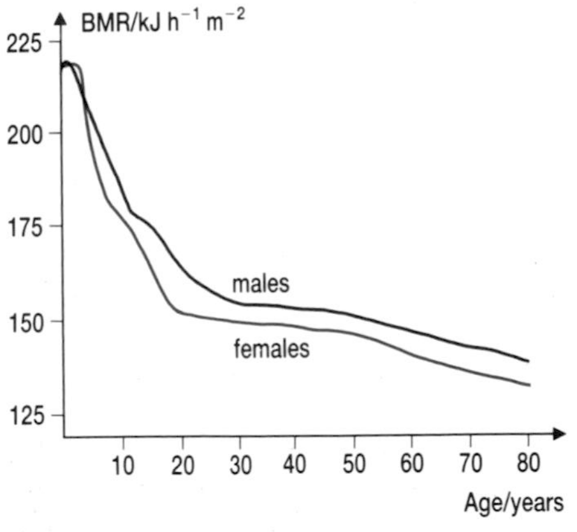

Fig 1.16 Basal metabolic rate variation with age and sex.

The body is continuously performing involuntary functions which require energy. These include the beating of the heart, breathing, digestion and growth. This maintenance expenditure of energy is called the **basal metabolic rate** (BMR). In children BMR is high because of the energy used in the biosynthesis of components for growth. The rate levels off as we mature, falling again in old age as all processes slow down (Fig 1.16). From childhood onwards, males have a higher BMR than females, on average. This is because of the lower proportion of fat in the male body. If this is excluded from the calculations, both sexes have similar BMRs.

The energy expenditure can be calculated indirectly from temperature or respiration measurements as described below. Alternatively we could calculate the work done in the various processes. In Chapter 6 for example the power of the heart is calculated from a knowledge of its volume, pulse rate and the pressure of the blood leaving it.

Getting warm and keeping cool

The values shown in the graph (Fig 1.16) are averages, individuals show a considerable variation. One of the factors which causes this variation is the maintenance of constant body temperature. The human body operates best at around 37 °C, and will cease to function if the temperature changes by more than about 6 °C from this, as described in Table 4.2 in chapter 4. In temperate climates, such as Britain, the body will need to expend energy to keep itself at a higher temperature than the surroundings. In tropical climates, or during periods of intense exercise, the body will need to expend energy to cool itself, like a refrigerator. So the amount of energy expenditure will depend on the ambient temperature, the work being done by the body, the surface area of the body and what clothes are worn.

Table 1.1 Energy loss processes

(a) Process	Determining factor	Values
conduction	temperature difference, area of surface, conductivity	very low due to insulating properties of fat, hair, clothing (skin 0.042 W m^{-1} K^{-1})
convection	temperature difference air speed around body	low, except in draughts
radiation	absolute temperature, area and nature of surface (see Chapter 4 for details)	100 W for an unclothed body at 22 °C
evaporation	temperature difference, area of skin exposed, humidity, air movement	625 W maximum
respiration	air temperature and humidity	20 W resting at 22 °C

(b)	studying at 22 °C	sunbathing at 32 °C	walking at –18 °C
% of body clothed	85	15	95
heat loss rate /W	170	400	400
% loss by:			
1 radiation	21	8	8
2 conduction and convection	67	10	50
3 evaporation	10	80	2
4 respiration and excretion	2	2	40

The body can lose energy by conduction, convection, radiation, evaporation (of perspiration), respiration (breathing out warm air and water vapour) and excretion. The relative importance of each depends on the condition of the body and its surroundings. Table 1.1 summarises the processes and gives some typical values.

The amount of energy released in body processes depends mainly on the **volume** V, of the tissues. The rate of loss of this energy depends on the **surface area** A, of the body. The smaller the body, the higher is the **surface to volume ratio**, A/V. This means that for example babies are much more at risk of hypothermia (low temperature), than adults. It also follows that smaller people need to eat more, relative to their mass, than larger people. Typical values of these quantities are given in Table 1.2.

Table 1.2 Body sizes and surface area to volume ratios

	surface area A/m^2	volume V/dm^3	surface area to volume ratio $\frac{A}{V}/\text{m}^2\,\text{dm}^{-3}$
Smaller person	1.53	56	0.0273
Larger person	1.97	80	0.0246

QUESTIONS

1.7 With reference to the data in Table 1.1, answer the following:

(a) The specific latent heat of sweat is about 2425 kJ kg^{-1}. What is the rate of its excretion from the skin to produce the maximum rate of energy loss quoted?

(b) Why is the percentage of energy loss by evaporation (i) high when sunbathing and (ii) low when walking in cold conditions?

(c) Why is a higher proportion of the energy lost while a person is studying, by conduction and convection?

1.8 Because of the importance of surface area in determining BMR, the values are expressed in Fig 1.16 as kJ h^{-1} *per square metre*. From the figure and Table 1.2, calculate the range of power dissipation, in watts, for males and females, in their seventeenth year.

Total energy requirement

This is the BMR value added to the work done in activities. It can be measured directly by putting the subject in an isolated room, (called an Atwater chamber, after its designer), and monitoring all energy exchanges. For example temperature changes are measured by water circulating in pipes through the chamber. Any work done which may be stored (such as raising weights), is added to this. Work which is not stored, (such as exercise movements) will produce temperature rise so need not be separately considered. Typical ranges of values are given in Table 1.3. The lower values in the range are for young or old subjects, for females and for less vigorous activity. The highest values are usually for heavier fit adult males – and even fitter females!

These figures lead to estimates of total energy expenditures of about 10 MJ for an average woman, between the ages of 15 and 50, with a moderately active life style and about 13 MJ for a man of similar age and lifestyle. The differences are mainly due to different BMR and body weights. A woman in the latter half of her pregnancy, and while lactating to feed her baby, requires an extra 1.5 MJ.

Table 1.3 Energy conversion

Activity	Conversion rate/W
Resting (BMR)	60 – 100
Sitting or standing still	80 – 150
Moving around classroom or home	100 – 300
Walking or 'light' sports, housework	150 – 450
Moderate sports, e.g. swimming, gymnastics	300 – 550
Climbing stairs, carrying heavy loads	400 – 850
'Heavy' sports, e.g. running, rowing, squash	400 – 1400

QUESTION

1.9 From the data in Table 1.3 calculate the energy expenditure for yourself, on an active and on an inactive day. How do your answers compare with the totals given above?

1.5 ENERGY INCOME

Input

The body's energy is supplied from the chemical energy of the food we eat.

There are three types of fuel foods: carbohydrates, fats and proteins. The digestive system breaks these substances down into chemically simpler ones: (a) carbohydrates to sugars such as glucose, (b) fats (triglycerides) to glycerol and fatty acids, and (c) proteins to amino acids. These simpler molecules are absorbed into the bloodstream and transferred to the cells, where they are oxidised, by the oxygen inhaled by the respiratory system. Enzymes are required to catalyse the reaction, but, as when any fuel is burnt, energy is released. The process of releasing energy by oxidation in the body is called **aerobic combustion** or **respiration** and its general reaction is:

fuel + oxygen = carbon dioxide + water + energy

and the particular equation for glucose is:

$$C_6H_{12}O_6 + 6O_2 = 6CO_2 + 6H_2O + 4 \times 10^{-18}\,J.$$

The respiration reaction is the reverse of photosynthesis in which plants 'fix' energy from sunlight. The energy released by combustion can be measured by calorimetry, typical values (of standard enthalpies) are given in Table 1.4. In the body the amount of energy released will normally be less than this, due to incomplete digestion of the foods and incomplete absorption of substances into the blood, giving the available energy values shown.

Thus the energy used at any time can be measured by the consumption of oxygen. A **spirometer** is used to monitor the conversion of oxygen to carbon dioxide. The average energy liberated from the combustion of carbohydrates, fats and proteins is 20.17 kJ per litre of oxygen. The rate of oxygen intake is a useful measurement to make on people who are in training for sports or under treatment for obesity.

An alternative breakdown of glucose to lactic acid can also release energy. This does not require oxygen, it is **anaerobic respiration**. It occurs therefore when energy is required in the absence of oxygen, for example during extreme exercise. The process is inefficient, liberating only about one twentieth of the aerobic process. It also causes feelings of fatigue and pain as lactic acid is poisonous. It is this which sets a limit on an athlete's performance. Recovery consists of the oxidation of the lactic acid and the making up of the oxygen shortage in the muscles.

The food normally used as a fuel is carbohydrate, with fat being drawn on as a reserve, protein being used only in extreme cases of starvation or overeating. In the former the protein is not then available for growth or maintenance of the body. In the latter the body needs to remove the excess consumption. Usually excess fuel food can be stored in the body.

Table 1.4 Foods as fuels

	Standard enthalpy of combustion /MJ kg^{-1}	Available energy /MJ kg^{-1}
carbohydrate	17.2	16.5
fat	39.4	37.5
protein	23.4	16.5

QUESTIONS

1.10 What is the BMR of a person who, when resting, consumes 1.5 dm^3 oxygen in 5 minutes? Give your answer in (**a**) kJ h^{-1} (**b**) W.

1.11 Explain how an increase in pulse rate and panting (deep, fast breathing), helps an athlete perform better. If an athlete is fatigued, how does massaging the muscles help?

Storage

Fat is the major energy store of the human body. Typically 12 per cent of a man's weight and 25 per cent of a woman's weight is made up of two types of fat, to be drawn on when energy is required. Fat is stored in adipose tissue, distributed throughout the body, especially under the skin where it provides good insulation. Adipose tissue is the white fat, and more is found in overweight people. Brown fat is usually found elsewhere in the body, and particularly in lighter people. As the brown fat has a higher metabolic rate it is thought that this can provide a less bulky energy store. Table 1.4 shows that fat provides double the energy produced by carbohydrate for a given mass. **Carbohydrate** breakdown products are stored in different parts of the body. Glycogen can be stored in the muscles or the liver, where it is broken down by enzymes to glucose, when required. It can also be converted to the high energy store substance adenosine triphosphate (ATP), which is again held in the muscle. The major supply of energy from carbohydrates is the glucose in the bloodstream.

Excess intake of food will result in excess body weight, or obesity. This is known to cause many common ailments; physical such as heart, circulatory, and respiratory malfunction, and psychological, for example poor self image. Despite the impression you may have gained from the above sections, that it seems a simple matter for a person to decide what energy intake is required, obesity is a major problem in Western societies, as witness the

Research into the causes and prevention of obesity receives considerably less funding than the estimated £300 M yr^{-1} profits generated by the UK sales of slimming aids and advice.

size of the 'slimming industry'. This consequence of Western affluence is thrown into sharper relief, when seen against the widespread malnutrition and frequent starvation experienced by people in poorer parts of the world.

SUMMARY ASSIGNMENTS

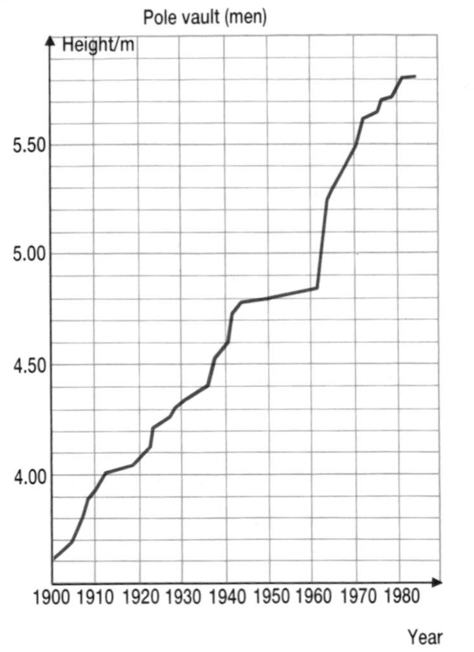

Fig 1.17 Effect of pole material on vault height.

1.12 Obtain details of a current slimming programme which gives some information about energy inputs and outputs. Compose a short appraisal of its likely effectiveness from what you have learnt in this chapter.

1.13 Summarise the factors which need to be taken into account in deciding an individual's energy requirements. Which of these will be difficult to quantify?

1.14 **(a)** Explain how, in terms of energy transfer, a pole vaulter has more in common with a long jumper than a high jumper.

(b) The importance of the pole is illustrated in Fig 1.17. What do you think happened to the type of pole available, in the 1960s?

(c) A pole vaulter has a take-off speed of 9.5 ms^{-1}, and a centre of gravity height above the ground of 1 m. Assuming all the kinetic energy is converted to potential energy, calculate the maximum vault height.

(d) How does this compare with actual performances?

1.15 Hypothermia may be combatted by vigorous exercise, external heating of the body, insulation such as duvets or thermal underwear, and reflecting blankets. Explain how each is appropriate to an at risk group such as elderly people, mountaineers or exhausted runners.

Further reading

The body report, *Observer Magazine*, 1988.

Chapter 2

THE EYE AND VISION

The human body is sensitive to most electromagnetic radiations. Microwave and infrared produce the sensation of warmth. Ultraviolet and ionising radiations can produce chemical changes and cause biological damage such as skin cancer. The eye is the one organ of the body designed to respond specifically, to receive (or perceive), a part of the electromagnetic spectrum, visible light, having wavelengths between about 380 and 760 nanometres. Of all the sense organs the eye has the greatest number of nerve cells dedicated to it, emphasising the importance of vision in our lives. There are two parts to the eye's perception; detection by the sensitive cells of the retina and image formation by the optical system. In this chapter we will consider how each of these work, and how to correct them when they do not.

LEARNING OBJECTIVES

After studying this chapter you should be able to:

1. give an outline description of the structure and function of the eye;
2. explain how the retina responds to light, including the visual effects of colour, acuity or resolution, adaptation to illumination levels and persistence;
3. explain how the optical system produces images, including the optical features of depth of focus, depth perception and resolution (diffraction);
4. explain how optical defects of vision may be corrected with lenses.

2.1 STRUCTURE AND FUNCTION OF THE EYE

The eye is often referred to as the living camera. If we consider the whole visual system of eye and brain, then a better comparison is with the closed circuit television system. Both receive light as a stimulus, form an optical image on a sensitive screen, transmit information about this image in the form of an electric current, process this electrically and generate an image in (at least) two dimensions (Fig 2.1). In this and later chapters of this book each of these stages is described in some detail. Answer the following question on the basis of what you know now; when you have finished the book you might like to return to the question to see if your answer would change.

QUESTION

2.1 List those features of the structure and functions of the TV system that are shared by the eye. In which is the eye superior and which the TV?

A summary of the structure of the eye is given in Fig 2.2. The following are detailed descriptions of the functions of the more important parts.

The cornea is the transparent bump on the front of the eye which is responsible for about two thirds of the **refraction** in image formation, most of it at the front interface. Variations in its shape account for most of the optical defects of vision that people have; contact lenses which sit on the front of the cornea, are one way of correcting these. The cornea is supplied with

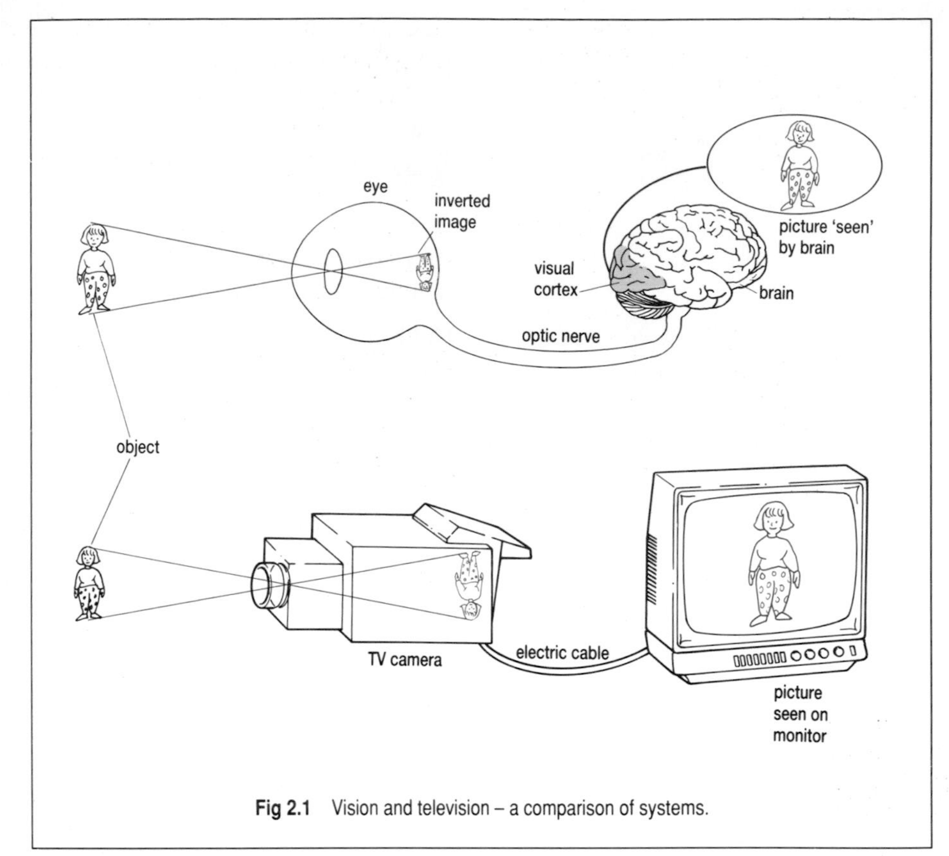

Fig 2.1 Vision and television – a comparison of systems.

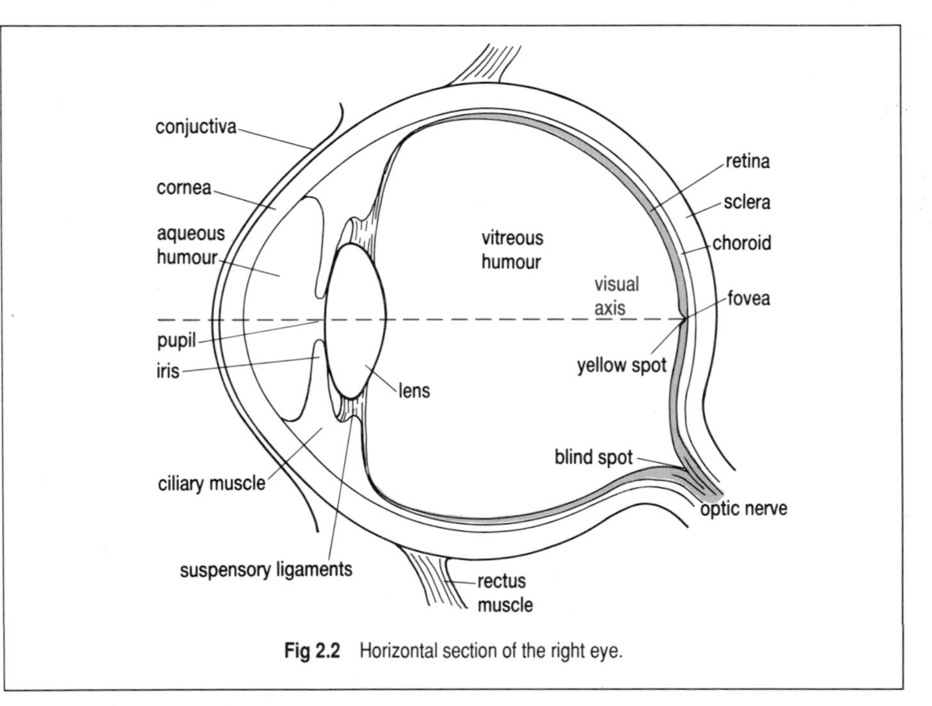

Fig 2.2 Horizontal section of the right eye.

oxygen from the air as blood cells are impervious to light. It is self healing if scratched, but can become opaque as a result of disease or exposure to ionising radiation; corneal transplants are a successful treatment.

The aqueous and vitreous humours are clear liquids which provide nutrients to the parts of the eye, without interfering with the passage of light. They also provide an excess pressure to keep the eyeball firmly in shape.
The lens is biconvex and divides the eye into two chambers, containing the two fluids; refraction occurs at both lens surfaces. It is made of a transparent fibrous material which is added throughout life to produce a layered

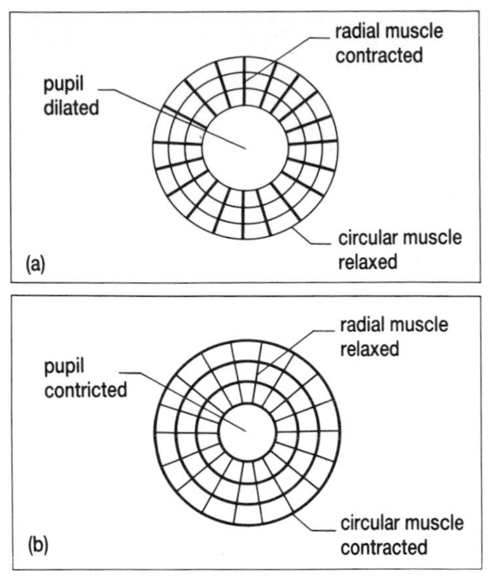

Fig 2.3 Operation of the iris.

structure with a varying refractive index. The cells in the centre die through oxygen starvation, so the lens gets gradually stiffer as we age. It is attached by the suspensory ligaments to the ciliary muscle and these together can extend or compress the lens. The refraction of the lens is changed by this so that it acts as the **fine focus control** of the eye. This automatic process is called **accommodation** and it is necessary to form clear images of objects at different distances. The lens like the cornea can become opaque, and develop cataracts. These can be surgically removed, and bifocal spectacles provided to cope with the loss of accommodation.

The iris is the eye's adjustable **brightness control**, allowing in more or less light through the aperture, the **pupil** of the eye. It consists of radial muscles which contract to dilate the pupil when illumination falls, and circular muscles which contract to constrict the pupil when the light becomes brighter (Fig 2.3). The retina can pass a message direct to the iris – a so called feedback loop, to ensure that this response is very rapid.

The retina is a layered structure on the rear inner surface of the eye which is shown schematically in Fig 2.4. It consists of two types of light-sensitive cells, the **rods and cones**, so called because of the shape of their outer ends. The rods and cones are, perhaps surprisingly, situated at the back of the retina behind the jumble of connecting nerves that carry the electrical signals to the brain. There are about 120 million rods in the retina and about 6 million cones, with the cones predominating around the centre, the optical axis, and the rods being more frequent at the periphery. On the optical axis a **yellow spot** of about 1 mm in diameter can be observed with a depression in the centre called the **fovea**. This is the position of most acute vision since it contains only cones, and there are no nerve fibres obstructing the light as they are displaced radially from it. A white spot can be observed about 10° from this, towards the nose. This is where the nerve

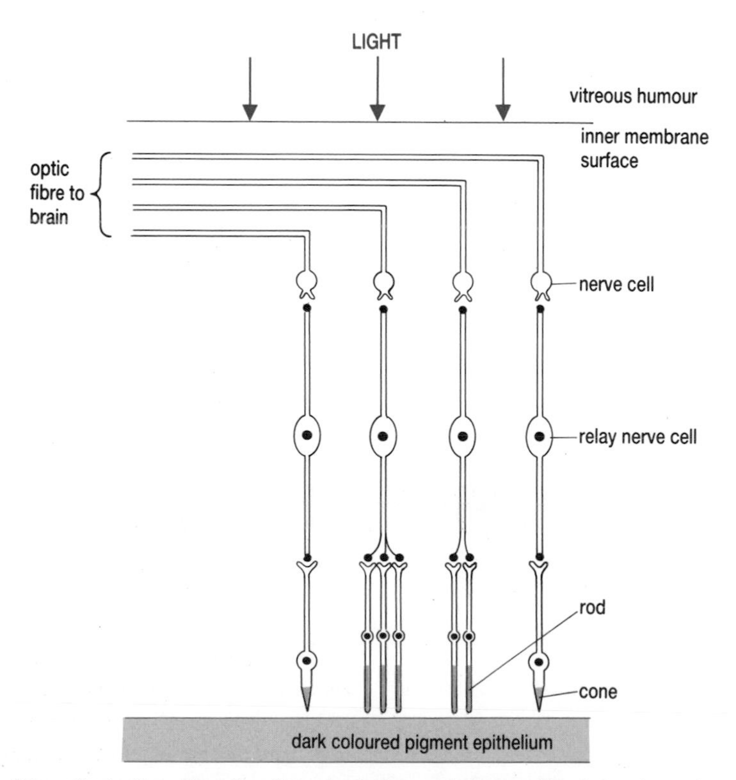

Fig 2.4 Schematic structure of the retina. Several rods share a single nerve fibre, but each cone is connected to a single nerve fibre.

Trauma is the medical expression for a sudden shock, in this case physical. One of the most common causes of detached retina is to be hit in the eye with a squash ball – if you are a player, take care!

fibres collect to form the **optic nerve** to the brain. There are no receptors here so it is called the **blind spot**, and is responsible for a small field of blindness if only one eye is used. The retina can become detached from the outer layers of the eye as a result of trauma or disease. One of the successful uses of the laser is in the reattachment by local heating. This is described in Chapter 7.

Sclera is the external covering of the eye which is made of a tough fibrous tissue to protect the contents. Inside this is the **choroid** or pigment epithelium layer which provides blood to the retina, and is heavily pigmented to prevent reflection of light within the eye, which would cause a blurring of the image.

Rectus muscles hold the eyeball in the skull socket called the **orbit**. There are three pairs of muscles, one pair for each of the three directions of rotation to permit free movement. The aim is to have the **visual axis** through the fovea directed towards the object viewed, for this gives most acute vision. If the object moves towards the eyes, these muscles will automatically pivot the eyes to follow it, so the visual axes **converge**. This process can be defective, so that the images from the two eyes are not exactly superimposed, causing double vision or diplopia. This can occur especially when the eyes have to follow close movement, which may, for example, cause reading difficulties. Temporary diplopia can be caused by excess alcohol.

2.2 THE RESPONSE SYSTEM

Photodetection

A photon is a quantum of light energy, the smallest amount of light that can exist.

Each rod or cone acts as a photocell, releasing electric charge when light falls on it. Experiment has indicated that they can respond to a single photon. Rods and cones are responsible for different parts of our vision.

Rods do not register colour but are more sensitive to light than cones so can operate at lower light intensities. This is as a result of several of them converging on a single nerve fibre (Fig 2.4), which results in rods having poorer visual acuity. Rods are therefore used for **scotopic** (dim light or night time) vision.

Cones differentiate colour as they contain three visual pigments. They produce more acute vision because they do not share nerve fibres, but they need higher levels of light than rods. Cones are therefore used for **photopic** (bright or daylight) vision.

Rods contain the light sensitive pigment, **rhodopsin** (often called **visual purple** because of its appearance). This is made of a protein and a derivative of vitamin A called **retinene**, which is changed into a different form by light, in a process called **bleaching**. This reaction results in the generation of an electrical potential, and this passes down the nerve to the brain. The bleached rhodopsin is rapidly changed back into its light sensitive form by vitamin A supplied in the bloodstream by the choroid, so that it can again respond to light. Shortage of vitamin A (retinol) makes it difficult to see in dim light. It can also cause the cornea to become thick and dry. Good sources of the vitamin are fish liver oil (ugh!) and carrots.

Eating lots of carrots enables you to see in the dark. This was the claim made by the British during the Second World War to explain why they were able to detect German night-flying planes. This exaggeration of the truth was to conceal from the enemy the successful development of radar detectors.

Sensitivity

Full sunshine is about 10^6 times more intense than full moonlight.

The maximum sensitivity of the eye has been estimated as a response to a single photon. What is also remarkable is the eye's ability to distinguish differences in intensity, over an intensity range of about 10^9.

You will notice this if you try to take photographs using a camera which needs you to estimate the exposure, you will inevitably underestimate the

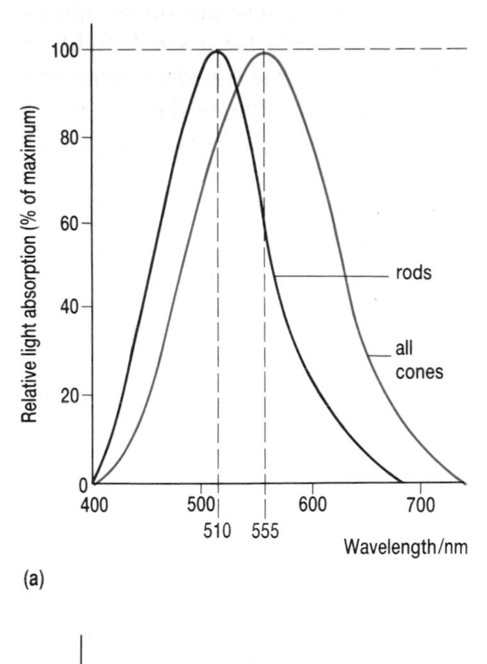

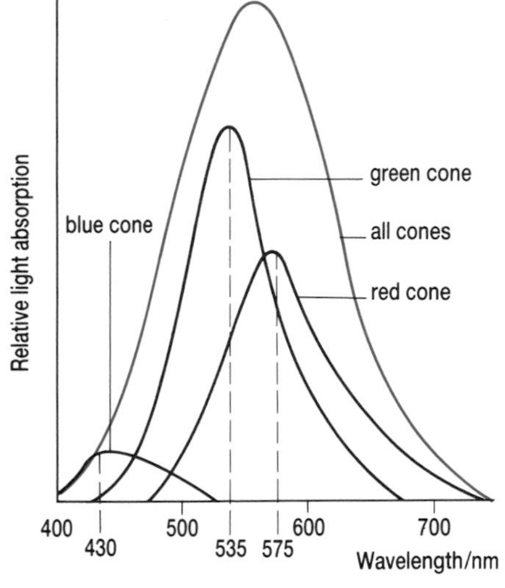

Fig 2.5 Receptor absorption **(a)** spectral absorption of rods and cones **(b)** contributions to total cone absorption.

exposure time needed in dim light and vice versa. The eye has this automatic control by a combination of the iris (total illumination), and dark adaptation.

Dark adaptation is the mechanism by which the eye increases its sensitivity by up to 10^6 as intensity falls. There are two parts to this process, the photochemical and the neural. Light on the retina causes the cells to be depleted of their photosensitive pigments in the process of response. In the absence of light, the concentration of the photosensitive chemicals in the rods and cones gradually builds up over a period of about 30 minutes, which increases the sensitivity to any dim light which is received. This occurs first for the cones and then for the rods. The second effect is the collection by the rods of energy from a greater area, by more sharing of nerve fibres. This means that smaller intensities produce a signal, at the cost of the ability to distinguish fine detail (a loss of visual acuity – see below). There is also a collection of light over a longer time. Thus the playing of ball games in dim light is difficult because of loss of accurate position and time information, though the ball can still be seen.

Spectral response

The minimum or threshold intensity needed to see a flash of light is very wavelength dependent. The cornea is opaque to wavelengths shorter than 300 nm, and the lens to wavelengths below 380 nm. Rods and cones do not detect wavelengths above about 700 nm.

Although the rods do not give any colour information, they are most sensitive to light in the green part of the spectrum, with wavelengths of about 510 nm – see Fig 2.5(a).

Experimental evidence has now confirmed the **trichromatic** theory of colour vision. This is that there are three types of cones containing three different pigments. These are:

- a 'red' cone containing the red sensitive pigment **erythrolabe**,
- a 'green' cone containing the green sensitive pigment **chlorolabe**,
- a 'blue' cone containing the blue sensitive pigment **cyanolabe**.

In responding to light, each is believed to undergo a bleaching process similar to that which occurs in rods. As Fig 2.5(b) shows, the green cones are most sensitive and the blue least, giving an overall maximum sensitivity in the yellow wavelength range.

The perception of colour is a result of the response of certain cones (Fig 2.5(b)). Monochromatic light of wavelength 430 nm will stimulate only the blue cones and so be seen as blue, whereas light of wavelength 550 nm will stimulate green and red cones and so appear yellow. Light with a particular wavelength will be detected by up to three types of cone in different proportions to give the specific spectral colours. This is strictly only one attribute of colour, the **hue**. The others are **brightness**, which depends on the intensity of the light and **saturation**. This is a measure of the purity of a colour, that is how much of the intensity is from the hue, or particular wavelength, and how much from a broad spectrum of a mixing colour such as white or grey. Thus a red hue may be reduced in saturation by mixing it with white to produce pink. The actual perception of colour is also affected by the material nature of the object viewed and its surroundings (Fig 2.6).

Colour blindness, or more usually, defects in colour vision are caused by the shortage of a particular kind of cone. The failure to distinguish between red and green is one of the more common ones, especially in males. Certain defects are not satisfactorily explained by the trichromatic theory, nor is the occurrence of negative afterimages. This is the perception of a complementary colour following the removal of a colour after viewing it for some time.

There are clearly complexities in the perception of colour that will not be fully understood until we have a much greater knowledge of the brain.

Use the information of this section to answer the following questions.

QUESTIONS

2.2 Why do you get sharper vision at night by looking out of the side of your eye?

2.3 Explain light adaptation, the opposite effect to dark adaptation. How does the time taken compare with dark adaptation?

2.4 Explain why an object emitting equal intensities of light of wavelengths 535 nm and 575 nm appears yellow.

2.5 Explain why yellow is a good colour to use for tennis balls or pedestrian warning clothing.

2.6 Explain the differences in appearance of objects when viewed in
(a) bright sunlight out of doors,
(b) dull natural light in a heavily furnished room,
(c) fluorescent light illuminated, white-painted office.

2.7 If you view a well-illuminated red object against a white background for about one minute, you may see a cyan (blue-green) after image when it is removed. Can you think of a possible explanation of this? How could you test the explanation ?

Fig 2.6 How the surroundings affect perception – all three figures have a centre of the same colour.

Resolution

The ability of the eye to discern detail is a very important factor in vision. **Resolving power** is defined as the minimum separation angle θ (in radians) of two objects that can be separately distinguished. It can be shown that $\theta \approx \sin\theta = \lambda/D$, where D is the diameter of the eye lens aperture.

Visual acuity is defined as the inverse of the angle θ (in minutes) by which the two nearest points are separated, that is

$$\text{visual acuity} = 1/\theta \text{ per minute}$$

This separation will depend on the structure of the retina. If a distinction between two points of light is to be made, then there must be at least one unstimulated receptor cell between those stimulated. Hence the minimum spacing will be twice the diameter of the receptor cell. The cells are most closely packed at the fovea and the cones have their own nerve fibre. Their diameter is about 1.5 μm, giving a minimum spacing of 3 μm, a visual angle of half a minute, giving a visual acuity of 2 per minute (Fig 2.7). This is in fact similar to the limit of resolution which is set by the diffraction of light as it passes into the eye (see the next section for details).

Elsewhere in the retina the cones are less frequent and the rods share nerve fibres, so a single signal is produced by an increased area of receptors. At the periphery of the retina about 600 rods are connected to one nerve, giving an acuity of about 0.1 per minute.

Angle in radians = $\dfrac{\text{arc length}}{\text{radius}}$

Response to illumination changes over time

When viewing an object under steady illumination, producing a fixed image on the retina, the eye executes a continuous **scanning** movement. This ensures that there are always some new nerves being stimulated to produce a signal, and avoids the fading of the image as the receptors cease to produce the chemical change necessary to generate the electrical potential.

By contrast, if the illumination changes rapidly, the excitation of the receptors takes a finite time to respond. This delay is called **persistence of vision** because you will see the light for longer than it was actually reaching

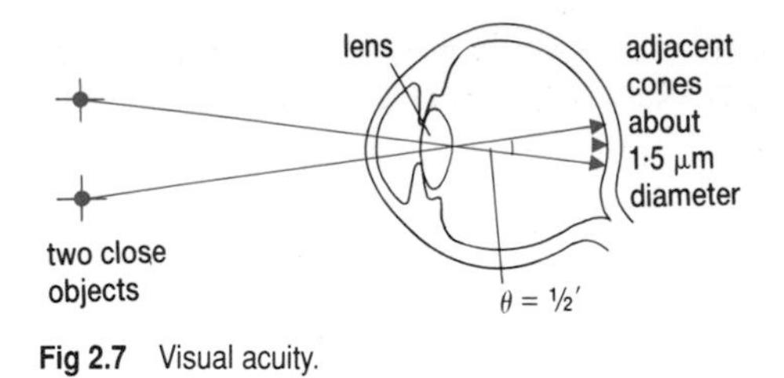

Fig 2.7 Visual acuity.

the eye. The delay period depends on the intensity of the light. Bright flashes will 'flicker fuse' into a continuous brightness at a frequency of about 50 Hz, in dim light this will occur at a flicker rate of only 5 Hz, because the rods respond less quickly to the changes than the cones. This ability (or disability), of the eye is the basis of movies, in which pictures are flashed on the screen at 24 frames a second, so that the gaps in the action are filled in by the eye to produce a smooth motion. Television screens use a similar principle in the scanning of the screen by the electron beam. This is described in Chapter 4.

2.3 THE OPTICAL SYSTEM

The ability of the eye to form images is as a result of refraction at the curved interfaces between the various media that the light passes through on its way to the retina. The curvature can be described as **refracting power.**

Refracting power

The refractive power of a surface is defined as:

power (in dioptres, D) = 1/ focal length (in metres),

$$P = 1/f \qquad (1)$$

Powers of a series of surfaces can be added algebraically, which makes the unit more useful than focal length. The amount of refraction is dependent on the change in the relative refractive index at the interface, and as Fig 2.8 shows this is greatest at the front of cornea. There is much less refraction (and it is in the opposite sense) at the rear, because the interface is with the aqueous humour, which has a similar refractive index. The lens produces less refraction than the cornea, but because its shape is variable it acts as

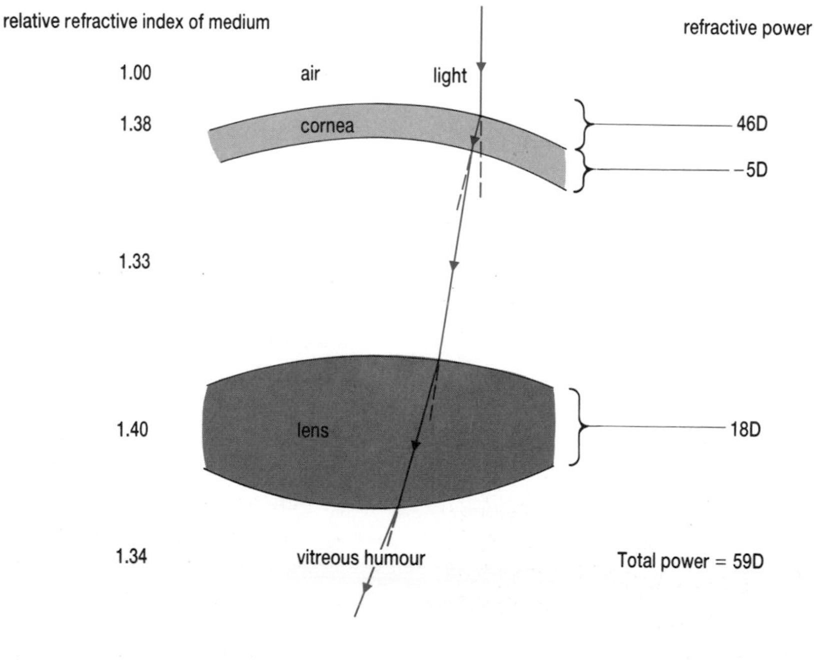

Fig 2.8 Refraction in the eye.

the adjustable focus. When the lens is at its flattest it has a power of about 18 D. This is for producing an image on the retina of distant objects. This is called the **far point**, and for the normal eye it is at infinity. To focus on closer objects the lens needs to increase its power, to converge the beam. In young people the maximum power is about 29 D, to produce a **near point**, the nearest position of clear vision, of less than 0.25 m. Measure your own near point. The range of focus available is the eye's **accommodation** (Fig 2.9). The power of accommodation can be calculated from the difference in power, in this case,

$$P_n - P_f = 29 - 18 = 11 \text{ D}.$$

In young children the power of accommodation may be as high as 14 D, in elderly people it can fall to zero.

QUESTION

2.8 Why is it impossible to focus your eyes under water? (*Hint* – refer to Fig 2.8)

2.9 Why do people lose the power of accommodation during their lives?

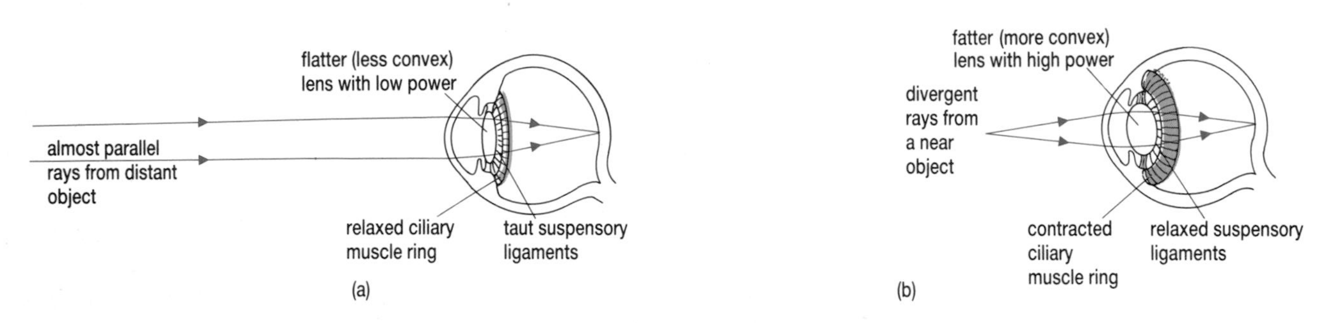

Fig 2.9 Accommodation – range of focus **(a)** distant object **(b)** near object.

Depth of field

The distance from lens to retina is the image distance and is fixed by the dimensions of the eye. For any particular lens focal length, objects at only one distance will form a sharp image. In practice there is a range of object distances which can produce an acceptably clear image. This range of distance is called the **depth of field**. It is much greater for distant objects than for near objects because the difference in divergence of the light is much greater for nearer objects (Fig 2.10). The lens has to be more powerful to produce the image. You will notice this if you use a camera with a powerful lens or look through a telescope. The shorter the focal length, the smaller is the depth of field.

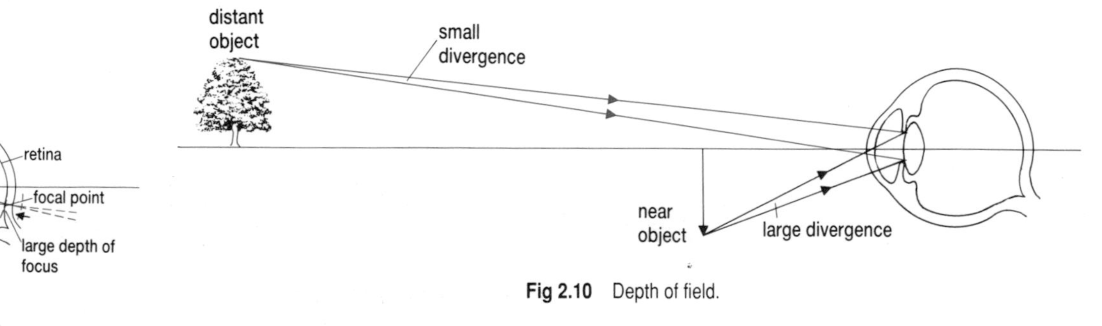

Fig 2.10 Depth of field.

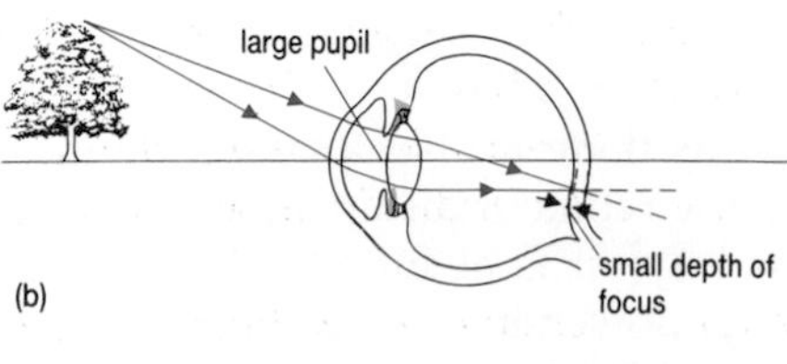

Fig 2.11 Depth of focus **(a)** small aperture **(b)** large aperture.

Depth of focus

The divergence of rays from an object can be reduced by reducing the size of the pupil (Fig 2.11). This increases the range of image distances that can be accepted by the retina as a focused image. This is called the **depth of focus**. You will notice this using a camera if you change the f number or aperture size. There is a depth of focus range shown for each f number which increases as the f number increases and the aperture reduces.

INVESTIGATION

Monocular and binocular vision

Here are some simple activities to examine the differences in the images from your two eyes.

1. Look at stationary objects whilst moving your head from side-to-side. How does this effect, called parallax, help in the judgement of distance?
2. Close one eye and explore your immediate surroundings, by reaching to grasp objects or by moving amongst them. Do you have a preferred eye? Do you find any difficulty in this task?
3. Repeat this with both eyes open. Can you explain why there was less difference between the two experiences than might have been expected?
4. With one eye open, try bringing together two fingers from different hands, or two pencils. How does this compare with your previous actions, why?
5. Roll a piece of paper into a tube and look with one eye through it at a plain, well-illuminated wall. Hold your other hand alongside and look at it with your other eye (Fig 2.12). Why can you see through your hand?!

Fig 2.12 Student carrying out investigation.

Depth perception

Perception is a complex process using not only the image on the retina, but also the brain's memory to interpret a scene. Familiarity with the scene or object viewed, and with the effects of perspective can help avoid the limitations of monocular sight. You will notice these more in new situations where there are fewer clues for the brain.

In binocular vision, each eye has a slightly different view, because of its position. The brain is able to compare the two images to perceive the third dimension, depth. This stereoscopic effect is used in photography, and can be

Fig 2.13 Stereoscopic scanning electron images of a cat flea. (To view, look beyond the picture to relax the eyes, then focus on the right-hand mouth part of both pictures.)

a useful diagnostic technique in internal medical examinations, for example using X-rays. In each of these examples, pictures are taken from two adjacent positions as two eyes would view the scene. The pictures are then placed in a mount so that each eye sees one picture, and the 3-D construction is carried out by the brain. It is also used routinely in the scanning electron microscope to give a fuller picture of what is being examined (Fig 2.13).

As the distance from the eye to the object increases, the visual angle between objects at different distances will decrease. The smallest that can be detected is, as was discussed above, about a half minute. This means that stereoscopic vision vanishes at about 60 m, and depth perception beyond this is due to the brain's interpretation, for which one eye would be satisfactory.

Other advantages of binocular vision are the increased field of view – over 180° horizontally, avoidance of a blind spot and a reduction in the impairment of vision if one eye is damaged.

Optical acuity

In addition to the perceptual limitations of the retina discussed above, the sharpness of the image is limited by optical factors.

Chromatic aberration is the blurring of the image due to light of different wavelength (colour) being brought to a focus in slightly different positions because the refractive index depends on wavelength.

Spherical aberration is the result of a wide beam of light entering the eye and being brought to a focus in different positions, not all of which will be on the retina. This will clearly be more serious when the pupil is larger.

Diffraction occurs as the light passes through the pupil, being more pronounced as the aperture becomes smaller.

The result is that the image of a point source of light is a spot with a maximum intensity at the centre and a blurred edge. The optimum size for the pupil is about 5 mm diameter, a compromise between the competing requirements of aberration and diffraction. This results in a spot of about 10μm diameter. The images of two object points may then overlap.

Rayleigh's criterion states that these two images are just resolved as separate points if the maximum of one intensity curve coincides with the first minimum of the other (Fig 2.14(a)).

The width of the spot, and therefore the resolution will depend on the wavelength of the light. It can be shown that the resolving power $\theta = 1.22\lambda/D$ where D is the diameter of a circular aperture. This occurs under optimum conditions when the centres of the two spots are 2 μm apart on the retina. This corresponds to the best resolution possible from the arrangement of the receptors in the retina (the cones in the fovea) so the two processes are perfectly matched.

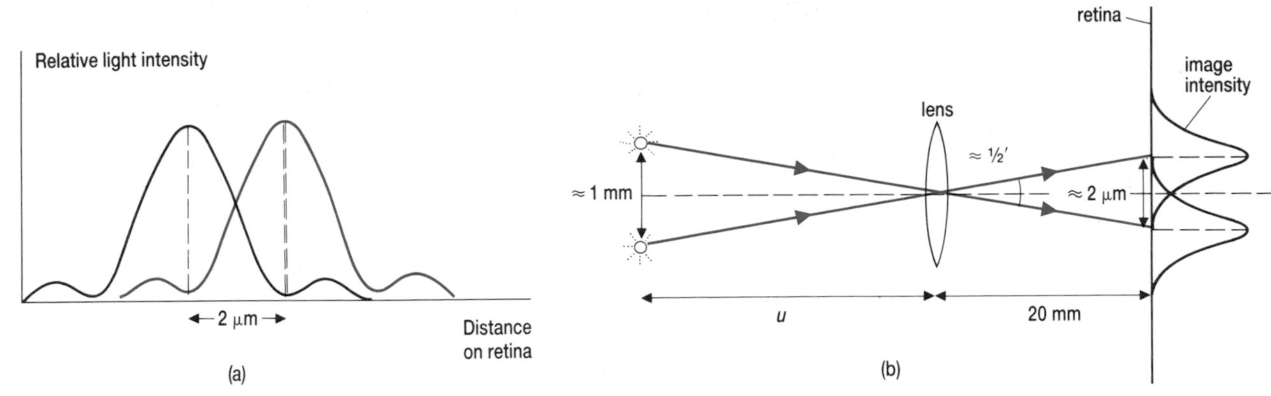

Fig 2.14 Diffraction in the eye **(a)** the Rayleigh criterion **(b)** the eye's resolution.

QUESTIONS

2.10 By considering similar triangles in Fig 2.14(b) find the maximum distance that this analysis predicts the eye could distinguish two points placed 1 mm apart. How does this compare with what you can actually see?

2.4 OPTICAL DEFECTS AND THEIR CORRECTION

There are four common types of defect in image formation by the eye, which can be corrected by spectacles or contact lenses. These are the failure to produce a sharp image of a distant object (myopia), or a near object (hypermetropia), failure to accommodate (presbyopia), and failure to focus in certain planes (astigmatism).

Myopia (short sight)

Near objects are clearly seen but the image of a distant object is formed in front of the retina, causing blurring of distant vision. The far point has moved from infinity to somewhere quite close to the eye so the range of vision has moved in towards the eye. It is a problem which often arises at adolescence, when the body grows quickly, so the eyeball is too large for the lens. In effect the power of the eye is too great and the problem is corrected with a diverging lens with a negative power (Fig 2.15(a)). This is calculated by finding the power of the eye at its far point and then selecting a lens with a power to move the far point to infinity. The thin lens formula

$$\frac{1}{f} = \frac{1}{u} + \frac{1}{v}$$

is used, and combined with the expression for power P, equation 1 hence

$$P = \frac{1}{f} = \frac{1}{u} + \frac{1}{d} \qquad (2)$$

where d is the distance from the eye lens to the retina, ~0.02 m.

(a) Consider a man whose far point is 0.5 m. The power of his eye when fully relaxed is:

$$P_f = \frac{1}{0.5} + \frac{1}{0.02} = 52 \text{ dioptres}$$

To have his far point at infinity, the man needs the power to be:

$$P_i = \frac{1}{\infty} + \frac{1}{0.02} = 50 \text{ dioptres}$$

When wearing glasses, the effective power is the algebraic sum of the powers

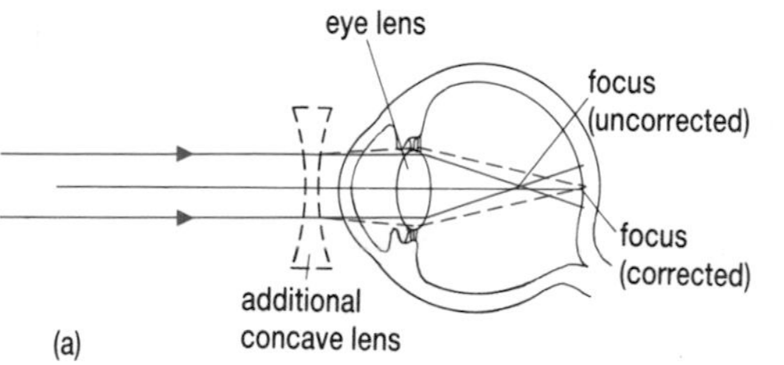

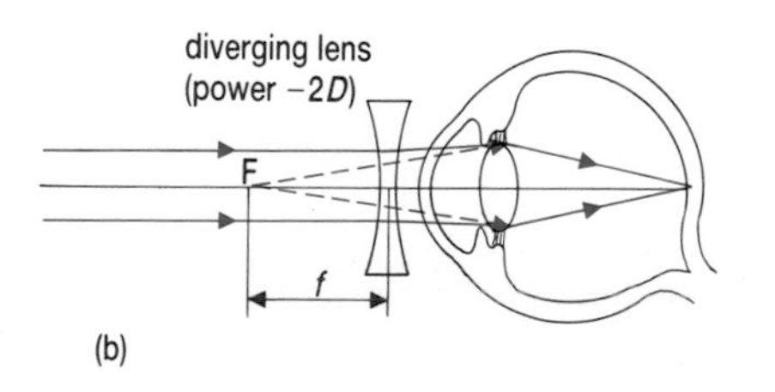

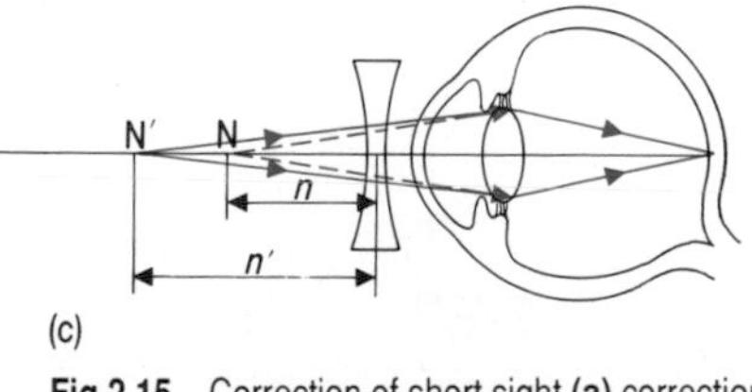

Fig 2.15 Correction of short sight **(a)** correction by concave lens **(b)** effect on far point (F) **(c)** effect on near point (N).

of his eyes and his lenses, assuming the lenses are close to the eyes so the power of the lenses needs to be

$$50 - 52 = -2 \text{ dioptres,}$$

so that when his eyes are relaxed he will have a net power of 50 dioptres, and see distant objects clearly (Fig 2.15(b)).

(b) A shortsighted person will typically have the near point closer than for the normal eye. Suppose the man has an accommodation power of 5 dioptres.

$$P_n - P_f = 5$$

So his power at the near point P_n is 52 + 5 = 57 dioptres, and he will focus at a point N, Fig 2.15(c) given by:

$$P_n = \frac{1}{n} + \frac{1}{d} \qquad \text{so } 57 = \frac{1}{n} + \frac{1}{0.02\text{m}} = \frac{1}{n} + 50 \text{ dioptres}$$

Therefore the near point is at 0.14 m from the eye.

(c) Wearing the lenses will move the near point further away to N′. The power is now that of the eye at maximum power combined with the lens,

$$P_{n'} = P_n + (-2) = 57 - 2 = 55 \text{ dioptres}$$

so the new near point is given by:

$$P_{n'} = \frac{1}{n'} + \frac{1}{0.02} \qquad \text{that is, } 55 = \frac{1}{n'} + 50 \text{ dioptres,}$$

Which gives the new near point 0.2 m from the eye. The man may wish to remove his spectacles for close viewing.

QUESTION

2.11 **(a)** What is the focal length of the corrective lens used in the above example?

(b) (i) What power of lens is required to correct the short sight of a woman whose far point is 0.2 m?

(ii) What is her near point without the spectacles?

(iii) What is her near point with the spectacles?

Hypermetropia (long sight)

This is the opposite of the previous defect. Distant objects can be seen clearly but the eye is insufficiently powerful to produce an image of nearer objects on the retina (Fig 2.16(a)). The range of vision has moved out from the eye. It is a common problem in infancy when the eyeball is too small for the optical system. It is corrected by an additional converging lens, with a power which will bring the near point to that of the normal eye, 0.25 m (Fig 2.16(b)).

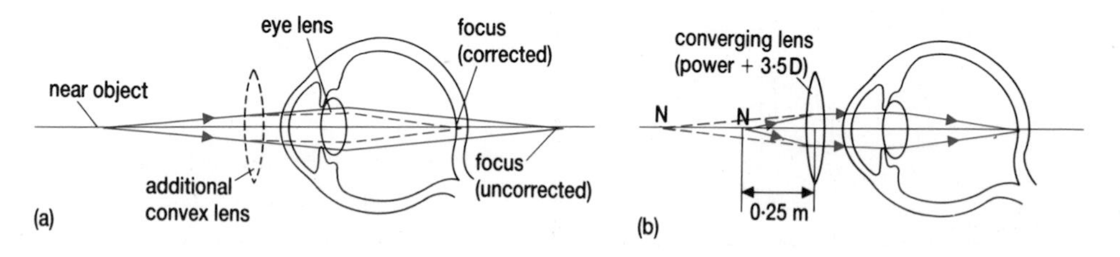

Fig 2.16 Correction of long sight **(a)** correction by convex lens **(b)** effect on near point (N).

A radical approach to optical defects has been pioneered in the USSR and is becoming available in Western countries. This is to change the shape of the retina surgically, using the laser. Other examples of laser surgery are given in Chapter 7.

Consider a woman whose near point is 2 m from her eyes. Using equation 2 the power of her eyes is:

$$P_n = \frac{1}{2} + \frac{1}{0.02} = 50.5 \text{ dioptres}$$

To focus at 0.25 m she will need a power

$$P_n' = \frac{1}{0.25} + \frac{1}{0.02} = 54 \text{ dioptres}$$

so the lens must provide a power of +3.5 dioptres, to give her a near point of 0.25 m.

Presbyopia ('old sight')

With age, the lens loses its flexibility and power of accommodation. There is also a weakening of the ciliary muscles that control the lens. Older people may require a lens for both distant and close viewing. This can be provided as bifocals in which the upper part of the lens is used for distant vision and the lower part for near vision. For example an elderly myopic person may require a strongly diverging lens in the upper part and a weaker one in the lower part.

Astigmatism

This is usually caused by the cornea not being spherically curved, so that it has different curvatures in different directions. Images are seen in sharper focus in one direction, e.g. vertical, than in others. It is corrected by lenses that have a cylindrical curvature in the correct orientation to compensate for the cornea. This is often combined with a curvature to correct short or long sight.

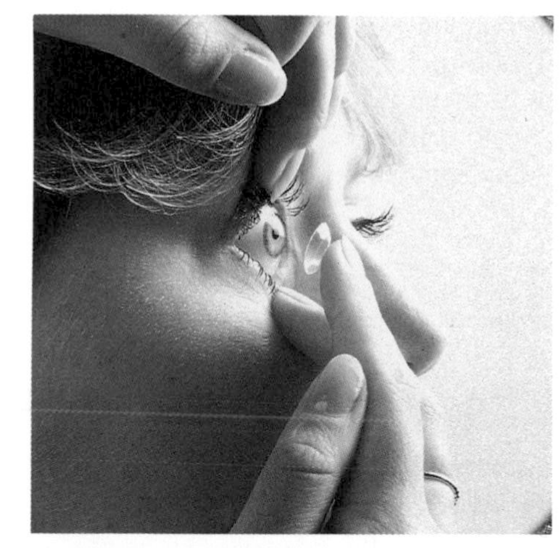

Fig 2.17 A contact lens being inserted.

Contact lenses

These are lenses placed on the surface of the eye to perform the same function as spectacle lenses. There are two main kinds, hard and soft. Hard lenses are made of rigid plastic, about 1mm thick and 1 cm in diameter. They are separated from the cornea by a layer of tears. This enables the lens to take up an orientation which can compensate for astigmatism. Soft lenses are rather larger and more comfortable to wear. They conform to the shape of the cornea so they do not correct for astigmatism. They are also more easily damaged. A newer, more expensive soft lens is the toric type which can maintain its shape and so correct for astigmatism, though at the cost of some loss of comfort, (Fig 2.17).

SUMMARY ASSIGNMENTS

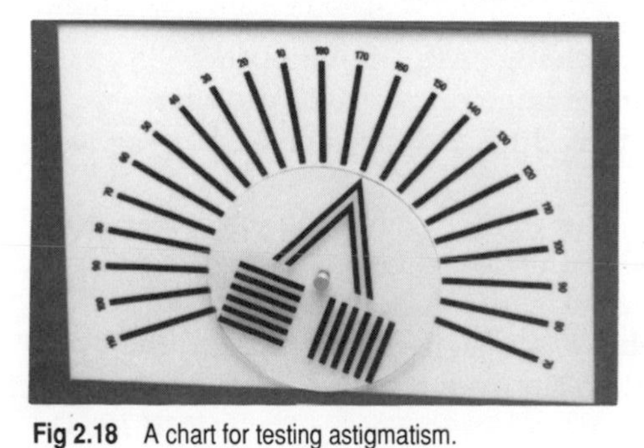

Fig 2.18 A chart for testing astigmatism.

2.12 Tabulate the differences between rods and cones in terms of their structures, functions and contributions to vision.

2.13 Diagnose the following defects of vision and suggest, where possible, suitable corrections, including the power of the lens and any advantage of a contact or spectacle type:

(a) Far point at infinity, near point at 1 m, power of accommodation 10 D.

(b) Far point at 5 m, near point at 0.5 m, accommodation 1.8 D.

(c) When viewing the chart shown in Fig 2.18, the vertical and near vertical lines are clearer than the horizontal lines.

(d) Normal near and far points with each eye, double vision for objects closer than 1 m.

2.14 Explain the following observations (you might try them out first!)

(a) When the red filament of a lamp is viewed through a blue filter, two images, red and blue, are produced side by side.

(b) Two objects about 10 cm apart are viewed by the right eye looking directly at the left object. When the eye is about 0.5 m away, the right hand object disappears from view.

(c) A printed page can be read clearly when it is placed nearer than the near point, if it is viewed through a small hole. This clarity is lost if the hole has a diameter of less than about 1 mm.

2.15 (a) Compare what is seen by an observer, with normal eyesight, when a coloured object is illuminated by white light of (i) high intensity and (ii) low intensity. Give an explanation for your answer in terms of the behaviour of the eye. Sketch a graph showing the spectral response of the eye.

(b) Two neighbouring, independent point sources of light are just resolved visually under optimum viewing conditions when they subtend 0.3 milliradians at the eye. If the sources are 0.5 m from the eye calculate their separation. Explain, in terms of the structure of the retina, how the two retinal images must be positioned to be seen as separate.

(JMB 1986 part)

2.16 (a) With the aid of a diagram give a brief account of the optical system of the eye. Explain how images of objects at different distances are focused on the retina, and how the amount of light entering the eye is controlled.

(b) Distinguish between *long sight* and *short sight*.

(c) Person A cannot obtain focused images of objects further away than 0.5 m. Person B cannot obtain focused images of objects nearer than 0.5 m. For each person, identify the defect and suggest the type of lens which would be required. Calculate the strength of each lens. Assume that the near point is 250 mm from the eye.

(Cambridge specimen)

2.17 (a) Explain what is meant by *accommodation* of the eye and describe how it is achieved.

(b) Opticians specify lenses by their power P and when an observer with normal eyesight looks at an object at infinity the eye lens has a certain value of power. The table shows the change ΔP *from that power* which is required by the eye in order to focus clearly on objects at different distances d from the eye.

ΔP /dioptre	0.1	1.0	9.8	19.8
d /cm	1000	100	10	5

By plotting $\log_{10} \Delta P$ against $\log_{10} d$ find the closest distance at which **each** of the following can see an object clearly with the unaided eye:

(i) a person whose maximum change ΔP_{max} is 4 dioptre.

(ii) a person whose maximum change ΔP_{max} is 0.65 dioptre.

(c) To read a book held at normal reading distance one of the people in (a) needs spectacles.

(i) State which person it is, the defect which the spectacle lens will correct, and the type of thin lens needed, providing a ray diagram to show how the correction is achieved.

(ii) Calculate the power of the lens which, placed close to the eye, will produce a near point 25 cm from the lens.

(JMB 1984)

Further reading

Eden, J. *The eye book*, Penguin 1981

A simply written book describing the eye's function, how you should look after your eyes, and what can go wrong. There are also chapters on eye examination and treatment.

Mason, P. *Light fantastic*, Pelican 1982

A stimulating and well written book which tells the story of 'a science that has transformed human life'. There are chapters on the eye, colour and X-rays. The book handles well the social and humanitarian issues arising from the application of science.

Chapter 3

THE EAR AND HEARING

Which sense is most valuable to you, seeing or hearing? Many people who are totally deaf claim that they would rather be blind. What they have lost is our basic means of communication; in a silent world it is difficult to learn to speak and language itself is at risk.

The ear is a remarkable organ. It has a sensitivity to sound vibrations with a loudness range of more than 10^{12} (1000 times greater than the range of light intensities that the eye can see). It has a frequency range which varies by a factor of 10^3, compared to the eye's range of a factor 2. For a mechanical analogy this might be compared with a weighing machine which could weigh just as accurately a flea or an elephant, whether they were jumping on and off the scales 20 or 20 000 times a second! Amongst its other impressive functions are the selectivity, which enables a person to hear a quiet word in a noisy disco; the recording of the complexity of sound so that the brain can interpret a false note in a musical performance; and the instant feedback that enables a person to monitor their own speech. In this chapter you will find out how the ear works, and what can impair its function.

LEARNING OBJECTIVES

After studying this chapter you should be able to:

1. give an outline description of the structure of the ear and how it transmits and detects vibration;
2. describe how the ear responds to the characteristic properties of sound waves, including frequency, intensity and loudness;
3. describe the common defects of hearing and their possible remedies;
4. explain the deterioration of hearing as a result of age and exposure to excessive noise.

3.1 STRUCTURE AND FUNCTION OF THE EAR

The ear is designed to convert weak mechanical vibrations of air into electrical pulses that can be sent to the brain. It consists of a mechanical collection and amplification system, (in the outer and middle ear) and transducers to produce electrical potentials in nerves (in the inner ear). The auditory nerves lead to the auditory cortex, the part of the brain which interprets the signals. We will now follow the sequence of events involved in hearing, in the structures shown in Fig 3.1.

The outer ear

The visible part of the ear, the **auricle** or **pinna**, is not strictly part of the outer ear as it plays very little part in the process. In some animals it has a role in collecting the sound vibrations, but in humans it can be removed without noticeable loss of hearing. This is because of the small size. Hearing can be improved by cupping your hands behind your ears, or by use of an ear trumpet, to make up for this deficiency.

The outer ear consists of the **external auditory canal** or **meatus**, which leads from the auricle to the ear drum or **tympanic membrane**. It is about

One of the first medical physicists to study the ear and hearing was Hermann von Helmholtz (1821–1894). His theory was developed by Georg Von Bekesey (1900–1970), a communications engineer who received the Nobel prize in 1961 for his contribution. There is still much that is unknown, particularly about the later stages of the process of hearing.

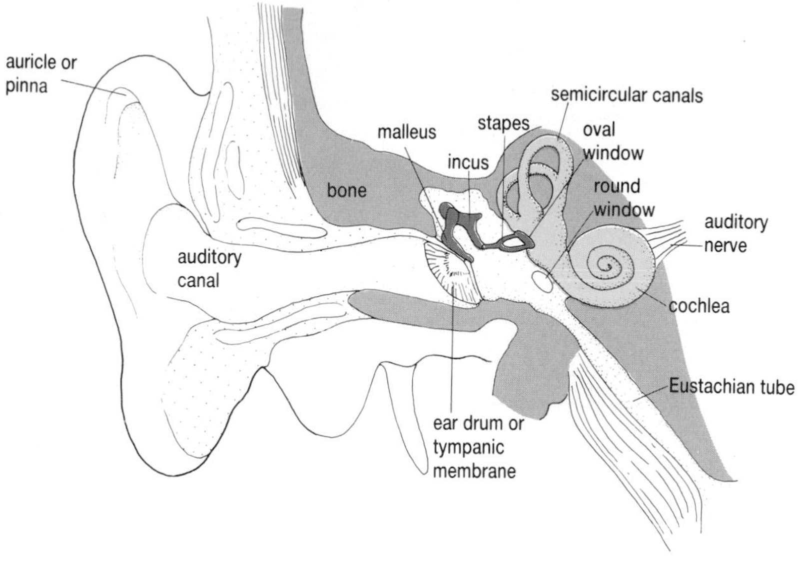

Fig 3.1 Vertical section through the ear.

2.4 cm long and about 7 mm in diameter. It acts as an organ pipe closed at one end, so that the air in it can vibrate. This vibration is passed to the tympanic membrane, which is paper thin (~0.1 mm), with an area of about 65 mm^2. This behaves, as its name indicates, like a drum, though the size of the vibrations can be as small as 10^{-11} m. Although its fibrous structure is relatively strong, it can be ruptured by loud sounds – or pencil poking!

The middle ear

There is a mechanical linkage of three small bones, **the ossicles**, between the ear drum and a smaller membrane, called the **oval window**. The bones are called the **malleus, incus** and **stapes**, or, reflecting their shapes, the hammer, anvil, and stirrup. They act as combined lever and pistons in the air-filled cavity between the membranes. The lever has a mechanical advantage of about 1.5 and the ratio of the areas of the eardrum and the oval window is about 15 to 1. This results in the pressure applied to the oval window being increased by a factor of more than 20, (Fig 3.2).

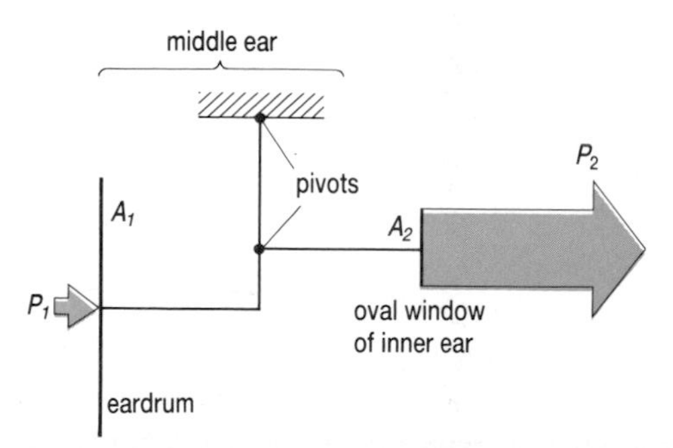

Fig 3.2 Schematic diagram of the lever and piston action of the ear. A_1/A_2 is about 15, P_1/P_2 about 20.

When sound is transmitted into a different medium a proportion of the sound is reflected, depending on a property of the medium called the **impedance**. If the impedance is similar, or matched then this avoids the loss of energy by reflection. The middle ear is designed to produce good matching.

The ossicles also help to protect the ear from damage by loud sounds, by a movement sideways which is not transmitted to the inner ear. This is too slow a reaction for a sudden increase in intensity. Another type of protection is given by the **Eustachian tube**, which connects the middle ear to the mouth. This enables the pressures on each side of the eardrum to be kept equal. Differences in pressure impair the hearing and can cause pain. The tube is usually closed and is opened by swallowing, yawning or chewing. You will have experienced the relief this brings when you are subjected to a sudden change in air pressure, perhaps in a plane, lift or car in hilly country.

The inner ear

This is where things get rather complicated!

The inner ear is a cavity deep within the skull which contains two organs. The **semicircular canals** are narrow, fluid-filled tubes which are not for hearing but for the sense of balance and body movement. The **cochlea** is also a fluid-filled tube of three chambers, coiled and about the size of the little finger nail (Fig 3.3(a)). If it were uncoiled the tube would be about 3 cm long (Fig 3.3(b)). The pressure wave from the oval window passes through fluid called **perilymph**, down the spiral in the **vestibular chamber** to the end and returns via the **tympanic chamber**. The pressure variation is absorbed at the return end by the round window membrane. Between these two chambers is the **cochlear duct** or **median chamber**, filled with **endolymph** fluid. This contains the sensors in which pressure waves generate electrical signals.

This is a complex process which is not yet fully understood, involving the hair-like structure of the **organ of Corti** (Fig 3.3(c)) which is attached to two membranes running the length of the spiral. The **basilar membrane** consists of fibres which are short (~0.04 mm) and tense at the base near the oval window, increasing in size (up to 0.5 mm) and becoming slacker at the narrow end. As the pressure wave travels along the chambers from the oval to the round windows it sets the basilar membrane into vibration, if the frequency is greater than 20 Hz. This movement is transmitted to the hair cells which brush against the **tectorial membrane**. The liquids on either side of these membranes, the **perilymph** and the **endolymph** have different ionic compositions which results in a potential difference across the membrane of about 150 mV. The shearing action of the hairs and their interaction with this membrane changes this electrical potential and causes a current to flow in the nerve fibres at the base of the hair cells, which is transmitted as information to the brain. A more detailed account of the generation of potentials in nerve fibres is given in Chapter 5, if you wish to pursue the story further.

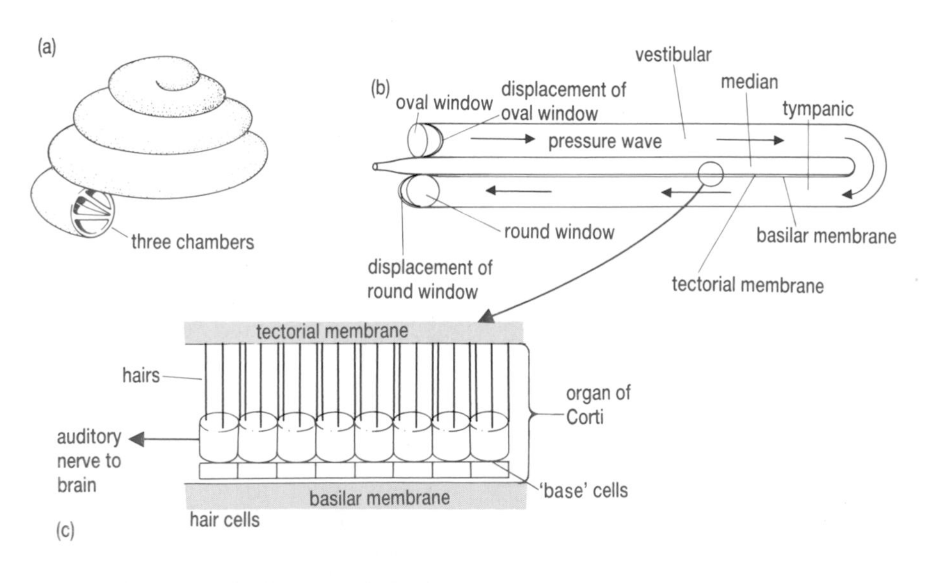

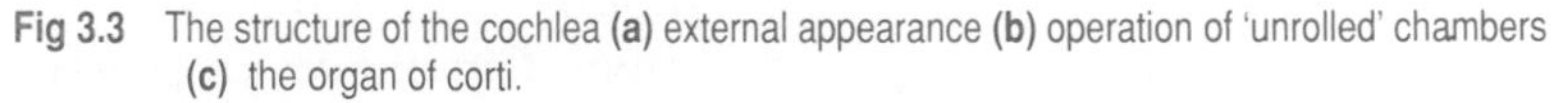

Fig 3.3 The structure of the cochlea **(a)** external appearance **(b)** operation of 'unrolled' chambers **(c)** the organ of corti.

QUESTIONS

3.1 Explain why you may find your hearing is impaired when you have a cold. It can also add to your discomfort to fly in an aeroplane.

3.2 It often helps in remembering a complex and detailed account like the one just given, if you can devise a simple summary, in the form of a diagram, showing the stages in the process without all the structural details. Try this now.

3.2 THE TRANSMISSION AND MEASUREMENT OF SOUND

Sound is a longitudinal pressure wave which is propagated by the oscillations of particles of the medium through which it is transmitted. The ear's response to sound is determined by certain characteristics of the sound, which depend on properties of the wave motion. These are:

- pitch, which depends on the frequency,
- loudness, which depends on intensity and on frequency,
- quality (or timbre), which depends on the range of frequencies and their relative amplitudes.

Musical notes are usually mixtures of related frequencies such as a fundamental and harmonics. Noise is a random mixture of many unrelated frequencies.

Transmission of sound in a medium

The velocity c of a sound wave in a medium depends on the density ρ, of the medium and its bulk modulus E. The relationship is given by the equation:

$$c = \sqrt{E/\rho}$$

For a gas $E = \gamma p$ where γ is the ratio of specific heats at constant pressure and constant volume, and p is the pressure, so

$$c = \sqrt{\frac{\gamma p}{\rho}}$$

There is a resistance to the passage of a sound wave through a medium which is analogous to electrical resistance. This is called the **acoustic impedance** Z, which is given by:

$$Z = \rho c$$

This is an important quantity in determining how much sound is transmitted from one medium to another. If the impedances of the two media are very different, for example a gas and a liquid, then much of the sound will be reflected, not transmitted at the interface. The media are said to be acoustically mismatched. The ear has to arrange for acoustic matching when the sound passes from air to liquid at the middle – inner ear interface. Matching is also very important in ultrasonics as described in Chapter 8.

The **intensity** I of a wave is defined as the power P per unit area of the wavefront:

$$I = P/A$$

It can be shown that the intensity is related to the amplitude a, and the impedance:

$$I = a^2/2Z$$

That is, the intensity is proportional to the square of the amplitude, or the maximum change of pressure.

As a wave passes through a medium it will gradually lose intensity by **attenuation**. This is the result of a number of processes of interaction between the wave and the medium, including absorption, diffraction, scattering and others. As with any wave transmission, the intensity falls by a constant fraction for each unit of distance travelled, which leads to an exponential fall of intensity with distance. Intensity at a distance x from the original intensity I_0 is given by

$$I_x = I_0 e^{-\mu x}$$

where μ is called the attenuation coefficient. Similarly an **absorption coefficient** k can be defined to describe the effect of that part of the attenuation.

QUESTION

3.3 Write down the expression for the intensity at depth x of a wave of incident intensity I_0 in a medium of absorption coefficient k.

Intensity levels

Atmospheric pressure is approximately 10^5 pascals, so the ear can detect a change in atmospheric pressure of two parts in ten thousand million.

The lowest intensity of sound that the ear can detect is called the **threshold.** It is typically about 10^{-12} W m^{-2}, corresponding to a pressure change of about 2×10^{-5} pascals. The highest intensity, beyond which the ear may be damaged is reckoned to be about 100 W m^{-2}. This is an immense range, and it is observed that the ear's perception of a sound of intensity I, the loudness L, depends on the relationship between this intensity and the threshold intensity I_0 according to the expression:

$$L \propto \log_{10}(I/I_0)$$

This is because the ear responds to differences in intensity rather than total intensity, as explained in the next section.

The **intensity level** is defined in order to quantify sounds in a way which relates to the ear's response. The threshold intensity I_0 is taken as 10^{-12} W m^{-2}, and the intensity level of a sound of intensity I W m^{-2} is given in bels (B), as

$$\text{intensity level} = \log_{10}(I/I_0)\ \text{B}$$

For example, if the intensity is ten times the threshold,

$$\text{intensity level} = \log_{10}(10^{-11}/10^{-12})\ \text{B} = 1\ \text{B}$$

The bel (named after Alexander Graham Bell, not the ringing device), is therefore rather a large unit, so it is more common to use one tenth, the **decibel** (dB), as the unit, so that

$$\text{intensity level} = 10 \log_{10}(I/I_0)\ \text{dB} \qquad (1)$$

The sound intensity of normal conversation (speakers about 1 m apart) is about 10^{-6} W m^{-2}, so the intensity level is

$$10 \log_{10}(10^{-6}/10^{-12})\text{dB} = 60\ \text{dB}$$

Doubling of the intensity of a sound increases the intensity level by ~3 dB. To combine two sounds, for example an alarm clock at 1 m (level 80 dB) with conversation, it is necessary to convert to intensities:

Conversation: $I_c = 10^{-6}$ W m^{-2} or $10^6 \times I_0$

Alarm clock: I_a is given by $80 = 10 \log_{10}(I_a/I_0)$

$I_a = 10^8 \times I_0$

so the total intensity $I_t = I_c + I_a = 1.01 \times 10^8 \times I_0$

and the intensity *level* $= 10 \log_{10}(1.01 \times 10^8)$ dB

$= (10 \log_{10}1.01 + 80)$ dB

that is very little more than 80 dB. The alarm clock completely drowns the conversation, as you would expect from the intensities, though this may be masked by the logarithmic nature of the decibel scale, for intensity levels.

The *difference* in intensity *levels* between two sounds of intensities I_1 and I_2 is found by subtraction:

$$10\ \{\log_{10}(I_2/I_0) - (I_1/I_0)\}\ \text{dB}$$

which simplifies to

$$10 \log_{10}(I_2/I_1)\ \text{dB} \qquad (2)$$

because the scale is based on comparisons.

Thus the difference in intensity level between a conversation and a door slamming $(I = 10^{-4}\ \mathrm{W\ m^{-2}})$ is

$$10 \log_{10}(10^{-4}/10^{-6})\ \mathrm{dB} = 20\ \mathrm{dB}$$

QUESTION

3.4 (a) The upper limit of acceptable noise, by law is 10^9 times the threshold intensity. What is its intensity level in dB?

(b) The intensity levels that can be found in a classroom range from 30 dB (rustle of paper during exams!) to 80 dB (everyone talks at once). Express this as a ratio of intensities.

3.3 THE EAR'S RESPONSE

Frequency response

The physical response of the ear to sound is essentially one of resonance. That is the vibrations of the sound match the natural frequencies of vibration of parts of the ear. The outer ear is a tube of length about 2.5 cm with one end closed. Its natural frequency is for a standing wave of length 10 cm to be produced (see a core physics text for details). Since

$$c = f\lambda$$

$$f = 330\ \mathrm{m\ s^{-1}}/0.1\ \mathrm{m} = 3300\ \mathrm{Hz}$$

The middle ear displays a broader range of resonance between about 700 Hz and 1500 Hz. Within the inner ear the pressure variation becomes a travelling wave within the fluid. The amplitude of the wave decays as it travels and this depends on the frequency of the wave. High frequencies are absorbed in a short distance, low frequencies in a longer distance. This causes different parts of the basilar membrane to be affected by particular frequencies. Thus the short fibres of the organ of Corti, near the oval window respond to high frequencies, while the longer fibres at the narrow end detect the low frequencies. Below 20 Hz, there is no stimulation of the hairs. The brain discriminates frequencies by distinguishing the places from which the impulses come. The combination of these various resonances gives the ear its **frequency range**. The high frequency limit for hearing is about 20 kHz, though this falls with age. Frequencies beyond this are called ultrasonic.

QUESTIONS

3.5 The data on the frequency response of the ear is summarised in Fig 3.4. Use this to write down:

(a) the intensity of the minimum threshold of hearing and the approximate range of frequencies over which the ear has this sensitivity. How does this relate to the structure of the ear?

(b) intensity levels of discomfort and pain; are these frequency dependent?

(c) the approximate audible frequency range of sound level 60 dB.

3.6 If you have access to a sound intensity level meter and an audio signal generator, measure your own audio response.

Frequency discrimination is the ability to distinguish one frequency from another. The ear's response in this is also frequency dependent. It is greatest at low frequencies; in the range 60–1000 Hz a difference of about 3 Hz can be distinguished. This falls away until at above 10 kHz it is very poor. This behaviour of the ear has resulted in the development of musical intervals, which depend on the ratio of frequencies rather than their differences. An octave is a ratio of upper and lower frequencies of 2:1. For example middle C has a frequency of 256 Hz, one octave above this is upper C with a frequency of 512 Hz. It is called an octave because there are 8 notes spanning

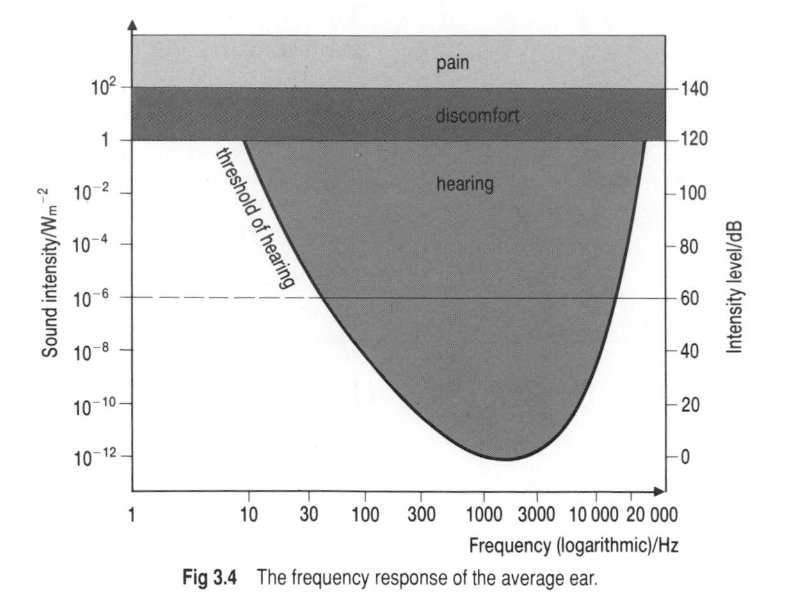

Fig 3.4 The frequency response of the average ear.

this frequency range, named A to G and an additional C. Each pair of notes are an interval in which the two frequencies are in a certain ratio.

If two notes close in frequency are sounded simultaneously, the waves will interfere with each other. This produces a regular variation in intensity which is heard as **beats**. The beat frequency is the number of intensity maxima which are heard each second. It is equal to the difference between the two individual frequencies. The ear can detect beat frequencies of 1–6 Hz, largely independent of the original frequencies.

Intensity response – loudness

The ear's response to an increase in intensity has at least three parts. In the inner ear the greater amplitude of the wave results in

(a) a greater movement of the basilar membrane, producing more stimulation of the nerve endings by the hair cells;

(b) additional hair cells are activated to stimulate nerve endings, in the particular location for that frequency of sound;

(c) nerves are stimulated beyond that part of the membrane as a result of its greater movement.

Each of these effects increases the extent of the nerve impulse which is sent to the brain. This increase is perceived as an increase in loudness L.

Thus loudness is dependent on intensity but is also a result of the energy transfer characteristics of the ear. It is measured in **phon**, and like the threshold of hearing, is strongly frequency dependent for a given intensity (Fig 3.5).

The perception of changes in loudness does not correspond directly to changes in intensity. Equal increases in loudness result from equal propor-

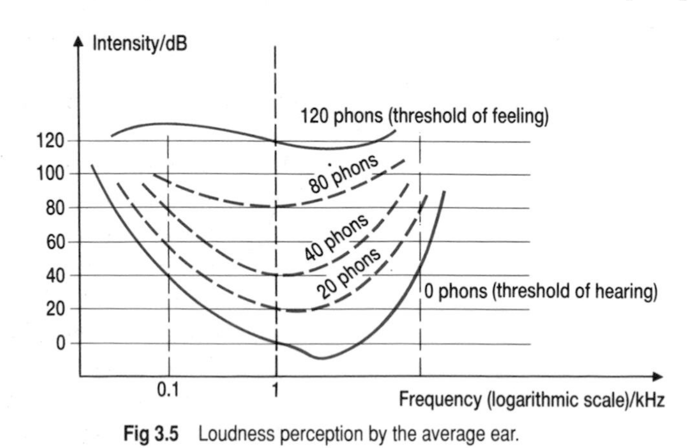

Fig 3.5 Loudness perception by the average ear.

tional changes in intensity. For example if the intensity is doubled say from 1×10^{-6} to 2×10^{-6}, and from 2×10^{-6} to 4×10^{-6}, the loudness will be heard to increase by the same amount each time. This can be expressed as

$$\text{loudness increase} \propto \frac{\text{intensity increase}}{\text{initial intensity}}$$

$$dL \propto \frac{dI}{I} \quad \text{or} \quad dL = k\frac{dI}{I}$$

where k is a constant. Integrating this gives

$$L = k \log_e I + C$$

where C is also a constant. It may be determined as follows:
when $I = I_0$ the threshold intensity, the loudness is zero, $L = 0$, so

$$C = -k \log_e I_0$$

$$\text{therefore } L = k \log_e I - k \log_e I_0$$

$$= k \log_e (I/I_0)$$

$$\text{or} \quad L \propto \log_{10}(I/I_0)$$

which expresses the ear's **logarithmic** response to intensity changes. This relationship is an example of the Weber-Fechner law, which states that the change in the perceived property is proportional to the *fractional* change in the stimulus. It is common to a number of stimuli including weight and brightness.

The **sensitivity** of the ear is its ability to detect the smallest fractional change in intensity dI/I. It is defined as

$$S = \log (I/dI)$$

Sensitivity, as you would expect, is strongly dependent on frequency and intensity, being a maximum for low intensities at about 2 kHz. The minimum change in intensity that can be detected is about 0.5 dB. This corresponds to a ratio of sound intensities I_1/I_2, given by equation (2) as

$$0.5 = 10 \log_{10}(I_1/I_2)$$
$$I_1/I_2 = \text{antilog}_{10}(1/20)$$
$$= 1.12$$

A variation on the "guess the weight of a cake" competition might be to find the largest slice that can be removed without it being noticeable to someone who is blindfold. By feeling the weight most people could tell when a tenth has been removed.

an increase in intensity of 12 per cent or one eighth. This is comparable with the sense of weight change at ~10 per cent but is considerably poorer than the sense of brightness change at ~1 per cent.

As previously described the intensity level scale is defined to be similar to the ear's response, that is logarithmic. This describes a physical quantity, however, which can be calculated from a knowledge of the amplitude or intensity of the wave. Loudness is a subjective sensation and so has to be measured subjectively. Since loudness is frequency dependent, measurement must take this into account. There are two ways in which this is done.

In **phon measurements** all sounds are referred to a standard frequency of 1 kHz. In order to measure the loudness of an unknown source the following procedure is followed. The source to be measured is placed near the standard source, whose intensity is adjusted until the two sources have equal loudness. If the standard source has an intensity level of l dB, then the loudness of the unknown source is l phons. This is not a very practical procedure in many situations so there is a more common method.

The **dBA scale** is an adaptation to the intensity level scale to take account of frequency dependence. Measurements are made with a sound level

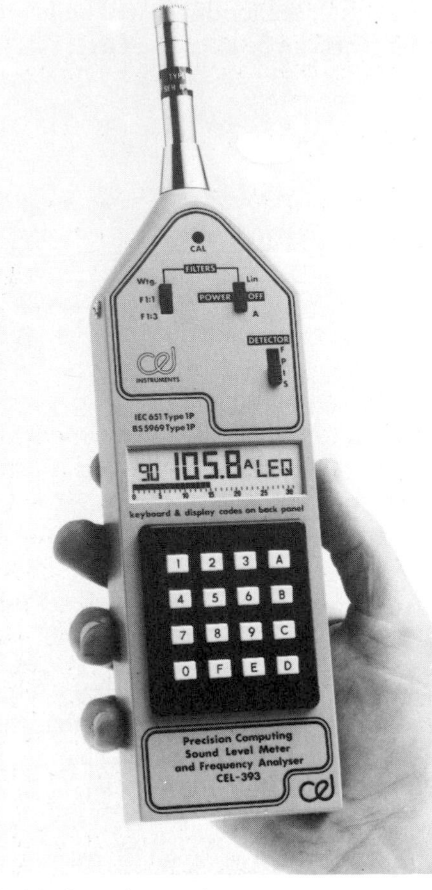

Fig 3.6 A sound meter or Leq meter.

meter (Fig 3.6) which is a microphone calibrated to register a 'weighted decibel' reading. This is done by means of circuits which suppress the contributions of high and low frequencies, so that the response is similar to the ear. This scale is now used almost universally in the measurement of noise, since it maintains the relationship between perceived loudness and measured value of 'weighted intensity', for example, an increase of 10 dBA doubles the loudness. Some values of typical noises are given in Table 3.1.

Table 3.1 Intensity levels of some typical noises

Level/dBA	Sound (and effect)
160	Rifle close to ear (eardrum ruptures)
140	Aircraft at 25 m (threshold of pain)
120	Discotheque, close to speakers (threshold of discomfort)
100	Very noisy factory, noisy food blender at 0.5 m
90	Road drill or heavy lorry at 7 m, underground train (upper limit of acceptable noise level, by law)
70	Busy street, vacuum cleaner at 3 m
50	Quiet street, 'normal' rooms at home
40	Quiet conversation
30	Whisper, tick of watch
20	Blood pulsing
0	Threshold of hearing

The damage that sounds cause depends on the time the ears are exposed as well as the intensity. Sounds or noise rarely continue for long at the same intensity, so a further quantity is defined to enable comparisons to be made of the effects of varying levels. The **equivalent continuous sound level** L_{eq} is that sound level which if constant for a defined period, would give an exposure to sound energy equivalent to that received. The defined period is usually 8 hours, as that is the nominal length of the working day. The quantity is thus called L_{eq} [8 h]. It is measured in dBA.

QUESTION

3.7 From Table 3.1, and values previously stated calculate:

(a) The ratio of intensities between the threshold of hearing and eardrum rupture.

(b) The intensity of sound to cause eardrum rupture, in W m^{-2}

(c) The change in intensity when an aircraft flies overhead at 25 m, when you are walking in the quiet countryside.

Stereophonic hearing

The localisation of a sound is possible because of the differences in the sound received by each ear. This is more difficult with low frequency sounds because of increased diffraction effects which cause the sound to be more diffuse. The brain can make use of three differences between the signals received by each ear.

(a) There is a time lag which is dependent on position, though there can be some confusion between locations giving the same time lag. This is particularly useful for short, sharp sounds and low frequencies.

(b) There is a phase difference (the two waves are at different stages in the compression–rarefaction cycle). This is important in continuous sounds and low frequencies.

(c) There is an intensity difference, because of distance attenuation and shadow effects of the head. This is most useful for high frequencies because diffraction effects are small.

QUESTION

3.8 **(a)** Why is diffraction less common in high frequency sound? How does it interfere with the location mechanisms?
(b) Speech is located using the first two mechanisms described above, why?
(c) Explain why the first method is useful for short sounds and the second for long ones.
(d) Find out which are the recommended positions for speakers for stereophonic recordings. How do these relate to the binaural hearing mechanisms?

3.4 DEFECTS OF HEARING

Causes

Hearing losses are usually divided into two categories. **Conductive** loss in which the sound vibrations do not reach the inner ear, and **nerve** loss where the cochlea does not pass impulses to the brain. The main causes are accident or trauma, disease, age and exposure to excessive noise. Young children are susceptible to temporary deafness as a result of cold-type infections in which fluid fills the middle ear. The eardrum can be ruptured by a sudden shock wave, the inner ear can be damaged by a blow on the head. Disease and age can reduce the ability of the bones of the middle ear or the oval window to respond to the pressure variations. Conductive hearing loss can be sometimes corrected by surgery, such as the replacement of solidified ossicles with plastic ones. An alternative is the use of a hearing aid which transmits the vibrations through the bones of the skull to the inner ear. Any hearing aid cannot of course replace the sensation. At present there is no cure for nerve hearing loss, because of the nature of the nerve conduction process. Electronic aids (and ear trumpets) amplify the vibrations to compensate for the loss of sensitivity of the ear. Usually the frequency range is very limited, for conversation 200–5000 Hz is adequate. This compares with the telephone's range of 30–3400 Hz.

There is another hearing defect whose cause is not well understood. **Tinnitus** is a constant ringing or hissing in the ears, which interferes with the reception of sound and can cause serious insomnia. It may be produced temporarily by a few hours exposure to a noisy factory or concert hall. But if this is regularly repeated the effects may not wear off and the sufferer may not notice the steady deterioration by getting used to the internal background noise.

QUESTION

3.9 Why would a person with an effective hearing aid suffer little loss in reception of telephone sound, but be wasting their money on Hi-Fi audio equipment?

Noise

Much is reported about the exposure of people to noise and the effects this has. This is because there is a concern that noise levels are increasing and also that the effects are not fully understood. The effects are certainly complex, ranging from the instantaneous physical damage of the eardrum or ossicles at levels of 190 dBA (close to a large gun at explosion), to the psychological effects of exposure to low frequency, low intensity vibrations such as that from a hand-held vacuum cleaner or engraving tool, or the interruption of sleep by occasional noise of moderate intensity.

The sources of noise are of course very widespread, those giving most concern include factory machines at work, overflying aircraft at home and firearms and pop concerts at leisure.

Measuring hearing

Ears are tested in a soundproofed room, usually with sound supplied by headphones to each ear in turn. The intensity is gradually increased until the subject indicates that it is heard. The procedure is repeated for selected frequencies to cover the normal range. These hearing thresholds are plotted on an audiogram on which normal hearing is taken to be 0 dB. Fig 3.7(a), shows typical results for people aged 40 and 65, whose loss is due only to ageing (presbycusis). Fig 3.7(b) shows the results for similar age groups who have been exposed to noise with a L_{eq} of 96 dBA for eight hours a day, five days a week since their eighteenth birthday. Note how the higher frequencies are most severely affected.

SUMMARY ASSIGNMENTS

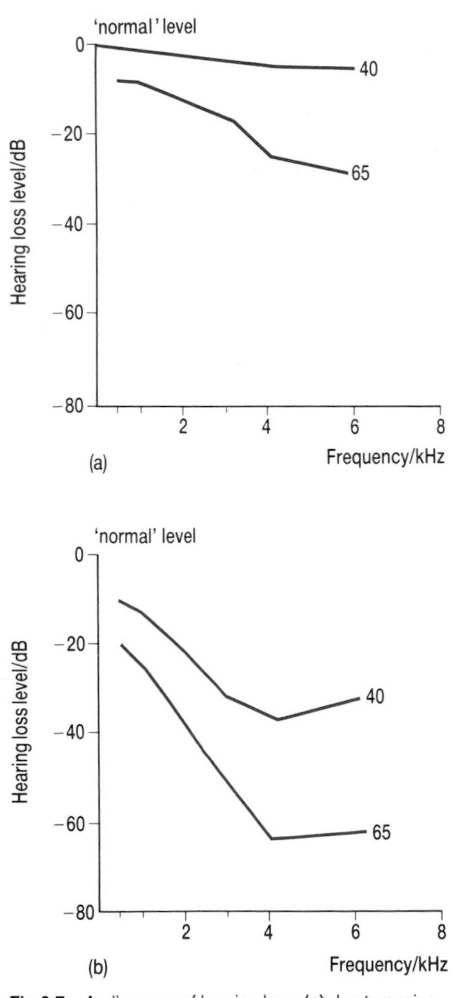

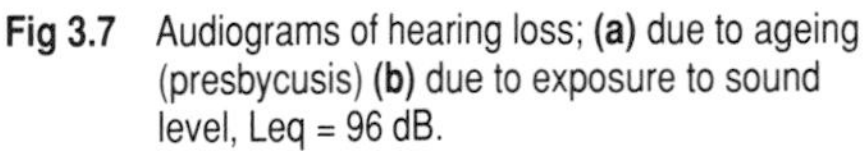

Fig 3.7 Audiograms of hearing loss; **(a)** due to ageing (presbycusis) **(b)** due to exposure to sound level, Leq = 96 dB.

3.10 Write a brief report on what you consider to be the main unnecessary sources of noise you and your family are exposed to. Give any values of noise levels you can obtain, explain the possible hazards, and state what actions could be taken to reduce these.

3.11 **(a)** Define *intensity of sound* and give a formula for relative intensity level measured in decibel (dB) units. Explain why it is convenient to compare intensity levels of sound in decibel units. The value of the *standard reference intensity* is 10^{-12} W m^{-2}. Explain what this value represents.

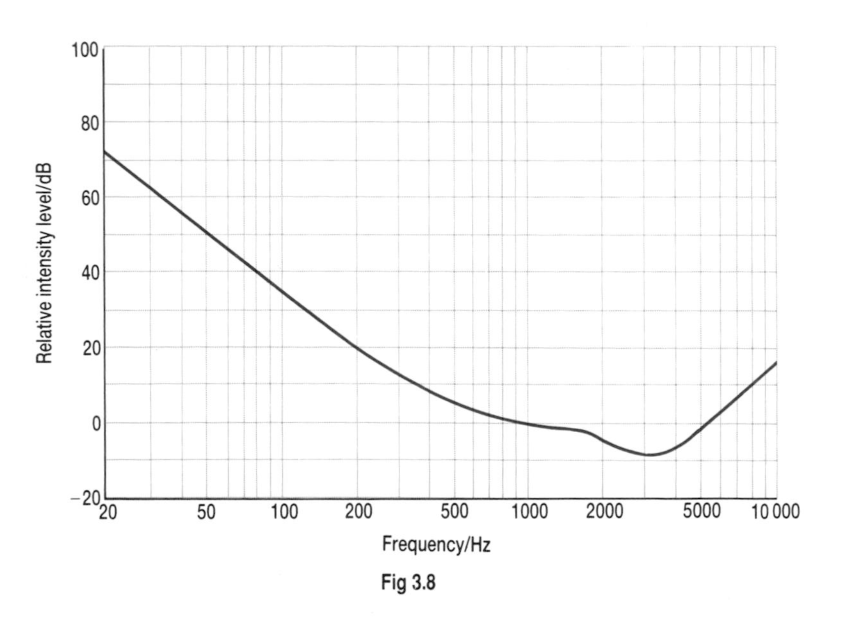

Fig 3.8

(b) The graph (Fig 3.8) shows the threshold of hearing for a young person with good hearing, as a function of frequency.

(i) Estimate the frequency at which the subject's ear is most sensitive.

(ii) Estimate the least intensity of sound the subject can detect at 100 Hz.

(iii) The external auditory canal of the subject behaves like a tube of effective length 28 mm closed at one end by the eardrum. Show, with the aid of a calculation, that this is consistent with your answer to (i) above.

The speed of sound in air = 340 m s^{-1}.

(JMB 1983)

3.12 **(a)** With the aid of a labelled diagram of the middle ear, explain how sound energy is transmitted across the tympanic cavity. Give a reason why the pressure changes due to sound are increased as a result of this transmission and state an approximate value for the increase.

(b) A meter, which measures the relative intensity level of sound referred to 1 pW m^{-2}, records a value of 97 dB when a pneumatic drill is switched on some distance away. Calculate the intensity of sound at the meter.

A second drill, identical to the first, is placed close to it. State the intensity of the sound at the meter when both drills are working and hence calculate the *increase* in the meter reading.

(JMB 1985)

3.13 (a) Give approximate values for the relative intensity levels in dB for:

(i) sound due to normal conversation,

(ii) the threshold of hearing for the ear at its most sensitive in a young person with normal hearing,

(iii) the threshold of pain for such a person.

Give the value of a typical frequency at which the normal ear is most sensitive.

By considering the differences between age-related hearing loss and loss due to exposure to excessive noise, explain how it is possible to estimate the hearing damage due to the latter in a person of a given age.

(b) At the site of a new machine in a factory the relative intensity level measured 70 dB before the machine was brought into operation and 79 dB when it was running. Find the relative intensity level due to the machine alone.

(c) Most sound level meters give readings on a modified dB scale. Give the name of the usual modified scale and explain why it is used instead of the relative intensity level in dB.

(JMB 1987)

Further reading

Noise Hobson's Science Support Series 1986

Useful background reading including the measurement and removal of unwanted sounds in industrial situations.

Kane, J.W. and Sternheim, M.M, *Physics* Chapter 22 Third edition John Wiley 1988

Theme 2

BIOMEDICAL MEASUREMENT

You may at some time have carried out a 'black box' investigation; trying to discover the functions of the box by measuring the outputs that it produces when subjected to various inputs. Biomedical measurement is such an activity where the black box is the human body. This makes its task very complex. The body contains mechanical, acoustic, thermal, electrical, optical, hydraulic, pneumatic, chemical and other types of system, all interacting with each other. It also contains a powerful computer, several communication systems and many control systems. In many systems there is no simple relationship between input and output, and often it is difficult to decide what is the input and which the output, since the body has many feedback loops. The variables to be measured are often inaccessible and often interact with other variables.

Even from this purely mechanistic view of the body, we can see that the task appears almost impossible. There are other complications which are of vital importance when the subject of the measurement is a conscious, feeling person. You may be able to identify some of these in, for example, the routine X-raying of all dental patients, or the trial of a new, uncomfortable investigation for a research project. The assumption that more measuring will necessarily lead to better understanding of the body is one which should not be accepted unthinkingly.

Prerequisites

Before you study this theme you should have some familiarity with:

- The principles of current electricity and its measurement.
- The principles of thermometry, and common thermometers.
- The wave nature of light and sound and their reflection and refraction.
- The functions of the hardware and software in a computer system.

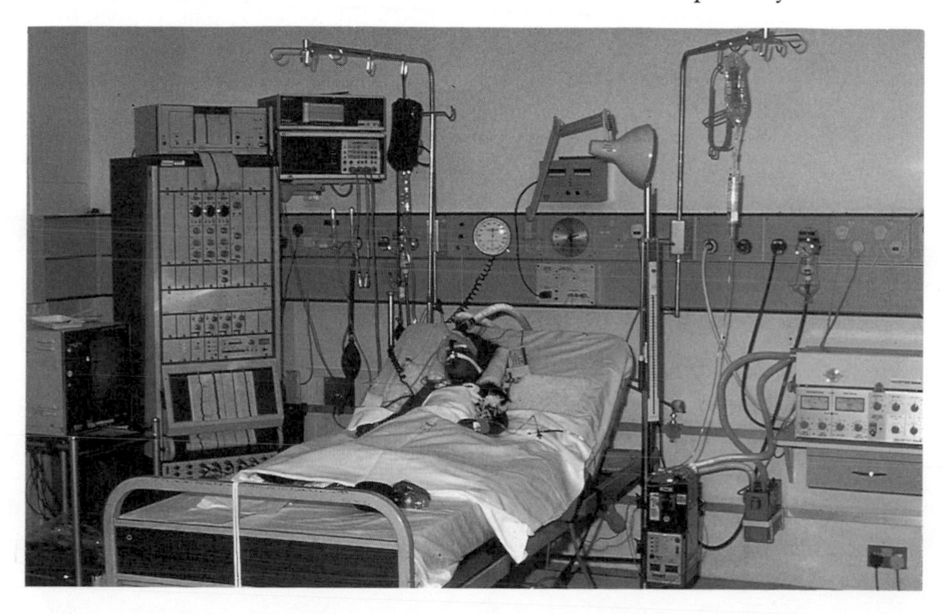

The comprehensive monitoring of a patient's condition in an intensive care unit.

Chapter 

INSTRUMENTATION

Biomedical instrumentation is the use of instruments and equipment to carry out the task (or part of it!) outlined in the theme introduction; methods of measuring the different systems in the body. In this theme we are concerned only with the most commonly made physical measurements. This chapter will illustrate some of the principles of the process of obtaining and presenting the measurement, using temperature as an example.

LEARNING OBJECTIVES

When you have studied this chapter you should be able to:

1. list the purposes of a biomedical instrumentation system and the type of components of which it is composed;
2. discuss the general problems involved in measuring human beings, and illustrate these with specific examples;
3. give examples of the uses of active and passive transducers in measuring and the ways in which electrical signals from the transducer can be processed;
4. give brief descriptions of the cathode-ray tube and thermograph, as display devices, and of the technique of telemetry;
5. give examples of the use of computer systems in the recording, data processing, transmission and control of measurements.

4.1 SYSTEMS

Purposes

The purposes of biomedical instrument systems can be classified as follows:

Information gathering will always be a part of the purpose of measurement. In research this may be the full extent of the objective. In any clinical situation other purposes will follow. Instruments for clinical use will typically be more fully developed than research instruments, enabling them to be used routinely by medical staff who are not experts in the equipment, only in the use of the information.

Diagnosis of the state of the body, or that part which has been measured, is the most important part of medical practice, as this book demonstrates. In many cases the information is more usefully presented as a qualitative image or picture, rather than a set of quantitative values. Much of the present development of diagnostic instrumentation is concerned with providing both.

Monitoring and evaluation. When the condition of the person is known, measurements are made to see if there is any change of this condition as a result of some input, or simply over a period of time. For example, the heartbeat rate and lung capacity of an athlete may be monitored during exercise in a training programme and used to evaluate either her body or the training programme.

Control. The measurement may be used to bring some change in the body's performance such as a heart pacemaker to ensure that the heart continues to pump effectively.

Performance characteristics

These are some of the criteria by which you could judge the suitability and effectiveness of an instrumentation system:

Range is the maximum and minimum inputs that the instrument can detect and produce a usable reading from. Inputs might be static, such as the amplitude of a current; or dynamic, such as the frequency of a sound wave.

Sensitivity is the smallest variation that can reliably be measured. This determines the resolution of the device, which is the smallest variation that can accurately be read. This is a very important parameter in imaging systems, where it is expressed as the shortest distance over which a difference can be accurately seen.

Linearity is the ability of an instrument to be equally sensitive throughout the range over which it operates. Direct current meters that you have used are usually linear; some types of alternating current meters are much less sensitive towards the limits of their range, than in the centre.

Frequency response. Many signals from the body are rapidly varying, so the instrument needs to be sensitive over the range of frequencies encountered.

Signal-to-noise ratio determines the ability to distinguish the required measurement (the signal), from any background or other effects (the noise). Clearly the sensitivity of an instrument is greater when the signal-to-noise ratio is low. The sources of noise are electrical interference, for example from 'mains hum', and to a lesser extent, thermal noise which arises from the nature of the electrical circuit itself.

Mains electricity in Britain is a.c. with a frequency of 50 Hz. As it flows in a cable it will generate an accompanying magnetic field. Any conductor near to a mains-carrying cable will have a current induced in it by this, which behaves as a noise to any signal in the conductor. The current changes direction 100 times per second, so the noise has a frequency of 100 Hz. This is known as mains hum.

Isolation. Often measurements must be made so that there is no direct electrical connection between the subject and the ground or other equipment. This may be for reasons of safety, convenience, or to avoid interference between instruments. **Telemetry** is one way of achieving this in which the measuring device in or on the body, transmits the information by radio waves.

Other important criteria which apply to any measuring instrument are, **accuracy** (which will include considering if the presence of the instrument affects what it is measuring), **stability**, and last but not least, **simplicity**.

INVESTIGATION

Measuring body temperature with a thermocouple

We will use this simple experimental procedure to examine aspects of the instrumentation system described above.

Procedure

1. Obtain or make a simple thermocouple by joining together dissimilar wires such as copper and constantan (details may be obtained from a standard physics practical book).
2. A reference junction should be kept at 0 °C and the other junction is placed where the temperature is to be measured. Choose suitable body sites.

3. The temperature difference of the junctions generates an e.m.f. of about one millivolt. This is measured by a sensitive voltmeter or potentiometer. Choose a suitable instrument.
4. The voltage reading represents a certain temperature. Tables are available to give this, or alternatively you can calibrate the instrument against another thermometer.

Suggestions for investigation

1. You may need to try out methods of improving thermal contact between thermocouple and skin, for example by interposing a copper block.
2. What is the effect of hot and cold water, how quickly does the skin respond?
3. What is the effect of a fan on wet and dry skin?
4. What is the effect of warm clothing?
5. Does body temperature vary with the time of day?

QUESTIONS

After you have carried out the above procedure and measured your own body temperature, answer these questions about the instrumentation system that you used.

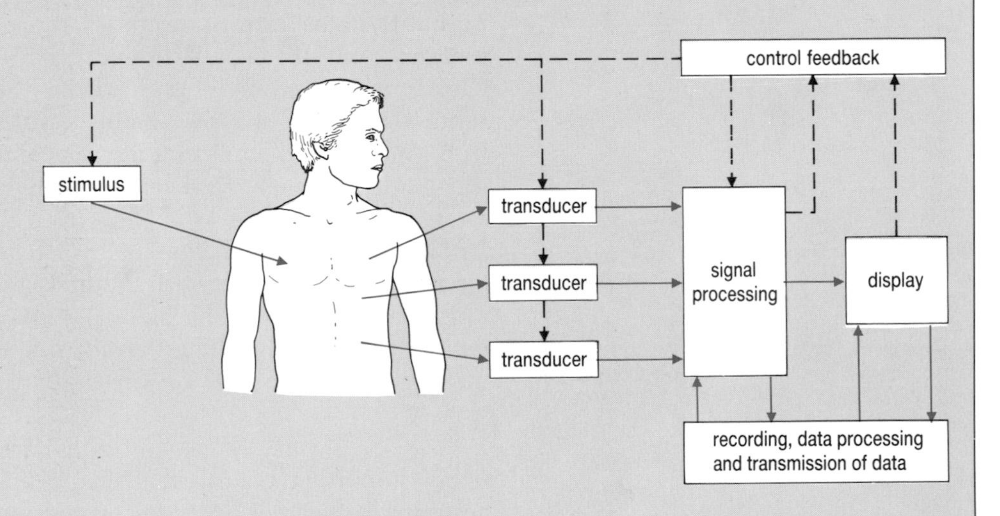

Fig 4.1 Components of an instrumentation system.

4.1 How good were the performance characteristics of your system? Comment on each of the headings listed above, which are relevant to your measurement. For example,

(a) were the range and sensitivity of the instrument suitable for this use? If not, how could these characteristics be improved?

(b) What are the sources of error in your measurement, with what accuracy do you give your result?

(c) What affects the stability of your instrument?

(d) Your instrument was, I expect, quite simple (otherwise you would have been unable to find the time or equipment to carry out the investigation). To what extent do your measurements rely on reference to a more complex or quality-controlled instrument?

4.2 (a) Draw a labelled block diagram similar to Fig 4.1, to show the components of your system. Write a brief description of each of the **components** you used, e.g. stimulus–ambient temperature and the particular conditions you investigated.

(b) How would you include a feedback loop, for example to prevent a patient with a fever from overheating?

The core temperature of the body, or systemic temperature, is maintained by careful control of the energy generated by active body tissue, and the energy lost to the environment. Its value measured by mouth is about 37 °C, or one degree lower under the armpit, or one degree higher in the rectum. Body temperature of even a healthy person, varies by about 1°C over a daily cycle, being lower earlier in the day. This temperature is affected temporarily by exercise, over longer periods by illness, but not by ambient temperature. The temperature control centre is located deep inside the brain. It monitors the blood temperature and controls various parts of the body in order to increase or decrease the rate of energy release. By contrast the skin temperature varies, usually between about 30°C and 35°C, as a result of local factors such as the exposure to ambient temperature, the amount of fat covering the capillaries and the blood circulation through them. Table 4.1 gives some indication of how intolerant the human body is, of changes in body temperature.

Table 4.1 Effect of temperature change on the body

Core temperature/°C	Change	Condition of subject
45 44 43	hyperthermia	Dead
42 41		Central nevous system affected; convulsions
40 39		Dilation of peripheral blood vessels, increased heart rate, reduced flow of blood to brain
38 37 36		Normal
35 34		Constriction of peripheral blood vessels, reduced heart rate; shivering
33 32 31	hypothermia	Depression of nervous system function; amnesia
30 29		Temperature regulating system fails, sleepiness, depression of respiration, heart beat affected, loss of consciousness
28 27 26		Dead

How a fun run meant meltdown for Mark Dorrity's body

By Sue-Ellen O'Grady

On February 27 this year, Mark Dorrity set off on what he expected to be an easy 8-kilometre fun run in Wagga, southern NSW. But near the finishing line, the fit 28-year-old collapsed, his body destroyed.

In less than an hour, his thigh muscles had overheated, liquefied and died. One leg has since had to be amputated at the buttock, because of gangrene.

Before Mark collapsed, his kidneys failed because the dying muscles had released toxic proteins into his blood, which thickened to a molasses-like consistency. Every organ in his body was affected.

He suffered brain damage. His lungs could not function unaided. His buttock and hamstring muscles also liquefied, but not as severely as his thigh muscles.

Mark's heart stopped at least once. When it started again, it hammered away at 150 beats a minute, compared to its normal 70. He was on a dialysis machine for eight weeks, and in a coma for three months. When he regained consciousness, he could not walk or talk.

Even now, five months later, Mark cannot turn over or get out of bed unaided. He faces months of intensive rehabilitation.

The devastating damage to Mark Dorrity's body was caused by heat exhaustion and dehydration resulting in a rare condition known as rhabdomyolysis, the extreme result of what every runner and athlete knows as muscle fatigue.

The director of research at the Sports Medicine Institute, Dr Tony Miller, says the condition usually affects runners taking on more than they are used to in training.

"In Mark's case it was caused by his body severely overheating – to 42.8 degrees. At the same time, he was extremely dehydrated. When someone has a temperature that high, they are delirious. They ignore the body's warnings to stop."

Mark Dorrity was no weekend jogger. When he graduated from the University of NSW with an honours degree in Science in 1974, he won a Blue for athletics. He moved to Melbourne to work as a wool exporter, and ran four kilometres through the Botanical Gardens every day, As well, he swam a kilometre three times a week. He had minimal body fat.

He travelled to Wagga in February with a group of friends, all planning to compete in a local event. When the temperature that day rose to 42 degrees, the locals cancelled their run.

But Mark and his friends, deceived by the dry heat, decided to hold their own race.

"It just didn't feel that hot," he remembers. "So we ran off."

He drank several glasses of water before beginning to run, but none during the race. That, say doctors, proved to be his near-fatal mistake.

When Mark collapsed, he was leading the race by a kilometre. Friends driving alongside rushed him to the local hospital, where he was packed in ice to lower his body temperature. He remained there for two weeks before being moved to St Vincent's for dialysis treatment.

Mark recalls nothing of this. "I don't recall collapsing. I remember waking up twice at Wagga Hospital, and then waking here."

"How do I feel? I'm very lucky to be alive. I know that. I'm a medical miracle. And it's a warning to other runners to be extremely careful."

The Sydney Morning Herald 3.8.1988

QUESTIONS

4.3 (a) What action might have enabled the runner, Mark Dorrity to have avoided this tragedy?

(b) What appears to have been the sequence of events which led to his body being damaged?

(c) What is his condition at the time of the article?

(d) Is further improvement expected?

4.4 The measurement of body temperature is a routine check to monitor the overall condition of someone who is unwell. It is simple but suffers from several limitations which originate from the subject rather than the instrument. Identify as many as you can, and comment on how serious they are.

Problems of measuring a living subject

The answer to the previous question indicates some of the difficulties in a very simple body measurement. Many of the variables that the clinician wants to know about are much less accessible than temperature, and subject to many unavoidable influences, both known and unknown. The presence of the transducer may physically affect what it is measuring, for example by obstructing the flow of blood. It may also have a psychological effect which can affect the physiology being measured, through fear. Many readings and images are altered just by the movement of the patient. This can give false results called **artifacts**. A general problem which is particularly serious in radiological measurement is the effect of the energy applied to the body during measurement. In the extreme this can cause more damage to the patient than the disease that is being investigated. The issue of safe practice is always of prime importance, and may impose severe restrictions on the use of high voltages and high intensity radiations.

Most of the measurements described in this book take place in the body, this is referred to as **in vivo**. This reflects medical practice, though there are tests which are done on samples taken from the body. These are called

in vitro, (in glass) tests, and in some cases this can overcome the problems of measuring the body directly.

4.2 TRANSDUCERS

These are the devices that change the physiological variables to be measured into electrical signals. They are part of a circuit which produces an electrical output from a non-electrical input. The relationship between input and output need not be linear but must be unambiguous; that is, for any value of the input there is only one value of the output. Some transducers actually convert the input energy into electrical energy, these are called **active** transducers and the thermocouple is one example, see Fig 4.2(a). In others there is simply an electrical change produced in the circuit of which they are a part, for example the resistance of a conductor will change if its dimensions change. This is made use of in detecting movement or pressure with a strain gauge. This is a **passive** transducer see Fig 4.2(b). In practice it is not significant to which group a transducer belongs, except that some active conversions are reversible, whereas passive effects are not. This can be important for example in ultrasound where the transducer both generates and detects the signal. Table 4.2 lists some basic transducers used in biomedical applications. A brief comment on each follows.

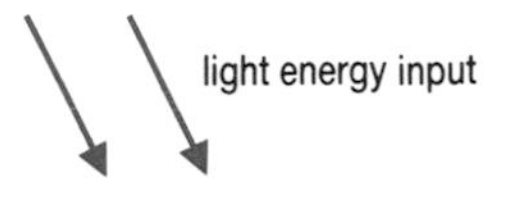

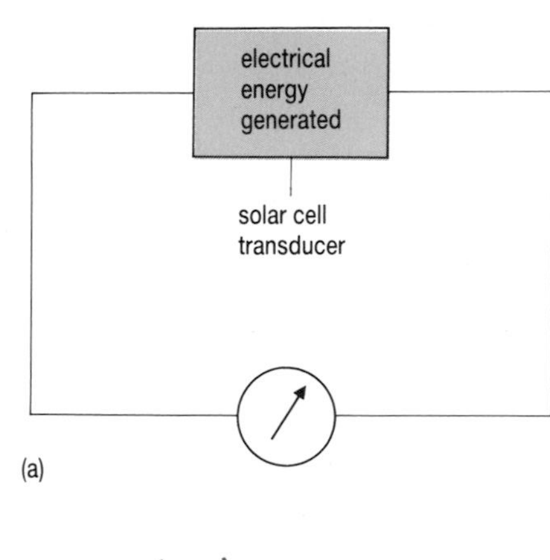

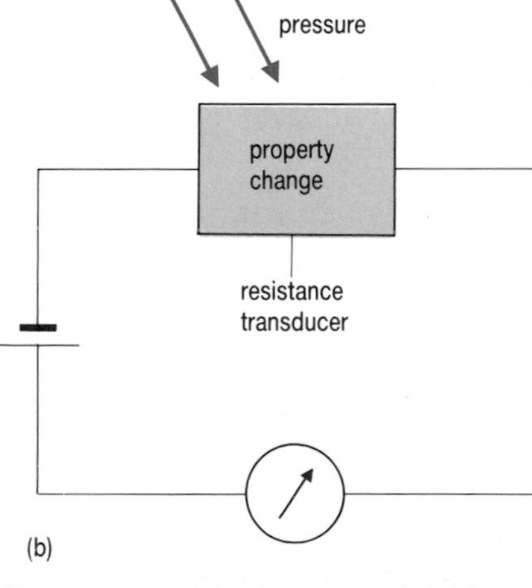

Fig 4.2 Transducer types **(a)** active **(b)** passive.

Table 4.2

Transducers for biomedical applications

Physical variable	**Type of transducer**
Force (or pressure)	Piezoelectric strain gauge
Displacement	Variable resistance Variable capacitance Variable inductance
Surface strain	Strain gauge
Velocity	Magnetic induction
Light	Photovoltaic Photoresistive Photodiode
Magnetic field	Hall effect
Temperature	Thermocouple Thermistor

The **piezoelectric effect** is the conversion of movement energy into an electric current, and vice versa. It is used in both directions in ultrasonics and is described in Chapter 8.

Movement or displacement can be detected by the movement of (a) a slide wire to change the length of **resistance** wire in a circuit, as in a potentiometer, (b) a magnetic core in a coil or a transformer to produce a change in **inductance**, (c) the plates or dielectric of a capacitor to change the **capacitance**. If the movement is more rapid it can be used to produce **magnetic induction** by the relative movement of a coil and a magnet. **Strain gauges** detect a range of mechanical changes and these are des-cribed in Chapter 6.

The **photoelectric effect** is used in the silicon **photovoltaic** cell or photocell which is the basis of the charge coupled device used for video endoscopy, described in Chapter 7. In semiconductor materials such as cadmium sulphide the conductivity is increased by the absorption of light. This is the **photoresistive** cell which is a sensitive transducer with a rather limited frequency response. The **photodiode** is a semiconductor transducer which is less sensitive but with a better frequency response.

QUESTION

4.5 Make a list of the body systems or functions that you can think of to measure, with the transducers listed in Table 4.1. (You may find it helpful to discuss this with a fellow student and then pool your ideas.)

Temperature transducers

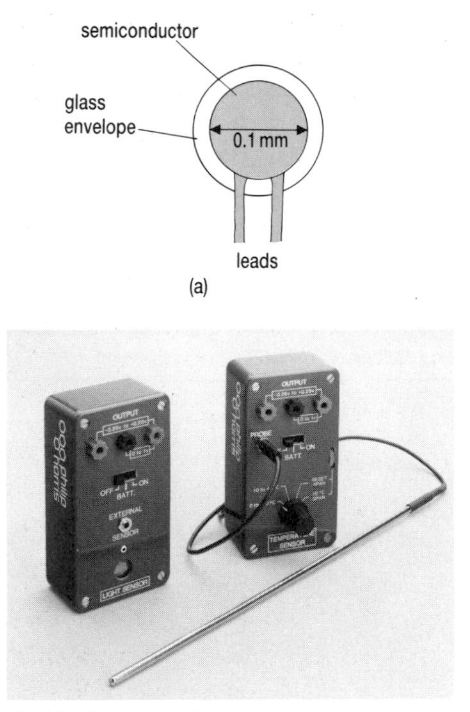

Fig 4.3 Temperature detectors **(a)** small thermister probe **(b)** laboratory thermister probe.

In most situations the clinical mercury thermometer is the standard instrument. This is an example of a transducer which does not produce an electrical output. If continuous recording of temperature is required then a **thermocouple** or a **thermistor** is used. The latter is now more common because it has a greater sensitivity and does not require a reference junction.

A typical thermistor for biomedical use is shown in Fig 4.3(a). It consists of a bead of semiconductor material, usually a mixture of nickel, cobalt and manganese oxides, enclosed in a thin glass envelope and attached to connecting wires. The sensitivity is typically ten times that of a thermocouple, enabling temperature differences of as little as 0.01 °C to be measured. The disadvantages are that thermistors are more expensive and have a shorter life than metal thermocouples as they are susceptible to overheating, and their temperature coefficient of resistance is exponential (and negative) which means they require complex circuitry to be used as signal measuring devices. An example for laboratory use is shown in Fig 4.3(b). There are thermistors with a positive coefficient of resistance, called positors, but these are not in common use.

Newer thermoelectric devices have been produced with very small dimensions and fast response times, and automatic cold junction compensation is incorporated in the circuit. A number of thermocouples can be connected together to produce a greater sensitivity, at the expense of increased size. Such a device is called a **thermopile** and it is used to detect the heat radiated from the skin. The relationship between energy radiated (E/joule) and temperature (T/kelvin) is given by **Stefan's Law:**

$$E \propto \sigma A T^4 \qquad (1)$$

where A is the surface area and σ is Stefan's constant. The thermopile is usually calibrated to read temperature directly. This last effect can be used to detect defects in the circulatory system by recording the skin radiation, as described in a later section.

4.3 DISPLAY DEVICES

A great variety of instruments are used to present the electrical outputs of transducers in a form that the operator can use. It is obviously important that these are suitably matched, for example it would have been no use trying to measure the output of your thermocouple with an ammeter. Processing circuits such as amplifiers, pulse counters and filters are introduced to provide an appropriate input to the display device, those specific to a certain measurement are described in the relevant chapter. We will not be describing electrical meters in this book – they are widely used in biomedical measurement, but in principle they are identical to the meters you will already have studied and used.

For medical diagnosis the most useful form of display is often a picture of the organ or system being investigated. This means that signals have to be located in a certain position and reproduced in that position on a screen. This may be a simple process of illuminating a film, as in the case of X-radiography, or it may involve complex scanning procedures, as with X-ray tomography. For immediate, so-called 'real-time' viewing the cathode-ray tube is invaluable, and very widely used.

Cathode ray tube (CRT)

The details and operation of the CRT will be found in a core physics text-book. As a display device the important features are its ability to reproduce electrical information in three dimensions, vertical, horizontal and brightness, and so build up a picture.

In the 625 television system used in Britain, the vertical time base moves the beam down the screen once every 1/50 seconds, while the horizontal time base moves the beam across the screen tracing out 312.5 lines. In the next 1/50 s another 312.5 lines are traced out, so that in 1/25 s 625 lines are traced out.

In the laboratory use of the CRT only two of these are used. A typical example is the display of a rapidly varying electrical voltage, say the output of a microphone. This signal is fed to the Y-deflection plates, and produces a vertical displacement in proportion to the amplitude of the signal. At the same time a regularly varying voltage called the **time base** is applied to the X-deflection plates, so that the change of the output is displayed with respect to time as the horizontal axis, see Fig 4.4(a). This type of display is common in medicine. You will find examples in Chapter 5 (the electrocardiogram, ECG), and Chapter 8 (the ultrasound A-scan).

The three dimensions are used in television, where time base voltages are applied to both the X- and Y-deflectors so that the electron beam scans the whole surface of the screen, at a rate which gives a continuous image.

The signal is fed to the electron gun so that it controls the intensity of the electron beam and hence the brightness of the image produced at a particular position on the screen. By linking the scanning of the screen with the scanning of the subject by the camera, the brightness distribution forms a picture of the subject, see Fig 4.4(b). Examples of this are to be found in all of the later chapters of this book, where the particular features of the system for that type of signal (ultrasound, X-ray, radioisotope etc.), are described, that for infrared is described below.

The use of computers for storage, processing and retrieval of information has extended the use of CRTs or monitors for display, and made obsolete the specialised CRT in which storage was achieved within the tube.

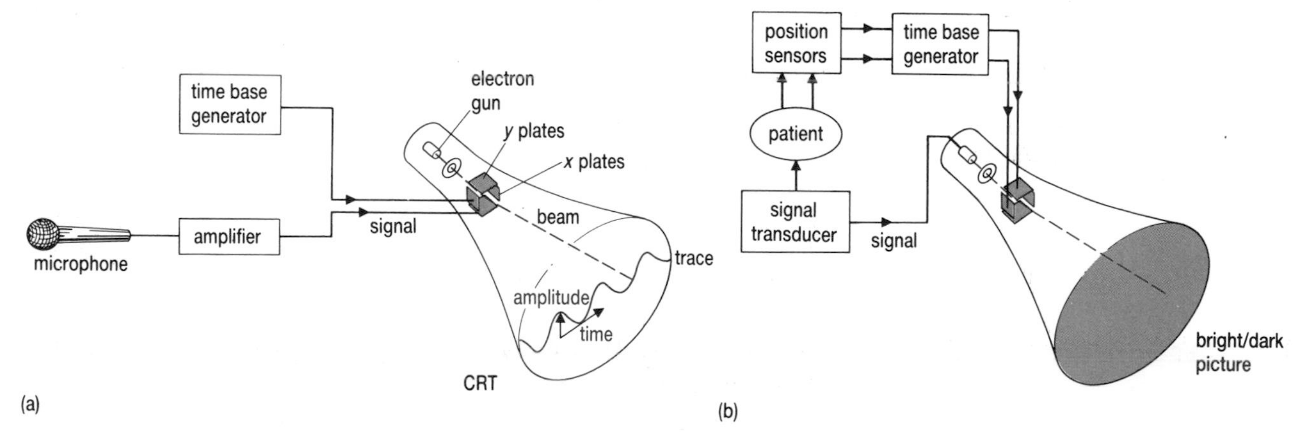

Fig 4.4 Cathode ray tube displays **(a)** amplitude/time **(b)** brightness/scan.

Thermography

This is a technique to display images of the infrared radiation from the skin. Stefan's law (equation 1) can be written as

$$E = e\,\sigma A T^4$$

where e is the **emissivity**, a constant which depends on the nature of the

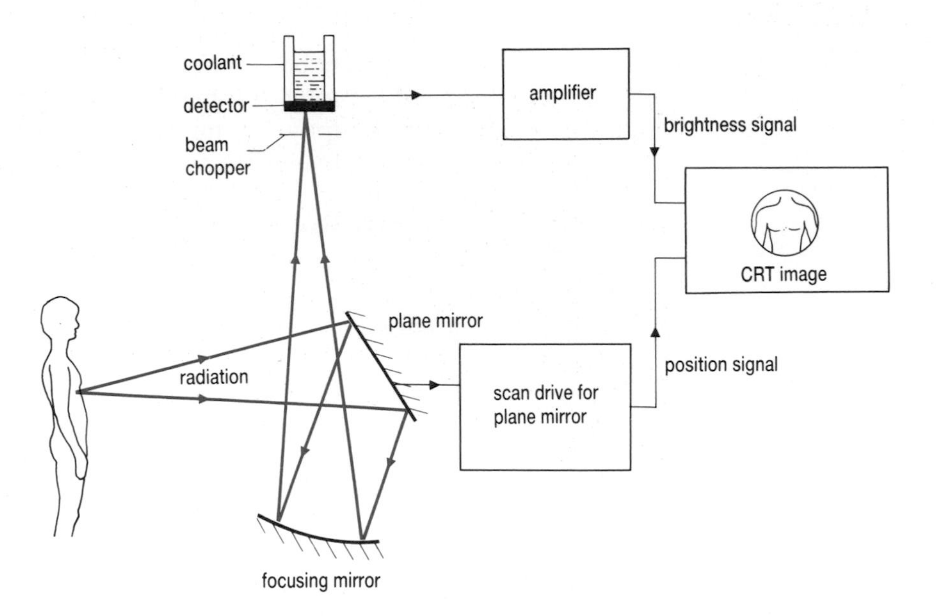

Fig 4.5 A schematic diagram of a thermographic unit.

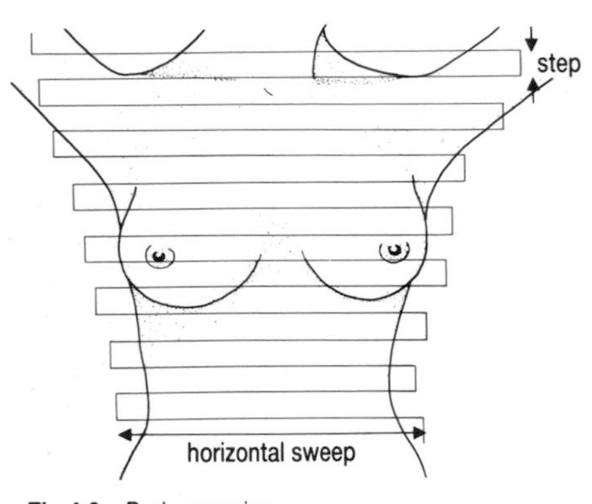

Fig 4.6 Body scanning.

surface. Black and matt surfaces which are good absorbers and emitters have $e \sim 1$, whereas shiny surfaces have a low value of e. A perfect radiator with $e = 1$ is called a black body. The emissivity varies somewhat with wavelength. In the visible range (e.g. sunlight) the darkest human skin has an emissivity of 0.82 and the lightest 0.65. Bodies at room temperature will radiate in the infrared region, for which all human skin has an emissivity of almost 1.

The basic system is shown in Fig 4.5. The radiation from a small area of the subject's skin is focused by the mirror arrangement shown, onto the detector. The detector is usually an indium-antimony photoconductor which increases its conductivity when infrared radiation falls on it. Thus a current is produced in proportion to the radiation, though as this is small there is a high noise-to-signal ratio, which is usually reduced by cooling the detector in liquid nitrogen to reduce the thermal noise. The radiation passes through a mechanical chopper to produce a pulsed signal which can more easily be amplified. This amplified signal then becomes the brightness control for the CRT screen. The scanning of the subject is carried out by moving the mirror in regular steps and sweeping across the body at each step (Fig 4.6), so that a two dimensional picture is built up on a CRT screen. The intensity or brightness of the visible image is related to the temperature of the radiating surface.

Typical dimensions are: the patient is positioned at 3 m from the camera, an area of about 0.1 by 0.1 m is scanned in four minutes and a temperature resolution of 0.2 °C is achieved. The resulting display can indicate abnormalities in the skin temperature, and therefore in the underlying tissue. Tumours produce a local change in the circulation system so it was expected that it could assist in the early diagnosis breast cancer. Unfortunately the resolution is not fine enough to detect the malignant cells at a sufficiently early stage to intervene successfully and so X-radiography is still the routine screening method. Thermography is used to detect direct circulation problems such as those occurring in diabetes and gangrene. This can enable preventative measures to be taken to avoid complete loss of circulation and the amputation of the leg.

QUESTION

4.6 Fig 4.7 shows that people get hot playing squash! What more can you tell from these thermograms?

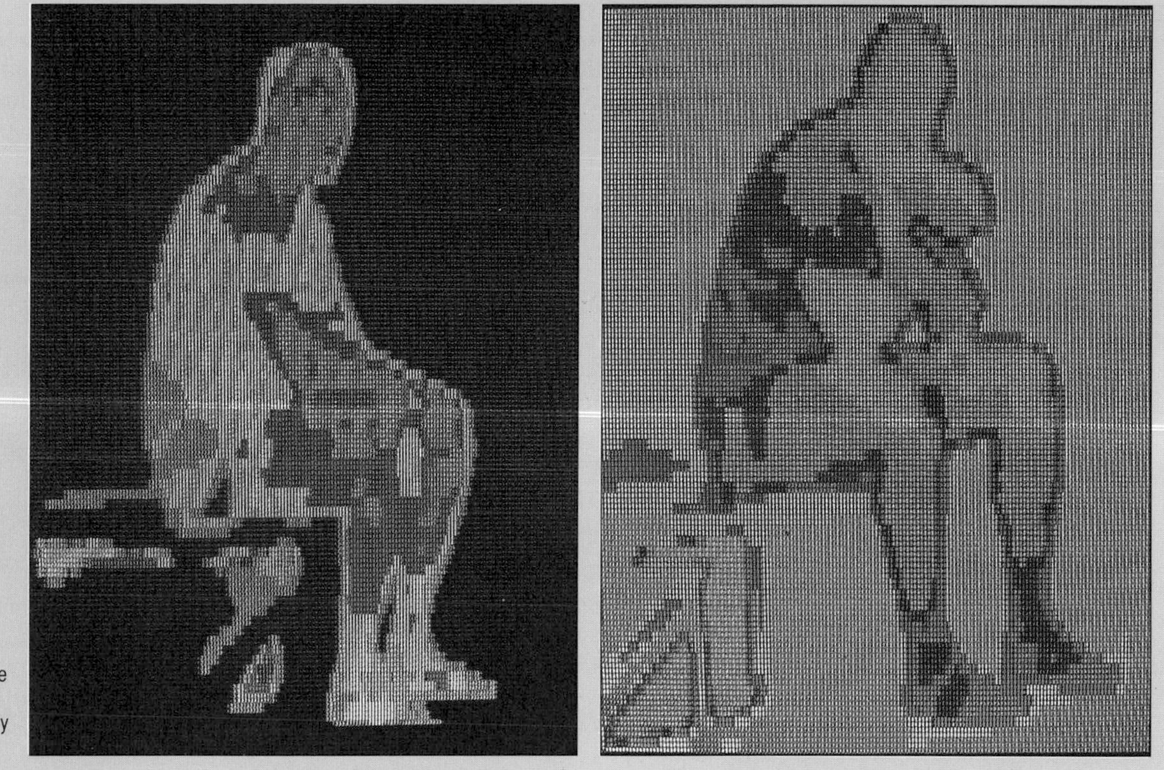

Fig 4.7 Thermogram of a man **(a)** before playing a game of squash and **(b)** immediately afterwards.

4.4 TRANSMISSION METHODS

The principal method of transmission in any biomedical measurement system is electrical. The transducer converts the signal to be measured into an electrical one which can then be amplified, digitised, or in other ways processed so that it is suitable for transmission, either to other parts of the system, or to a different place. There are however a number of situations in which direct electrical connections are either impossible or undesirable. Two examples are the monitoring of astronauts in space and the isolation of a specially electrically sensitive patient. In these circumstances a method of transmitting signals over distance is used. This is called **biotelemetry**.

QUESTION

4.7 List some of the reasons why readings may need to be taken at a distance.

Biotelemetry

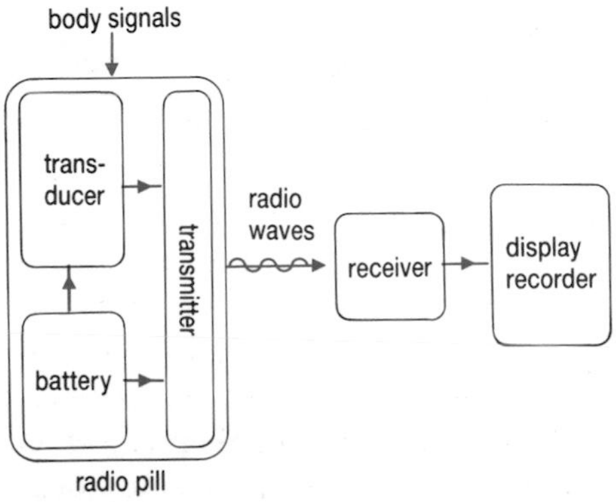

One of the simplest examples of taking readings at a distance is the stethoscope. The sound of the heartbeat is amplified acoustically and transmitted along a tube to the ear of the physician. The transmission of the electrical signal from the heart is probably the most widespread use of telemetry, at the present time. It was also one of the earliest, having been transmitted on the telephone system by the originator of the electrocardiogram in 1903, (see Chapter 5). Telemetry's common use nowadays is between the place of a cardiac emergency such as an ambulance and the hospital, where a cardiologist can interpret the signal and instruct the ambulance crew in resuscitation procedures. The method of communication is usually radio, and the methods are called **radiotelemetry**. Other frequent uses are in monitoring patients who have certain conditions which put them at risk. In such cases it may be necessary to implant the assembly of transducer, transmitter and power supply. This so-called 'radio pill' (Fig 4.8), is either

Fig 4.8 Radiotelemetry system.

swallowed or introduced surgically. The radio transmitter is essentially an electronic circuit which produces an alternating current (the oscillator) in a coil of wire (the aerial). This generates a radio wave which is detected in a corresponding coil, the receiver.

Most of the important parameters of the body have been measured and transmitted in this way, using appropriate transducers in small, encapsulated pills to protect them against body fluids. Examples include pressures and temperatures in various body cavities such as the vascular system, stomach and the intestines.

4.5 COMPUTERS

The availability of ever greater computing power at reducing cost in recent years has affected hospitals, as much as anywhere else. The use of computers in measurement, imaging and control is now very widespread, and has transformed the types of equipment now in use as the following chapters will show. The development of databases for patient records and so on is now changing the way hospitals are run, as you will find in the final chapter of this book.

Components of a computer

Essentially a computer is an electronic device which can automatically and rapidly perform a long and complicated sequence of operations as directed by instructions called a *program*. It can also store and retrieve large quantities of information. All computers comprise four elements: the *calculator* or *arithmetic unit*, the *memory*, to store data and instructions, *input–output* (I/O) *devices*, to allow the computer to communicate with the world, and the *control unit* (Fig 4.9).

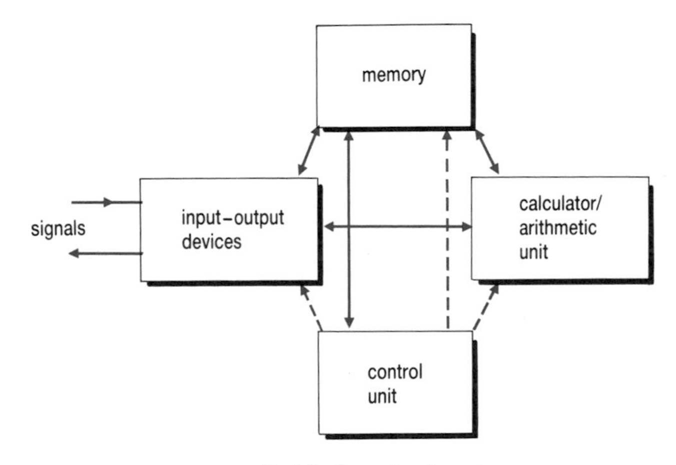

Fig 4.9 Computer elements.

The reason for the rapid growth of computing has been the successful miniaturisation of electronic circuits in semiconductors. This has led to the production of tiny integrated circuits, microprocessor units and memories of great complexity, at relatively small expense. The basic structure of all these components is like a series of switches, and the fundamental operation is that of switching 'on' or 'off'. All information processed by the computer must be reducible to this binary state, when it is called a *bit* of information. These characters are often manipulated in groups of 8 or 16 bit words, called *bytes*. Most information from biomedical measurement is produced as a continuously varying, *analogue* signal, so it must be changed to *digital* information, that is given a numerical value, before the computer can handle it. This is done by an *analogue-to-digital converter* (A/D), and its reverse, the *digital-to-analogue converter* (D/A), is used to display the product of the computer processing. This electronic circuitry and physical equipment is called the *hardware*; the programs of instructions, the computer *software*.

This is of two types, the *system* software which manages the operations of the computer and the *applications* software which enables specific functions to be carried out.

Biomedical applications

Examples of specific computer uses will be found throughout the remainder of this book. They include the following categories:

- data acquisition,
- data processing,
- mathematical calculations,
- pattern recognition and limit detection in monitoring,
- storage and retrieval,
- statistical analysis,
- data presentation,
- control.

INVESTIGATION

Temperature monitoring using a microcomputer

Consider how the simple measurements of body temperature that you made earlier in this chapter could be extended to investigate the dangers of athletes overheating, as in the news item about Mark Dorrity.

1. *Plan* an investigation incorporating the following features and any others that you consider important:
   - **(a)** continual measurement of temperature whilst the exercise is being performed;
   - **(b)** a method of warning the athlete when his/her temperature rises too high;
   - **(c)** the effect of water consumption on temperature.
2. *Discuss* your plans with your teacher and *carry out* as much as possible depending on the equipment, time and money available to you. A range of data collection devices are available from equipment suppliers (e.g. VELA from Educational Electronics, EMU from Philip Harris), which take data from a temperature sensor, see Fig 4.4(b), and output it in suitable forms, directly or to a computer.
3. *Record* your results and *communicate* any conclusions you may draw from the work, in an appropriate form, to the P.E. department of your school or college.

SUMMARY ASSIGNMENT

4.8. Write an essay on the increasing use of computers in biomedical measurement. Use the description of the components and the list of application categories, together with some background reading of current journals, books and articles from the bibliography, to ensure that your account is up-to-date and clearly explained.

(You may wish to leave this question until you have studied in more detail, some specific areas of measurement and imaging in later chapters.)

Further reading

Cochran, T. Medical electronics and physiological measurement, *Physics Education*, vol 24, no 4, July 1989 p 201.

Curnow, J. Computing applications in medicine, *Physics Education*, vol 24, no 4, July 1989 p 207.

Allen, J. *Telecommunications*, Macmillan 1990

Chapter 5

ELECTRICAL POTENTIAL

There are many important systems in the body which generate electrical signals in the course of carrying out their functions. These are the bioelectric potentials associated with nerve conduction, heartbeat, brain and muscle activity, and so on. By the use of electrodes these signals can be measured and used in medical diagnosis. This chapter is an account of the process, with particular reference to that most vital of organs, the heart.

LEARNING OBJECTIVES

When you have studied this chapter you should be able to:

1. give an illustrated account of the origin of biopotentials in the human nervous system;
2. describe the action of the heart as a pump to the circulation system;
3. describe the method of measuring an electrocardiogram (ECG), recall its main features, and explain its use in diagnosis;
4. give examples of other measurements of the body's electrical activity, and of the therapeutic use of electricity.

5.1 BIOPOTENTIALS

Luigi Galvani, an Italian anatomy professor, discovered the presence of electricity in living tissue, with his observations, in 1786, of the stimulation of a frog's leg, which you may yourself have seen repeated. It was not until 1903 when the Dutch physician Willem Einthoven found a way of measuring these potentials, that any practical use was made of them. The advent of microelectronics and the computer, as described in the previous chapter, have now made their measurement a simple routine, and led to important advances in knowledge of the functioning of the body.

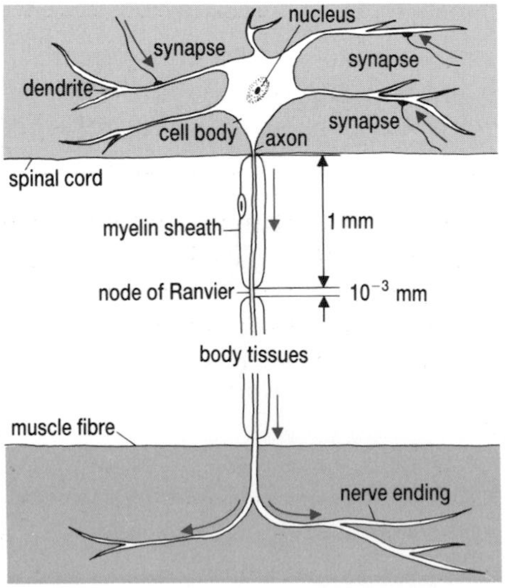

Nerve cells

The communication system of the body is the nervous system, which passes its messages along long, thin nerve fibres, called *axons*. A typical nerve which controls movement is shown in Fig 5.1. The main cell body, containing the nucleus is situated in the spinal column, and the axon runs from here to the muscle to be moved. So the axons to the foot for example, are up to 1 m long. The nerves pass messages between each other at the *synapses*, which are connected to the branching ends of the nerves, the *dendrites*. The axon is typically only a few micrometres in diameter, and covered with a sheath of *myelin*. This acts as an insulator between the axon and the surrounding tissue. There are gaps in this sheath every millimetre and at these, the *nodes of Ranvier*, charge can flow, by the passage of ions, to attenuate (reduce), or amplify (increase) the nerve signal. This flow will depend on the local biopotential.

Fig 5.1 A nerve cell from spinal cord to muscle.

Resting and action potentials

The axon wall is a membrane through which only certain things can pass under certain conditions. The small K^+, Na^+, and Cl^-, ions can permeate but

Fig 5.2 The source of biopotential in a nerve fibre.

not the larger organic ions such as proteins, so the membrane is said to be semipermeable. In its resting state, the fluid inside the axon, the axoplasm, contains an excess of K^+ and negatively charged organic ions, and the fluid outside, the interstitial fluid, contains an excess of Na^+ and Cl^- ions. The imbalance between the Na^+ and K^+ ions is maintained by an imperfectly understood process called the sodium–potassium pump. This transfers Na^+ out of the cell and K^+ into it. Because the membrane is relatively impermeable to Na^+ ions, they do not diffuse back with the concentration gradient, but the K^+ ions can. This results in an excess of positive charge on the outside and a potential difference between the inside and outside of about 90 mV, the inside being negative compared to the outside (Fig 5.2). This is the **resting potential** of the axon, which is said to be polarised.

When there is a stimulus to the nerve the membrane suddenly becomes much more permeable to the Na^+ ions, which therefore flow into the axon to reduce the potential difference. This process of depolarisation takes only a few milliseconds, and continues until there is a positive potential inside of about 20 mV (Fig 5.3). This is called the **action potential**. As soon as this potential is reached, the membrane again becomes less permeable to Na^+ and more permeable to K^+ which flow out of the axon to repolarise the cell. Over a longer period of time (about 50 ms) the Na^+ ions and the K^+ ions are exchanged across the cell wall, by the sodium–potassium pump. This ensures that the resting axon is in a state that can respond to the next stimulus.

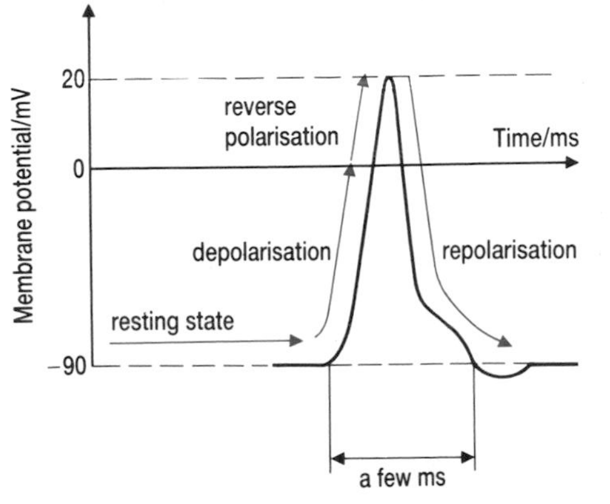

Fig 5.3 Resting and action potential of a nerve.

It is called a pump because it builds up an energy store in the imbalance of the ions across the membrane, like water being stored behind a dam. When the stimulus causes the permeability to change, the flow of charge is like the water flow when the floodgates are opened. The whole operation is designed to amplify the electrical stimulus signal as it passes along the nerve fibre at a speed of about 150 ms^{-1}, with the depolarised region triggering a change in the adjacent section of the axon.

The changes in the potential are a characteristic of the nerves concerned, those in the heart usually travel more slowly, last longer (150 to 300 milliseconds) and rise to a higher level (40 mV), (see Fig 5.6).

QUESTION

5.1 Draw a diagram to show the movement of ions to match the sequence of these changes: resting state – depolarisation – reverse polarisation – repolarisation.

5.2 THE CARDIOVASCULAR SYSTEM

The circulation of the blood by the heart is the most important system in the body, if this fails to work satisfactorily death occurs in a very short time. It is not surprising therefore that measurements of this system are routine. In this chapter we will cover the use of biopotentials in the operation of heart muscles, in the next, the measurement of blood flow.

Components of the system

The heart (*cardio-*) can be considered as a double pump, circulating the blood through two major tubular (*vascular*) systems, the *pulmonary* circulation to the lungs, and the *systemic* circulation to the rest of the body. A simplified anatomical diagram of the heart is shown in Fig 5.4 and a schematic illustration and description of the system in Fig 5.5.

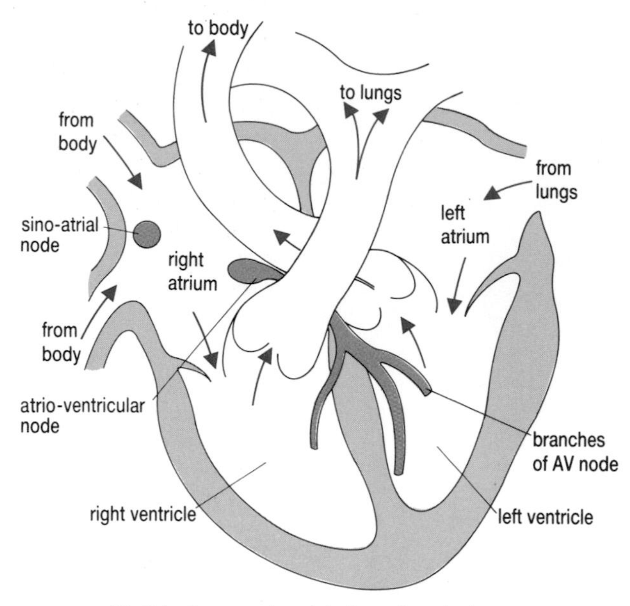

Fig 5.4 Cross-section of the heart, from the front.

QUESTION

5.2 **(a)** Make brief notes on the cardiovascular system from the above information, if you have not previously studied it (supplemented by a GCSE biology text if necessary).

(b) Why would you expect the walls of the left ventricle to be much stronger than those of the rest of the heart?

(c) Why is the blood that leaves the heart in the arteries at a much higher pressure than that which returns to it via the veins? Why is the pressure in the pulmonary artery less than that in the aorta?

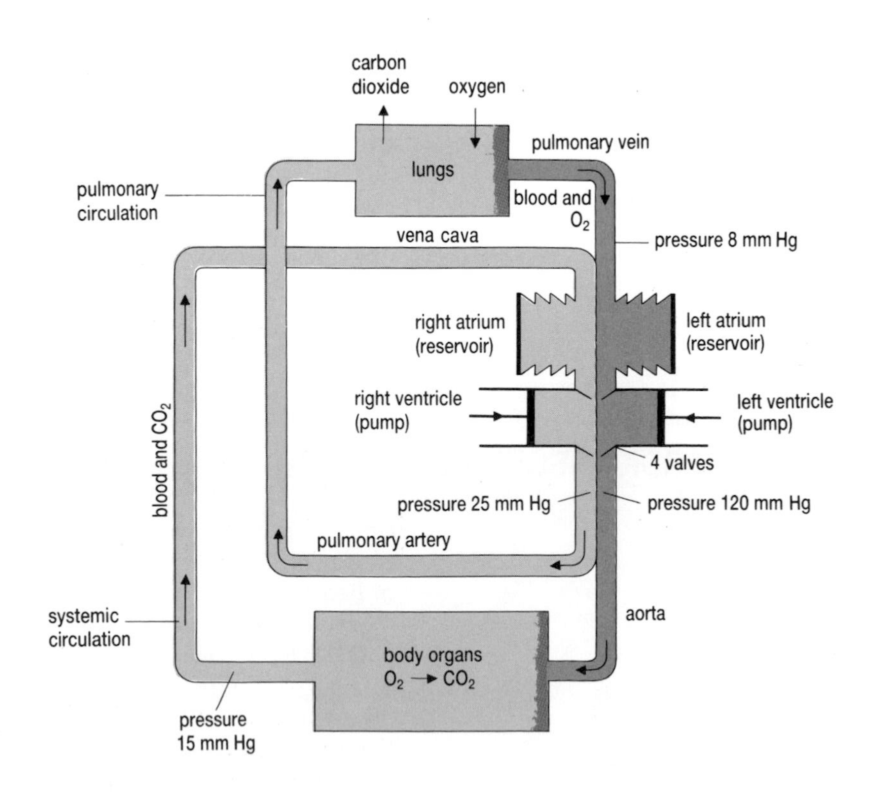

Fig 5.5 Schematic diagram of the circulatory system.

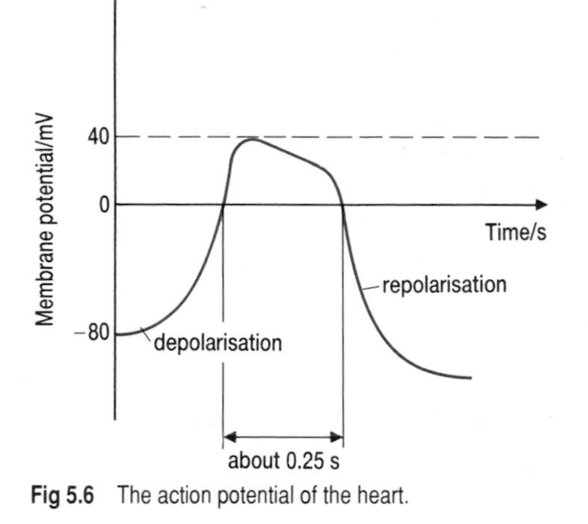

Fig 5.6 The action potential of the heart.

Controlling the heart's action

All muscular activity is associated with the changes in electrical potentials, which result from the migration of ions, as described above. The regular pumping action of the heart is controlled by special muscle cells, the *sino-atrial (SA) node* or *pacemaker*, located in the right atrium (Fig 5.4). This produces an electrical stimulus about 70 times a minute and initiates the depolarisation of the nerves and muscles of both atria. These contract and pump blood through one-way valves (tricuspid and mitral) into the ventricles. Repolarisation follows as the ions move to reduce the potential, and the muscles relax. The electrical signal then passes to the *atrio-venticular (AV) node*, which initiates the depolarisation of the two ventricles. They contract and pump blood through two more one-way valves (pulmonary and aortic) into the two circulatory systems. The ventricle nerves and muscles then repolarise and relax so that the sequence starts again. The action potential is shown in Fig 5.6, and the complete cycle represents a heartbeat. The rate will depend on the body's demands on the circulatory system, being greater for example at a time of greater physical exertion.

5.3 ELECTROCARDIOGRAPHY

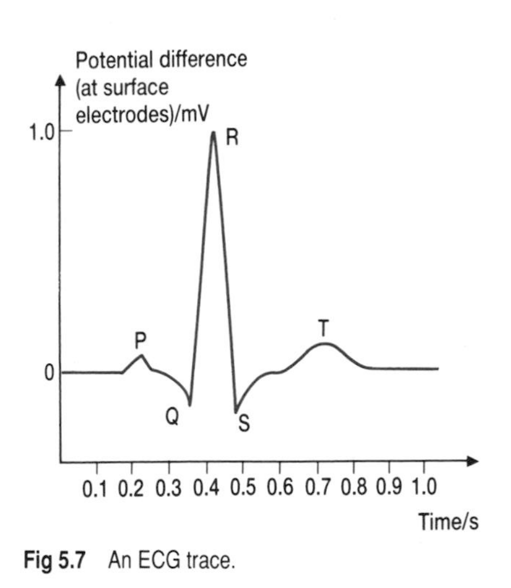

Fig 5.7 An ECG trace.

The electrocardiogram

During each heartbeat, the spread of the action potential through the heart, causes potential differences between depolarised and polarised cells. The conducting nature of body fluids transmits these potential differences so that they can be detected at the surface by electrodes. The signals are very much reduced in size by their passage through the body having typical values of 1 to 2 mV. They need amplifying before they can be recorded, electrodes must be placed in particular positions, and the patient must be relaxed so that no other signals interfere with the heartbeat. Their display over time is called an **electrocardiogram (ECG)**, and their study, electrocardiography, is a method of diagnosing various diseases and conditions of the heart.

A typical ECG is shown in Fig 5.7 and includes the following important features, which are usually referred to by letter labels:

- The P-wave due to the depolarisation and contraction of the atria.
- The QRS-wave due to the depolarisation and contraction of the ventricles.
- The T-wave due to the repolarisation and relaxation of the ventricles.

QUESTION

5.3 With reference to Fig 5.7,
  (a) Why is the wave for the repolarisation and relaxation of the atria not visible?
  (b) What percentage of the original membrane potential is detected at the surface (refer to Fig 5.6)?

Measuring the ECG

This was first carried out by Einthoven in 1903 and though the equipment has changed out of all recognition since that time (Fig 5.8), much of the terminology which originated then has become standard throughout the world.

The waveform measured at the surface depends not only on the subject, but also on the placing of the electrodes. At least two will be needed to measure a difference. These are called *bipolar leads*, and any two of three limbs are used. The right leg is not used as it is farthest from the heart, though it may be used as an earth connection to minimise interference from the equipment. In *unipolar leads* one or more electrodes are held at zero potential, while the other records differences. Thus one of these limb sites

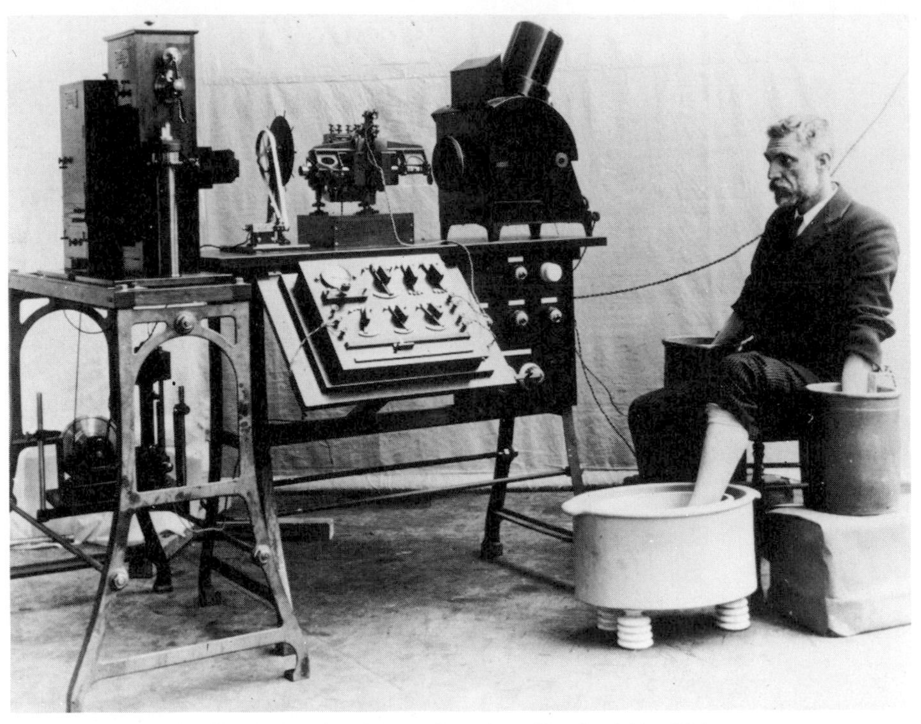

Fig 5.8 The first electrocardiograph machine, Cambridge 1908.

can be compared to the average potential of the other two, or all three limb sites can be connected together to give a neutral, with the 'active' electrode placed in one of six specified chest sites (Table 5.1). This gives a total of twelve alternative arrangements. It is obviously most important in any diagnostic use to know exactly where these were situated for a given ECG, so a labelling system of these standard sites is used, as shown in Table 5.1.

Table 5.1 Labelling of ECG leads

Type of load	Active electrode	Neutral electrode(s)	Name
Bipolar	Right arm	Left arm	Lead I
	Right arm	Left leg	Lead II
	Left arm	Left leg	Lead III
Unipolar limb	Right arm	Left arm and leg	aVR
	Left arm	Left leg and right arm	aVL
	Left leg	Right and left arm	aVF
Unipolar chest	One of six chest sites	Both arms and left leg	V_1 to V_6

INVESTIGATION

Measurement of ECG

There are a number of commercially available sets of equipment for measuring and displaying ECGs. If one is available to you, take this opportunity to carry out a simple investigation, following the manufacturers' instructions (e.g. Biological amplifiers from Unilab can produce a signal for display on a CRO for capture and processing by a microcomputer, using an appropriate interface).

1. Try the different electrode position and compare results.
2. What is the effect of exercise, the time of day, a stimulant such as coffee etc. on heart rate?

QUESTIONS

5.4 Describe the results of your investigation, and discuss the factors which were important in the production of a reliable, reproducible signal. (If you were unable to carry out a practical investigation you should still be able to make some predictions!)

5.5 What are the advantages of the following presentation styles of ECGs and when would each be most useful:
(a) continuous monitoring by display on a CRT screen;
(b) a permanent record (photographic or numerical)?

5.6 **(a)** Fig 5.9 shows a modern ECG recorder, what are the advantages of the various features given in the caption?
(b) Work out the pulse rates (in min^{-1}) from the ECG's of Fig 5.10.

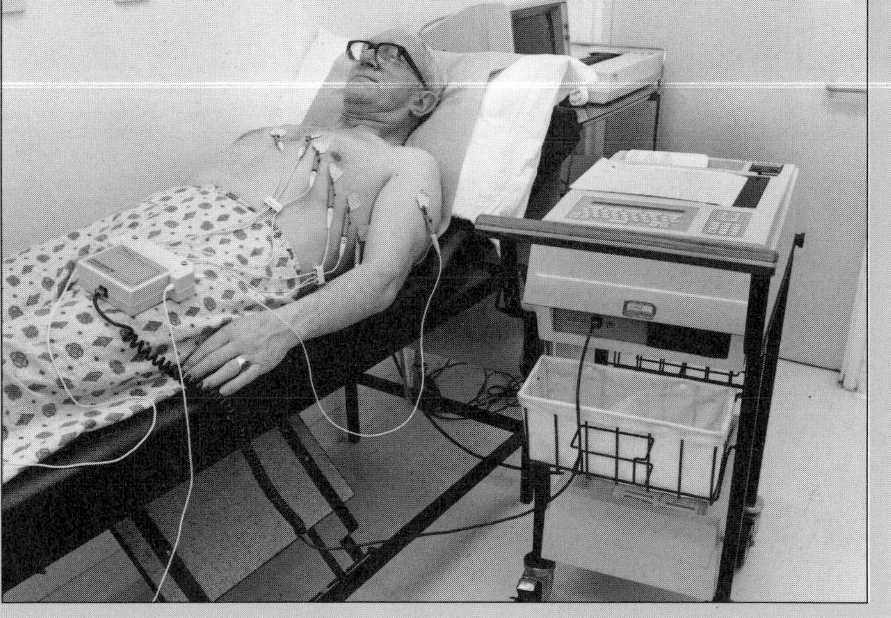

Fig 5.9 A modern three channel ECG recorder with store, playback and automatic diagnosis facilities.

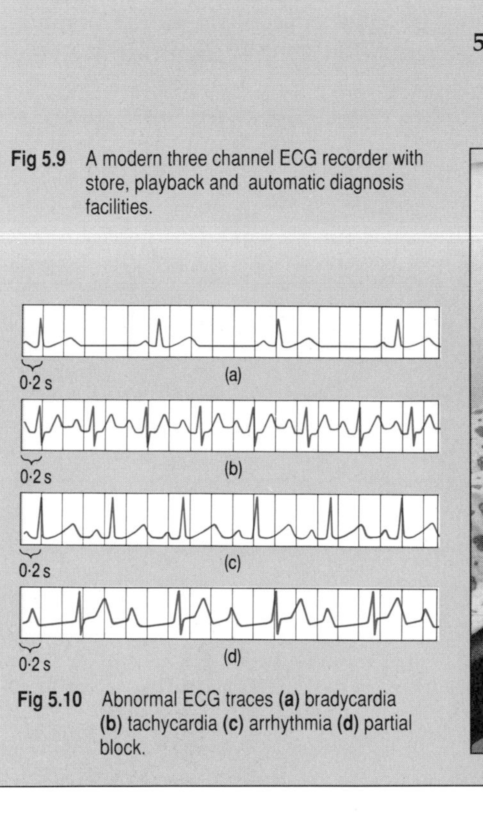

Fig 5.10 Abnormal ECG traces **(a)** bradycardia **(b)** tachycardia **(c)** arrhythmia **(d)** partial block.

Heart attack, myocardial infarct, and coronary thrombosis are names given to the blockage of a coronary artery, depriving the heart muscle of blood. The effect is a sudden collapse by the person who suffers severe chest pains. By the age of 65, one man in five in Britain has had a heart attack; about 50 per cent of them will have been fatal. Total deaths are about 150 000 a year, making this the most frequent cause of death.

Diagnosis

The following are examples of the indications of cardiac disorders that can be detected with the ECG :

- Normal heart rate is 60 to 100 beats per minute. A slower rate is called *bradycardia,* Fig 5.10(a), which may occur in athletes. A faster rate may be produced by sudden emotion, exercise or fever, and is called *tachycardia,* Fig 5.10(b).
- Irregular trace repetitions indicate irregular pumping, *arrhythmia,* Fig 5.10(c). This is quite common in young people. It is increased by deep breathing and reduced by exercise.
- A blockage of part of the heart, due to heart disease will result in a part of the trace not showing. If the heart muscle has been damaged, for example by a 'heart attack', the pumping action will be defective and the wave heights reduced, Fig 5.10(d).
- If the heart fails to pump properly, for example as a result of shock, this will show as a very irregular trace. The condition is called *ventricular fibrillation,* and it can lead to death in a very short time.

5.4 OTHER ELECTRICAL POTENTIAL MEASUREMENTS

Muscle activity

The nervous system and the muscles of the body can be investigated in a similar way to that described for the ECG. Muscle is made of motor units which consist of a single branching neuron from the brain or spinal cord and between 25 and 2000 muscle fibres (cells). The record of the electrical signals during movement is called the **electromyogram**, (EMG). Individual fibres cannot be investigated because of their small size, but a single motor unit can be measured by means of a needle electrode inserted into the muscle, or a small area can be measured with a surface electrode (Fig 5.11). The signals can be obtained either from the normal movement of the muscle, or more usually from direct electrical stimulation. Muscles can normally function for long periods without fatigue, provided that the frequency of restimulation allows a relaxation period of about 0.2 s. In the muscular disease *myasthenia gravis* this continued action is lost. The EMG can also be used to measure the speed of conduction of the action potentials in the motor and sensory nerves. A reduction in this, significantly below normal, indicates nerve damage.

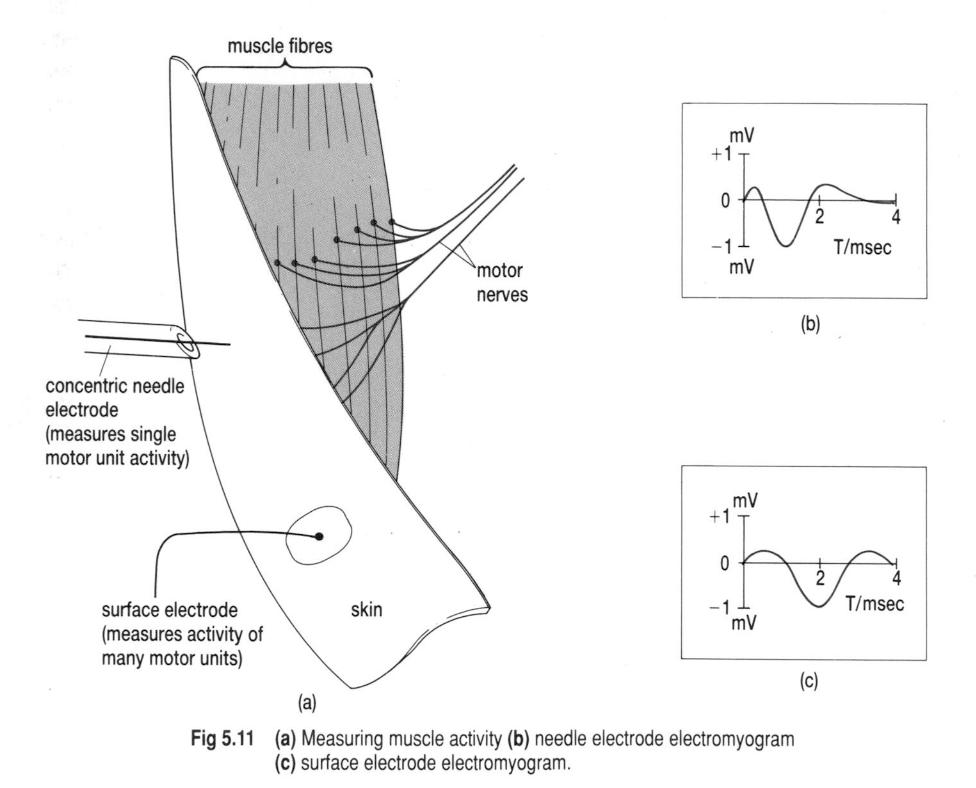

Fig 5.11 (**a**) Measuring muscle activity (**b**) needle electrode electromyogram (**c**) surface electrode electromyogram.

Brain activity

When you think of the concentration of neurons in the cortex of the brain, you will realise that the problem is to discern what is actually being measured, when electrodes are attached! These are usually small discs of chlorided silver in locations that, as with ECGs, are internationally agreed and labelled. The reference or neutral electrode is attached to the ear. The resulting **electroencephalogram (EEG)**, is a trace which has a very low amplitude (about 50 mV), and a frequency which seems to depend on the mental activity of the subject. The small size of the signal can cause problems with noise from, for example, the muscle potentials from eye movements. As the brain frequencies are low the signal is usually passed through a filter to remove the higher frequency EMG signals, as well as being amplified for display or recording.

The signals are thought to originate from groups of neurons whose activity is normally synchronised to produce regular bursts of charge from different sites over a period of time. A waveform contains variations of both amplitude and frequency, but certain ranges or 'rhythms' are distinguishable. The common frequency range for a relaxed or lightly sleeping subject is 8 to 13 Hz, the alpha brain wave. When a person is more alert the frequency increases to the beta wave range, above 13 Hz. In deep sleep the frequency falls to the theta (4 to 7 Hz) or delta (0.5 to 3.5 Hz) ranges, Fig 5.12 (a and b). Sometimes a sleeping subject shows a higher frequency pattern called paradoxical sleep or rapid eye movement (REM) sleep, because the eyes move during this period. REM sleep appears to be associated with dreaming.

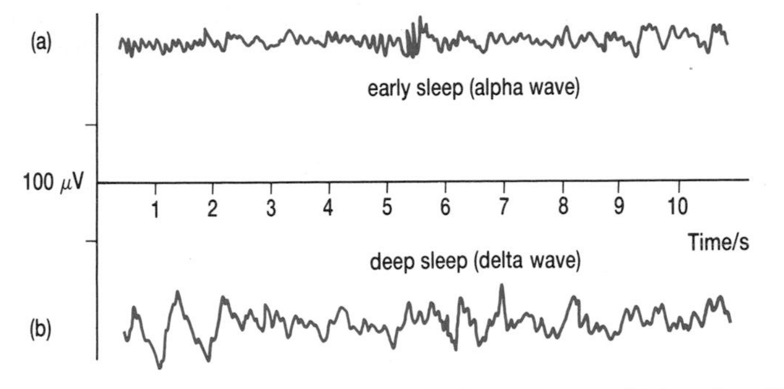

Fig 5.12 EEG traces for stages of sleep **(a)** early sleep (alpha wave) **(b)** deep sleep (delta wave).

EEGs are used in research on sleep and other investigations into normal brain function as well as for diagnosing brain disorders. One of the most distinctive is the high amplitude spikes which are associated with epilepsy (Fig 5.13). They can also be used in monitoring the condition of the patient during surgery to determine the degree of anaesthesia, for example.

Other uses of electrical potentials include the study of the retina and the movements of the eye.

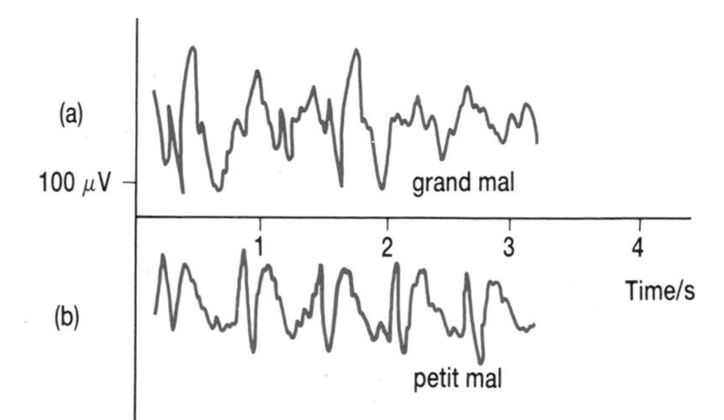

Fig 5.13 EEG traces for epileptic attacks **(a)** grand mal **(b)** petit mal

5.5 ELECTRICAL THERAPY

Electroconvulsive therapy (ECT)

This is a treatment of the brain with electrical signals, under general anaesthetic and has produced many improvements in the behaviour of patients suffering from mental illness, particularly severe depression. The way it works is not well understood, and it may have undesirable side effects, such as short-term memory loss. Though widely used, it remains a controversial technique.

Defibrillation

This is another dramatic intervention, the action that is taken when the cardiac monitor shows that the heart is failing to pump the blood; it is in fibrillation. A current of about 20 A at 3000 V is passed through the heart for about 5 ms. This produces a major contraction of all the muscles in the heart, which usually jolts the organ back into its proper rhythm. The 'shock' is applied by placing two large electrodes, called paddles, above and below the heart. As with all of the uses of electrodes, it is important that there is good electrical contact with the skin. This is achieved with the use of a conductive paste. In this case it is also important that the electrodes are insulated from the operator!

QUESTION

5.7 **(a)** How do you think the operator could be suitably insulated during defibrillation?
(b) How much energy (in joules) is passed through the heart in defibrillation?
(c) (i) What is the resistance of the circuit through the body in this case?
(ii) Where is most of this resistance located?
(d) Draw a defibrillation circuit which operates from the mains supply and incorporates a capacitor across the electrodes to prevent sparking.

Pacemakers

The pumping action of the ventricles is controlled by the atrioventricular node; if this is damaged pumping can go onto an autocontrol at a rate of about 30 beats per minute. At half the normal rate this is sufficient to maintain the life only of a very inactive person. To improve this situation the artificial pacemaker has been developed which produces a suitable electrical pulse 72 times a minute. It is placed in a cavity under the shoulder (Fig 5.14), by lifting a flap of skin, often under local anaesthetic. The wire to carry the signal is fed into a vein and advanced into the right ventricle by viewing the operation under X-ray fluoroscopy (details of fluoroscopic examination are given in Section 9.3). The pacemaker is powered with batteries that last several years and is enclosed in material that is impervious to body fluids. It has become a routine treatment in recent years.

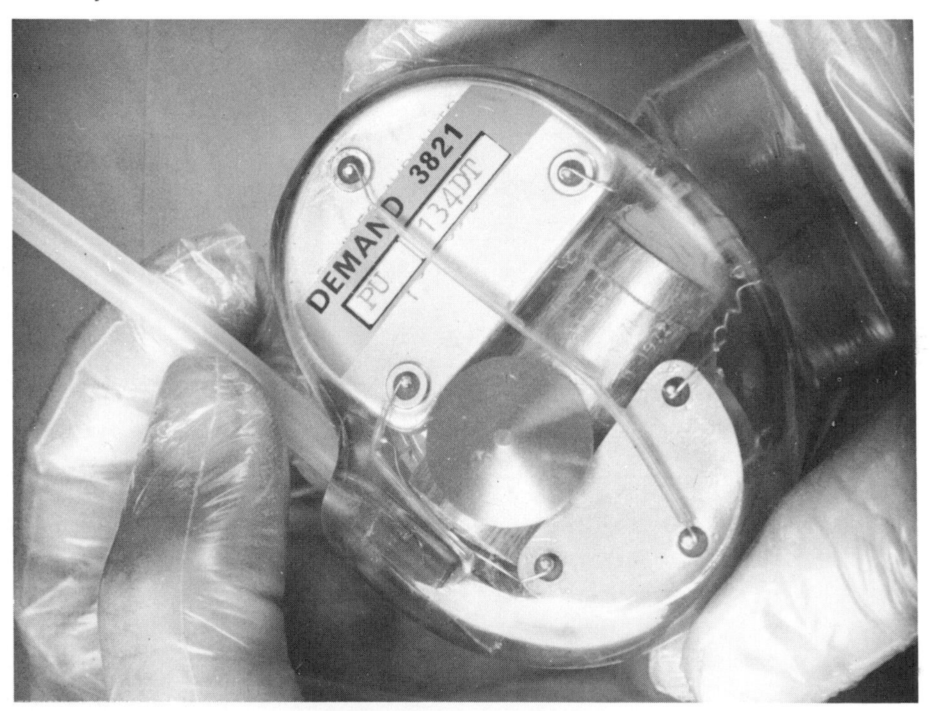

Fig 5.14 Heart pacemaker implant.

Diathermy

This is the technique of local heating of tissue by a high frequency source. Two methods are currently used, *short wave diathermy,* which uses an electric field oscillating at frequencies of around 30 MHz, and *microwave diathermy* which uses electromagnetic radiation of frequency 2450 MHz. In short wave diathermy the tissue to be heated is placed either between the plates of a capacitor or close to an inductor. Thus electrical heating is produced either by the changing field in the capacitor or the eddy currents around the inductor. Microwave radiation is absorbed directly into the tissue, and the frequency can be chosen to interact with particular molecules – in this case water. This interaction is thermal (it makes the molecules move more violently), rather than chemical as in the ionising radiations described in Theme 3; however, care must be taken not to cook the exposed tissue! Diathermy is used in the treatment of joints (in arthritis), muscles and tendons (in strains and other traumas). Many sports teams have a portable diathermy kit available to apply emergency treatment on the field.

Electrocautery and electrosurgery

More popularly known as burning and cutting (!) these common surgical tasks are often conveniently carried out with high frequency (>2 MHz), high voltage (~15 kV) electricity. This is supplied to a fine probe which is then used to stop bleeding during surgery by coagulating blood in vessels that are too small to tie (electrocautery), or to cut through tissue (electrosurgery).

Care must be taken that there is no burning at the other electrode (which connects the patient to the supply) and no shock hazard. The power density at the tip can be as high as 3300 W cm^{-3}. The cutting is produced by the extreme local heating (up to 800 °C) resulting in the physical rupturing of tissue by the rapid boiling of fluids. If the probe is moved rapidly the destruction can be limited to a depth of 1 mm from the probe. Electrosurgery is used in operations on the brain, bladder, prostate, spleen and cervix, often in conjunction with an endoscope (see Chapter 7).

SUMMARY ASSIGNMENTS

5.8 Draw up a table to compare the values of the electrical quantities mentioned in all sections of this chapter. These would include potential difference, current, frequency, power, time period, energy and perhaps others. Not all this data is given but you should be able to work some of it out. What conclusions (if any!) can you draw from these figures?

An alternative approach is to calculate the current from a knowledge of the ionic movement which generates the action potential, as in the following question.

An indication of the currents which flow in the body can be gained from laboratory safety regulations which include the following recommendations:
(i) In any work where there is a risk of the body coming into contact with mains supply, a circuit breaker should be fitted which will trip when 30 mA has flowed for 30 ms.
(ii) Equipment generating voltages above mains should be protected by a trip or fuse which operates at a current of 5 mA.

5.9 **(a)** Sketch a graph of the action potential as a function of time for a typical nerve axon, giving approximate scales on the axes. Air breaks down so that current flows when the electric field strength is 2.5×10^6 V m^{-1}. If a typical axon has a membrane thickness of 10 nm state, giving your reasoning, whether air or the membrane is the better insulator.

(b) 4.3×10^{-8} mol of sodium ions enter the core of an axon per square metre of membrane area during an action potential lasting one millisecond. Calculate the average electrical current density associated with this ionic flow and the average electrical current if the action potential involves a membrane area of 5.0×10^{-12} m^2.
charge of an electron = -1.6×10^{-19} C
the Avogadro constant = 6.0×10^{23} mol^{-1}

(c) Electrodes are placed on the surface of the body to record the cardiac waveform in a healthy person.

(i) Sketch a graph of potential difference between the electrodes as a function of time during a single beat of the heart, giving approximate scales on the axes.

(ii) Mark on the time axis the approximate points when the sino-atrial node is triggered and when ventricular stimulation occurs.

(iii) What change would you expect to find in the electrocardiogram of a patient suffering from poor ventricular contraction?

(iv) Why should the person under examination be as quiet and relaxed as possible?

(JMB 1987)

Further reading

Kane, J.W and Sternheim, M.M., *Physics* Chapter 18 Third edition John Wiley 1988

Chapter 6

PRESSURE

Living at the bottom of an ocean of air many kilometres deep, we are all subjected to the considerable pressure of the atmosphere. All parts of the body have to operate under this pressure. Fluid systems which require pressure to drive them must have pressure differences relative to this. When we breathe in or suck a drink through a straw, the pressure in the body will be less than atmospheric, the heart pumps blood at greater than atmospheric pressure. In this chapter we will be continuing the emphasis on the circulatory system, whose condition can be monitored in some detail by pressure measurements.

LEARNING OBJECTIVES

After studying this chapter you should be able to:

1. give a simple description of the human circulatory system, including typical blood pressures and flow rates, and give examples of the diagnostic importance of their measurement;
2. describe the use of the sphygmomanometer to measure blood pressure;
3. give examples of invasive methods of measuring blood pressure, including those using resistance, capacitance and inductance transducers;
4. explain the physical principles of resistance, capacitance and inductance transducers;
5. give examples of other fluid pressures within the body.

6.1 MEASURING PRESSURE

In biomedical measurement, pressures are usually given exclusive of atmospheric pressure. This is called the gauge pressure in contrast to the absolute pressure. Thus:

gauge pressure = absolute pressure – atmospheric pressure.

Units

Pressure is defined as the force per unit area, so its unit in SI is N m^{-2}, which is given the name **pascal (Pa)**. This is unfortunately a rather small unit; atmospheric pressure for example is about 10^5 Pa, so you will often find the unit kPa (kilopascal) used. 10^5 Pa is called a bar.

The pressure in a solid is referred to as the stress. In a liquid or a gas the pressure can be calculated from:

pressure (p) = density (ρ) × depth (h) × gravitational field strength (g)

(For a derivation of this formula see a core physics text.)

This has a practical importance, as the basic method of measuring pressures in the body is using a manometer. You may have measured your lung pressure by blowing into a Water U-tube manometer and reading off the height of the column. Thus the pressure is given in centimetres of water (cm H_2O), or alternatively in millimetres of mercury (mm Hg) because it

Direct measurement of the circulation system was first carried out by Rev. Stephen Hales in London, in 1733. Using a sharpened goose quill to pierce with, and a goose trachea to make a flexible connection, he connected a 9 foot vertical glass tube to the artery of a horse, and found that the blood rose to an average height of 8 feet above the heart.

has a particularly high density (though it is a health hazard to carry out this direct measurement with mercury in the manometer).

The measurement of blood pressure by the manometer is rather more difficult, involving the insertion of a catheter tube into the blood vessel, which is called an *invasive* technique. This is supplied with salt solution (saline) at pressure, containing heparin to prevent clotting (Fig 6.1).

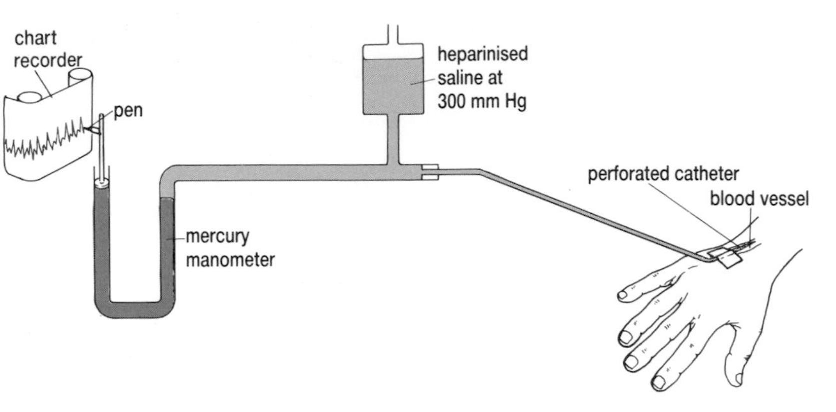

Fig 6.1 Schematic diagram of direct, invasive measurement of blood pressure.

QUESTION

6.1 What difficulties do you expect to be associated with this direct manometer method of measuring blood pressure? Which of these does the system illustrated in Fig 6.1 overcome, and which remain?

We will study a modification of this method, the sphygmomanometer, in which a manometer is used, but no direct contact with the blood is made, in the next section. Although many other methods are commonly used to measure blood pressure, especially when this is fluctuating, readings are still expressed in mm Hg. Equivalents in SI units and a range of other units still in common use, are given in Table 6.1

Table 6.1 Normal body pressures – typical values

	Typical Pressure /mm Hg
Arterial blood pressure – Maximum (systole)	100–140
– Minimum (diastole)	60–90
Venous blood pressure	3–7
Great veins	<1
Capillary blood pressure – Arterial end	20–35
– Venous end	10–20
Middle ear pressure	<1
Eye pressure – aqueous humour	~20
Cerebrospinal fluid pressure in brain (lying down)	5–12
Gastrointestinal pressure	10–20
Intrathoracic pressure (between lung and chest wall)	–10
The atmosphere	760
or 1.01×10^5 Nm^{-2}, 1.01×10^5 Pa, 1.01 bar	
1033 cm H_2O, 14.7 psi, 1 atm	

Values

Table 6.1 gives typical values of pressures within the body. You will note the highest fluid pressure is at the output of the heart, and that the pressure falls in stages through the circulation system.

The *cerebrospinal fluid* (CSF) cushions the brain within the skull. If the pressure builds up too high it can cause the condition of *hydrocephalus* ('water on the brain'). This is not uncommon in infants, due to a blockage in the outlet to the spinal cord. If this is detected soon enough it can often be corrected with an operation to install a bypass drainage system. CSF pressure is not usually measured directly.

The fluids in the eye are under pressure to keep the eyeball in a fixed size and shape. In Chapter 2 it was shown how the clarity of vision is very dependent on this. When you press on your eyelid gently you can feel this pressure. If you partially open your eye as you do this you may notice a loss of focus. Pressure build up in the eye can cause a condition called *glaucoma*, which results in tunnel vision and in some cases blindness. The pressure is usually measured with an instrument called a tonometer, which measures the indentation of the eye by a known force!

The pressures in the gastrointestinal tract can vary considerably as a result of the contents of the various parts and the action of various valves, which are designed to permit food to flow in only one direction. Blockages of these valves such as the pylorus between the stomach and the gut, of other parts of the gut will increase the pressure and this can have fatal consequences so surgery may be necessary. The pressure in the digestive system is linked to that of the lungs through the flexible diaphragm that separates the two organ systems. Thus the pressure of the digestive system can be increased voluntarily by taking a deep breath, closing the windpipe at the glottis, and contracting the abdomen (as in defaecation).

The highest pressures in the body are to be found in the load-bearing bone joints as described in Chapter 1, and the pressure would be even higher if the joints were not as large as they are. The frequent damage to cartilage suffered by people in sporting activities is a measure of the exacting conditions under which it lubricates the joints.

QUESTION

6.2 Calculate the values of some of the body fluid pressures listed in Table 6.1, in kPa.

6.2 BLOOD PRESSURE

Blood flow

The circulatory system has been described in Section 5.2, and illustrated in Fig 5.5 which shows how the pressure changes around the body. There is a corresponding change in the blood flow as the branching vessels take the blood into ever smaller arteries and then into veins of increasing size. A summary of the dimensions, blood flow velocities and blood pressures is given in Table 6.2. The values are averages, for easy comparison. You will notice that veins are wider than arteries. This is partly because arteries have thicker walls for the higher pressures they have to sustain. The veins hold about 80 per cent of the volume of the blood (about 5 litres for an adult) at any one time, so they serve as a kind of reservoir. The highest flow is at the aorta and pulmonary artery, where between 3.5 and 5 litres per minute flow in a normal adult at rest. This is called the *cardiac output*. In the capillaries by contrast the flow is so slow that the movement of individual blood cells can be observed under the microscope.

There is a direct relationship between pressure, flow rate and resistance in the blood vessels that is analogous to the electrical relationship. However, there are other factors that affect flow. The capillaries can be narrowed *(vasoconstriction)*, or widened *(vasodilation)*, to respond to body temperature control, as described in Chapter 4. Certain drugs such as nicotine can also produce vasoconstriction.

Nicotine is the habit forming substance in tobacco smoking. Its effect on the body is complex. Chemically it is a depressant producing a calming effect on the smoker. However it also has a stimulating effect which raises blood pressure. This puts an extra strain on the heart and so can contribute to heart disease, one of the dangers associated with smoking.

Table 6.2 Cardiovascular System – Typical values

Vessel	Number /1000	Diameter /mm	Length /mm	Mean velocity /mm s^{-1}	Pressure /mm Hg
Aorta	–	10.50	400	400	60–140
Terminal arteries	1.8	0.60	10	<100	35–50
Arterioles	40 000	0.02	2	5	40–25
Capillaries	>million	0.008	1	<1	25–12
Venules	80 000	0.03	2	<3	12–8
Terminal veins	1.8	1.50	100	1	<8
Vena cava	–	12.50	400	200	3–2

The flow is not constant throughout the cross section of a vessel, but increases as the distance from the wall increases. Detailed information of the flow profile can be obtained by Doppler ultrasound, which is described in Chapter 8. It can indicate the condition of the vessel walls. Obstructions can build up from fatty deposits on the walls (*atheroma*), when these form a complete blockage (clot or *thrombus*), the flow of blood to a part of the body is prevented. If this is in the brain the result is a *stroke,* in the heart muscle it is called a *coronary* or a heart attack. Clearly these are potentially fatal conditions, so the measurement of blood flow is very important.

Pulse

The pumping action of the heart causes a regular pulsation in the blood flow which can be felt in an artery near the surface, such as at the wrist. This is the pulse, and a typical rate is about 70 per minute. We will now consider in more detail how the pressure changes with time. As the ventricles contract, they increase the pressure in the arteries leaving the heart, until this back pressure closes the valves from the heart. The blood then runs away through the arteries, so the pressure falls in the aorta and pulmonary artery. A pressure wave then passes around the circulatory system, getting smaller as it progresses to bring the venous blood back to the heart. The cycle then begins again, with the rising pressure period of

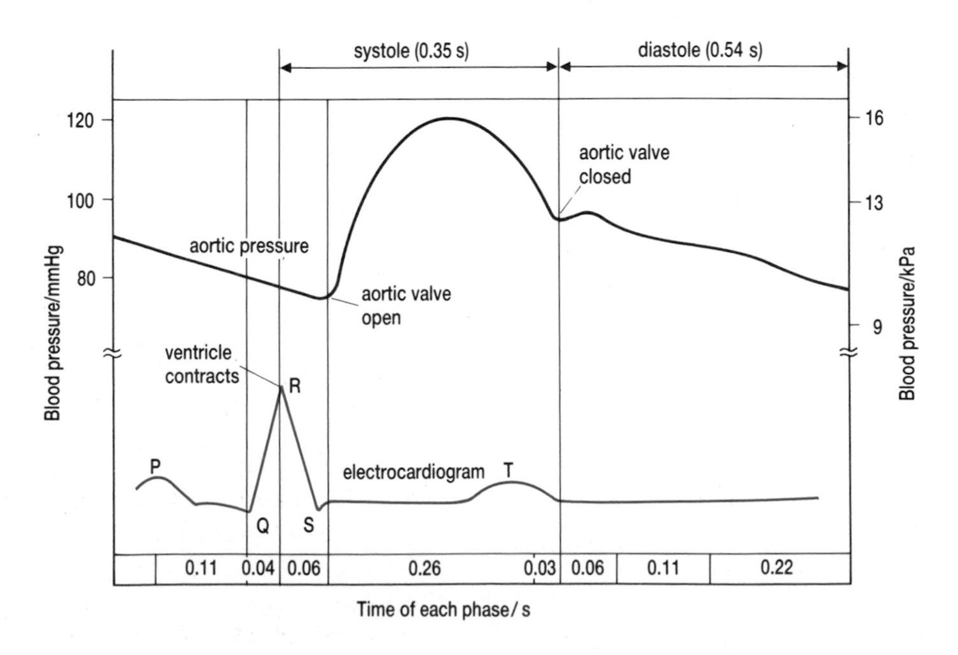

Fig 6.2 Pressure variation in the aorta, during one heart beat, and associated ECG.

It has been estimated that during the course of an average lifetime, the heart will have pumped a total volume of blood around the body, sufficient to fill the Royal Albert Hall in London. How many heartbeats would have occurred over this period?

about one third of a second, which is called the *systole*, being followed by the longer falling pressure period, the *diastole*. Fig 6.2 shows how this cycle relates to the electrical signals of the ECG. These changes are dependent on the physical condition of the heart and the blood vessels, so detailed measurement of blood pressure is a valuable diagnostic tool. For example, high blood pressure (*hypertension*), is a condition that can lead to heart failure, if it is not treated.

QUESTION

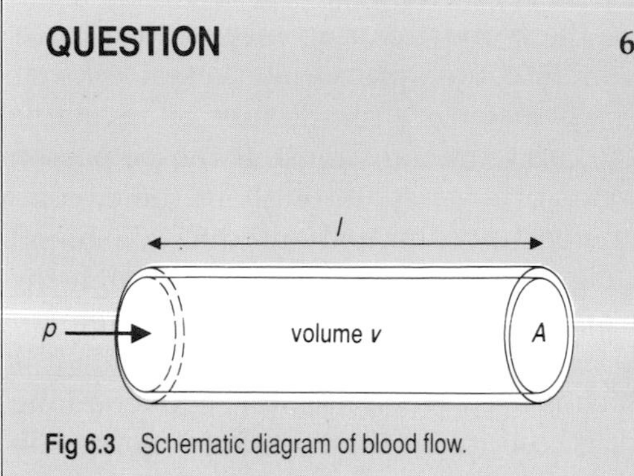

Fig 6.3 Schematic diagram of blood flow.

6.3 We can calculate the rate of working (the power) of the heart, from a knowledge of the flow rate and pressure of the blood.

Suppose blood flows with pressure p with flow rate v per second, through a major artery of area A (Fig 6.3):

(a) The work done by the blood per unit time is force (F) multiplied by distance travelled per unit time (l/t). Write down expressions for F and l/t in terms of p, A and v.

(b) Show that the power $P = p \times v$

(c) Calculate the power developed in an artery which has blood flowing at a rate of 5 $dm^3\ min^{-1}$ and a pressure of 125 mm Hg.

Sphygmomanometer

This instrument is used for routine blood pressure measurements. It is easy to use and involves no direct contact with the bloodstream. It is however less accurate than those methods that do, is not suitable for continuous monitoring, and gives only the systolic and diastolic pressures, with no indication of the shape of the waveform shown in Fig 6.2. The method often fails when the pressure is very low, for example, when the patient is in a state of shock.

The instrument consists of an inflatable cuff and a manometer. The cuff is strapped round in the arm at heart height. Air is pumped in manually until a pressure of 200 mm Hg is applied to the brachial artery, so that it obstructs the blood flow. When a stethoscope is placed on the artery below this point no sound is heard (Fig 6.4). The pressure is slowly reduced by releasing air until the blood begins to spurt through a small opening in the artery, when it is at its maximum pressure. This causes turbulence and the sound, called *Korotkoff sound*, can be heard in the stethoscope. This highest, systolic pressure can then be read on the manometer. These sounds continue as the pressure falls, until the blood vessel remains open throughout the pulse period. This means that the applied pressure has now fallen as low as the lowest, diastolic pressure. So a second manometer reading is taken when the Korotkoff sounds cease. The technique of listening for the sounds is called *auscultation*.

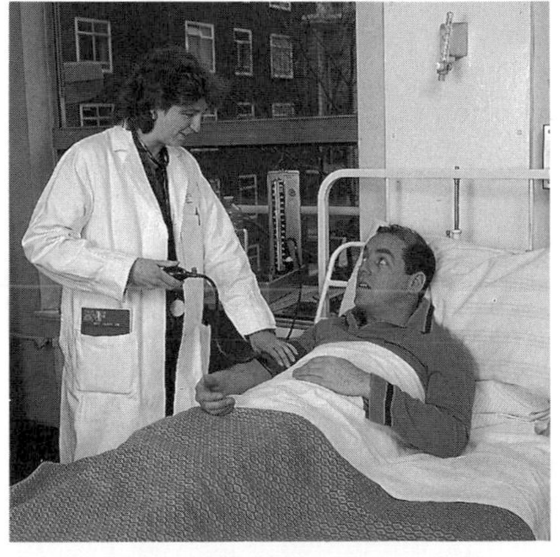
Fig 6.4 A doctor measuring a patient's blood pressure with a sphygmomanometer.

Electrical or automatic sphygmomanometers have been devised which incorporate a microphone or a pressure transducer to detect the onset and end of the blood turbulence as the pressure is reduced. These are obviously easier to use and are sold for educational and personal health use. Trained medical staff usually prefer to use the auscultatory method however as this gives more reliable results.

INVESTIGATION

Measurement of blood pressure

Use a manual or electronic sphygmomanometer to measure blood pressure. Try to identify factors which increase and decrease blood pressure. Things to try include exercise and caffeine to increase heart rate, and deep breathing or meditation to decrease.

QUESTION

6.4 (a) How reliable is the sphygmomanometer? Does operator practice improve this? What do you think are the major sources of inaccuracy?

(b) Why should the operator check that the position of measurement is at the same height above the ground as the heart?

6.3 INVASIVE MEASUREMENTS

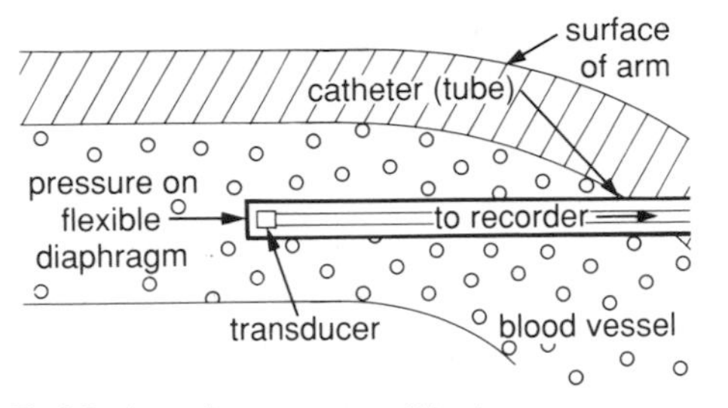

Fig 6.5 Internal measurement of blood pressure

Because of the shortcomings of the indirect methods it is often necessary to introduce a transducer directly into the bloodstream to measure pressure. This is done by inserting a **catheter** into the appropriate vessel *percutaneously*, that is through the skin. The catheter is a plastic tube of a suitable diameter to fit into the vessel and to carry the transducer. It can be passed via the superficial (near surface) veins or arteries, throughout the circulatory system and into the heart. Catheters are commonly used to take blood samples and to administer drugs. Fig 6.5 shows one inserted through the skin with an adaptor in place for all these functions. The transducer can either be ready mounted at the end of the catheter or introduced while it is in place. An alternative method which is used when monitoring over a long period, or at a distance is required, is to implant the transducer surgically at the site to be measured.

Transducer types

The principle of the pressure transducer is that it produces a change in an electrical circuit as a result of movement produced by the flow of blood. A thin, flexible **diaphragm** (of stainless steel, or phosphor bronze) is usually placed in the bloodstream, and the transducer attached to it. There are a number of different types available. The selection of instrument will depend on the measurement site and the size and range of the pressures to be measured. For example the pressure in the arteries is high (up to 120 mm Hg) and changes rapidly, so a high frequency response is important. In the veins the fluctuations are slower and the pressures much lower (from 0–10 mm Hg), so the sensitivity of the instrument is more important. The properties of resistance, capacitance and inductance are all used in pressure transducers. The **resistance** transducer is the most common, and is often called a **strain gauge**.

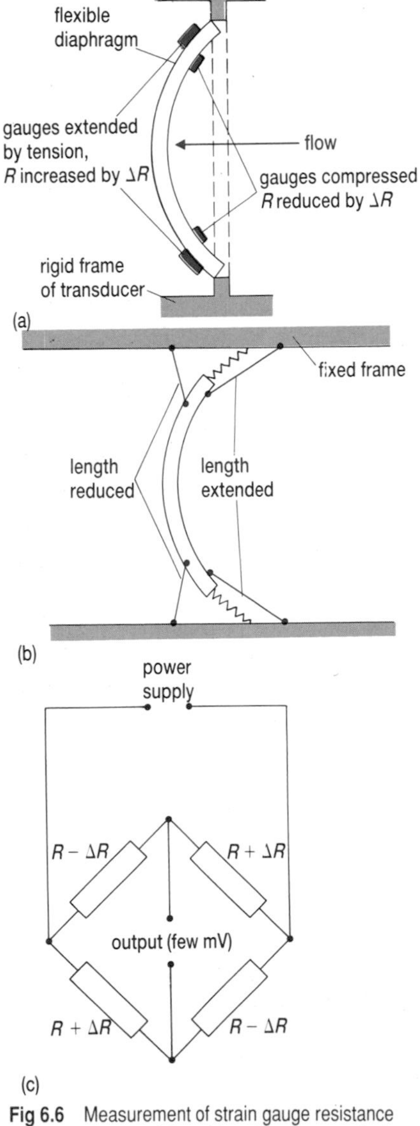

Fig 6.6 Measurement of strain gauge resistance changes **(a)** position of gauges on diaphragm (bonded type) **(b)** unbonded strain gauge **(c)** bridge circuit.

Resistance transducers

When the dimensions of a wire change because of stretching, its resistance shows a corresponding change, since

$$R = \rho L/A$$

The resistance elements are attached to a deflectable diaphragm, placed in the flow, to produce a distortion or strain. The devices are compressed or extended, Fig 6.6(a) to follow the strain of the diaphragm hence the name strain gauge.

The transducers may be of the **bonded** or **unbonded** type, (Fig 6.6). In the bonded type the wire is arranged in a zig-zag pattern and firmly bonded to the diaphragm so that it moves in the same way. In the unbonded type one end of the wires are connected to the diaphragm and the other to a membrane or frame which can move relative to the diaphragm. This causes two of the wires to stretch and the other pair to relax. With very thin wires this can provide a sensitive response.

In order to compensate for temperature fluctuations, the gauges are usually arranged in a bridge circuit, Fig 6.6(c). The four have the same resistance when unstrained and will be affected equally by any temperature changes. When the diaphragm moves the change ΔR is then the result only of the dimension change. The transducer inside the body is connected to the bridge circuit and its power supply, and can be calibrated to read pressure directly.

The alternative is to use a semiconductor. The silicon bonded gauge has been widely used for its sensitivity to very small displacements. For example with a supply voltage of 10 V it can produce a 30 mV signal from a 3 μm displacement, over a pressure range of 300 mm Hg. The sensitivity of a pressure transducer is usually expressed as microvolts per (applied) volt per mm Hg. Alternatively the **gauge factor** is used to express sensitivity, this is the ratio of the fractional change in the resistance to the fractional change in the length:

$$G = \frac{\Delta R/R}{\Delta L/L}$$

The gauge factor of a metal is typically 2, that of a semiconductor about 150. Semiconductor gauges are very temperature sensitive however, so they are not used where temperature fluctuations are large.

QUESTION

6.5 (a) What is the pressure sensitivity of the silicon bonded gauge quoted here?

(b) A semiconductor pressure transducer of length 3 mm is extended by 3 μm in use. What would be the extension of a metal wire transducer of length 8 mm, to produce the same fractional change in the resistance? (Use the gauge factors given above.)

Capacitance transducers

The capacitance of a parallel plate capacitor is given by:

$$C = \varepsilon A/d$$

Where d is the distance between the plates, A is the area of overlap and ε is the permittivity of the insulator between them, called the dielectric. If any of these is changed, by attaching one plate or the dielectric to the diaphragm, (Fig 6.7) then the capacitance will change. This is the principle of the capacitance transducer. The diaphragm itself may form one plate of the capacitor, separated from the fixed plate by a few hundredths of a millimetre. The capacitance is measured using an a.c. circuit. The transducer may form one arm of a bridge circuit so that the imbalance gives a measure of the capacitance. Alternatively, included in a tuned circuit it will vary the resonant frequency of the circuit. A third method is to measure the voltage across the capacitor transducer when it is in series with an inductor. Compensation for temperature changes is achieved in the same way as with resistance transducers, by using identical capacitors on opposite arms of an a.c. bridge, so that one increases its capacitance whilst the other reduces, when the pressure is applied. This arrangement is called a differential capacitance transducer.

The advantage of this type of transducer is that it can have a very linear shape and so not obstruct the flow of blood. The displacement is very small

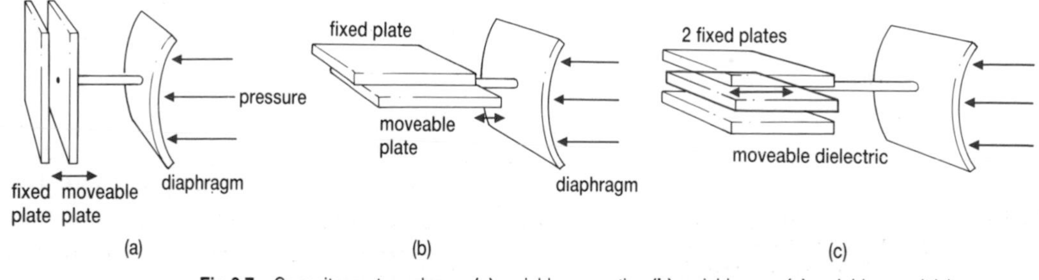

Fig 6.7 Capacitance transducers **(a)** variable separation **(b)** variable area **(c)** variable permittivity.

so they are very sensitive and have rapid response times. They are however more difficult to use than the resistance type and suffer from interference from the capacitance of the lead wires.

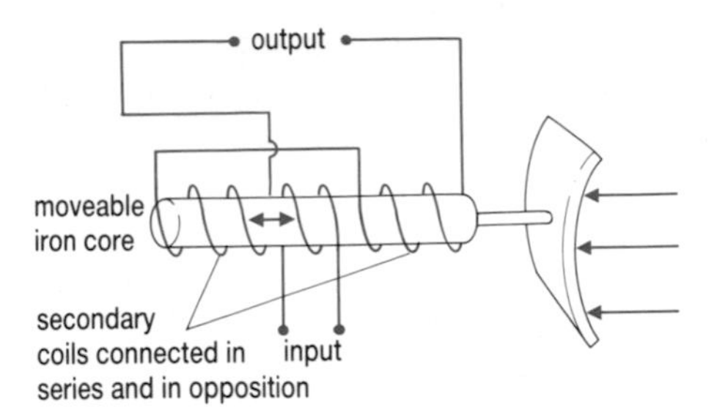

Fig 6.8 A linear variable differential transformer.

Inductance transducers

The inductance of a coil depends on the amount of permeable core within it. If this core is moved then the self-inductance of the coil will vary. Thus the inductance will depend on the pressure which moves a diaphragm connected to the core. Changes in inductance are measured by an a.c. bridge, a series circuit or an oscillator circuit. The core can link a primary and a secondary coil, so that the mutual inductance changes, but the most common arrangement is to use two identical secondary coils as shown in Fig 6.8. This is called a linear variable differential transformer (LVDT). The primary coil is given an audio frequency signal which induces voltages in the two secondary coils in opposition to each other, as they are wound in series opposition. All three coils are linked by the core of a high permeability ferromagnet, which is attached to the diaphragm. As the core moves from the balanced null position the voltage induced in one coil increases whilst the other decreases. This results in an output voltage which varies linearly with the position of the core and therefore the pressure. The core can be spring-loaded to accept pressure from one side, or it can accept pressure from both sides and measure the pressure difference between two points.

The LVDT is sensitive and has a rapid response. It is also insensitive to temperature changes and quite robust. It is however rather larger than the two preceding types.

Piezoelectric transducers

These are made of materials which generate a potential difference when they are subjected to pressure. They are active transducers, requiring no external power supply, as noted in Chapter 4, in contrast to the transducers previously described. They are not used directly in the detection of blood pressure as their sensitivity is limited at these pressure ranges. They have an important role in the measurement of blood flow rate in Doppler ultrasound. In this reflection technique they both generate and detect the ultrasound. This is described in Chapter 8.

SUMMARY ASSIGNMENTS

6.6 Explain which of the following types of transducer are, in their basic form, capable of measuring displacement and which measure pressure by measuring velocity: piezoelectric crystal, LVDT, capacitance transducer, bonded stain gauge, unbonded strain gauge.

6.7 List the factors which are important in choosing suitable methods of measuring blood pressure. What are the advantages of the methods described in this chapter, in relation to these factors?

6.8 **(a)** Name **two** types of bioelectric signals which are often measured at the surface of the body. Sketch the waveform you would expect to observe for **one** of the signals, indicating the type it represents. State, giving the reason, **one** precaution you would take in attaching electrodes to the surface of the skin to obtain satisfactory signals. The signal at the electrodes is often very small and requires considerable amplification. State **two** further requirements, apart from large amplification, which an amplifier should fulfil if it is to be suitable for such measurements, explaining why they are important. (No electrical circuit diagrams of amplifiers are required.)

(b) Describe, with the aid of a diagram if appropriate, **one** type of pressure measuring instrument or pressure transducer, mentioning how it is connected to the source of pressure. Give **one** application in medicine for which you consider it suitable.

6.9 The following passage is based on extracts from **Foundation of Biophysics**. Read the passage carefully and then answer the questions at the end.

The cardiac cycle of human heart

The sequence of events that occurs during one heartbeat constitutes the cardiac cycle, and is most easily described graphically. Fig 6.9 shows two pressures as functions of time during about two cardiac cycles.

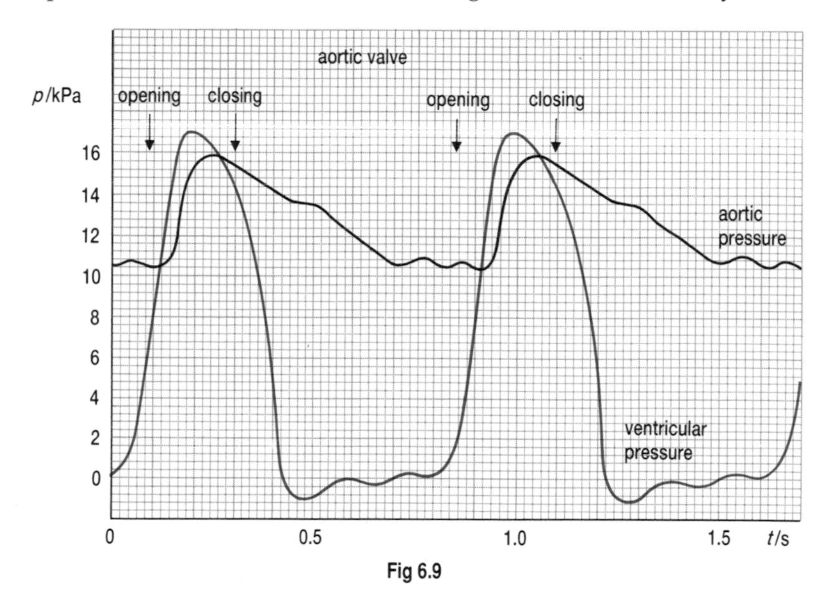

Fig 6.9

Energetics of the heart function

The heart is a transducer, converting chemical energy into mechanical energy of contraction, and finally transmitting that energy to the blood. The heart moves the blood, providing kinetic energy of amount

$$E_k = \frac{1}{2}\rho V v^2$$

where ρ is the density of blood, V the volume it occupies, and v its velocity.

Since we shall be considering the energy and power delivered by the heart to the circulatory system at the same level as the heart, changes in gravitational potential energy may be ignored. We must, however, think about the pressure energy as a form of mechanical energy. When a fluid is under pressure it has the capacity to do work by converting the pressure energy into other forms. For example, when a volume liquid is pumped to a height, h, above ground level, the pressure at ground level is

$$p = \rho g h$$

$$= \frac{m}{V} gh$$

$$= \frac{E_p}{V}$$

since E_p, the pressure energy, is equal to mgh, the increase in potential energy which it has produced.

The total mechanical energy generated by the heart in pumping into the aorta is thus

$$E = E_k + E_p$$

$$= (\tfrac{1}{2}\rho v_A^2 + p_A)V$$

For our purposes we may assume that blood is an incompressible fluid. This means that for the range of pressures encountered in the bloodstream, the volume of the blood will not change appreciably. Although arteries are elastic to some extent it is a good approximation to assume that the dimensions of the circulatory system are constant. Thus the volume flow rate, K, (the volume passing a plane in the system per unit time) must be the same everywhere in the body. We shall consider steady conditions of the circulatory system, i.e. assume that the blood's volume flow rate is constant.

The power, P, developed at any instant by the heart pumping blood into the aorta is given by

$$P = \frac{E}{V} \times K$$

$$= (\tfrac{1}{2}\rho v_A^2 + p_A)K$$

Of course, v_A and p_A vary considerably during the cardiac cycle. In addition to the aorta there is a second tube, the pulmonary artery, into which the heart pumps blood. We may therefore express the average power developed by the heart in terms of the velocities and pressures in the great arteries by

$$\overline{P} = (\tfrac{1}{2}\rho \overline{v_A^2} + \overline{p_A} + \rho \overline{v_A^2} + \overline{p_P})K \quad (1)$$

in which the bars represent average values.

We must now invoke experimental information in order to proceed. It has been determined that the velocities of the blood in the aorta and pulmonary artery are approximately the same, whereas the pressure inside the aorta is six times greater than that inside the pulmonary artery. The average velocity of blood, $\overline{v}$ can be simply related to the volume flow rate, K, and the cross-sectional area, A, of the aorta, Also the mean square velocity that appears in equation (1) is different from the square of the mean velocity, the two being related by

$$\overline{v_A^2} = 3.5(\overline{v_A})^2$$

Substituting this experimental information into equation (1) gives

$$\overline{P} = \frac{3.5\rho K^3}{A^2} + \frac{7\overline{p_A}K}{6} \quad (2)$$

Equation (2) represents the average power developed by the heart for a steady-state condition.

(a) (i) Explain what is meant by the phase steady conditions as used in the passage.

(ii) What evidence is there in the figure to justify the assumption that the circulatory system operates under steady conditions?

(b) (i) Show that $\overline{v} = K/A$

(ii) Show that Equation (2) follows from Equation (1) under the experimental conditions described in the passage.

(c) Use Fig 6.9 to estimate

(i) the pulse rate in pulses per minute,

(ii) the mean pressure of blood in the aorta,

(iii) the fraction of the cardiac cycle for which the aortic value is open.

(d) Reduce the units of the two terms $\frac{1}{2}\rho v^2 V$ and pV to their basic units of mass, length and time. Are the results what you would expect from the equations given in the passage? Explain your answer.

(e) For a person at rest the two terms on the right hand side of Equation (2) contribute 0.04 W and 1.3 W respectively to the total power developed by the heart. During strenuous physical activity K increases by a factor 7, with all other parameters remaining essentially unchanged. Calculate the power developed by the heart in the case, and comment on your answer.

(ULSEB 1983)

Further reading

The body report, *Observer Magazine* 1988

Chapter 7

OPTICS

The investigation of the human condition using light can hardly be avoided. All patients are seen by their doctors and external appearances are used in many diagnoses. Internal observation could obviously bring much additional information. Few opportunities are as simple as the mouth, all that is required here is to 'open wide and say "aah"'. Specially focused beams enable internal viewing of the eye (with an opthalmoscope) and the ear (with an auroscope). In most cases a tube (an endoscope) has to be inserted into an orifice, to see inside the body.

There were 2000 cases of skin cancer in the UK in 1987 and its incidence has doubled in each of the past five decades. Rona McKie, Professor of Dermatology at the University of Glasgow believes this cannot therefore be linked to the depletion of the ozone layer and increased levels of ultraviolet light, caused by chlorofluorocarbons released from aerosols and refrigerants. That could lead to an environmental catastrophe of skin cancers in the future, she warns.

The Observer 1988

Good health has traditionally been associated with exposure to light. During the industrial revolution in Britain in the nineteenth century, air pollution and working conditions reduced this to the extent that many people suffered diseases caused by lack of vitamin D. This is manufactured by the body by absorbing sunlight. Now there is more concern about the dangers of too much sunlight. An increase in the incidence of skin cancer has been associated with the greater exposure of people with light coloured skin to strong sunlight – for example on beach holidays. This is an example of the ionising properties of the ultraviolet radiation in sunlight, details of which are given in Theme 3.

Laser light provides the opportunity to expose the body to light in a very specific way (Fig 7.1). The intense beam is able to cut (incise), and burn (cauterise), tissue in a way which makes the laser a useful addition to the surgeon's toolkit.

LEARNING OBJECTIVES

After studying this chapter you should be able to:

1. use the following scientific terms correctly: total internal reflection, coherent, numerical aperture, resolution, laser;
2. explain the principles of the transmission of light by optical fibres;
3. describe how the fibre optic endoscope is constructed; compare the operation of the video endoscope which uses a charge coupled device;
4. give examples of the use of endoscopes in diagnosis and treatment;
5. explain the principles of operation of the laser and give examples of its medical application.

7.1 FIBRE OPTICS

Transmission

When light enters the end of a glass rod, it will be **totally internally reflected** if it strikes the surface of the rod at greater than the critical angle (Fig 7.2). These reflections can continue along the length of the rod so that the light emerges with little loss of intensity. If the glass is made into thin filaments it is surprisingly flexible, and can carry light to inaccessible places. A bundle of these filaments produces the **fibre optic light guide**.

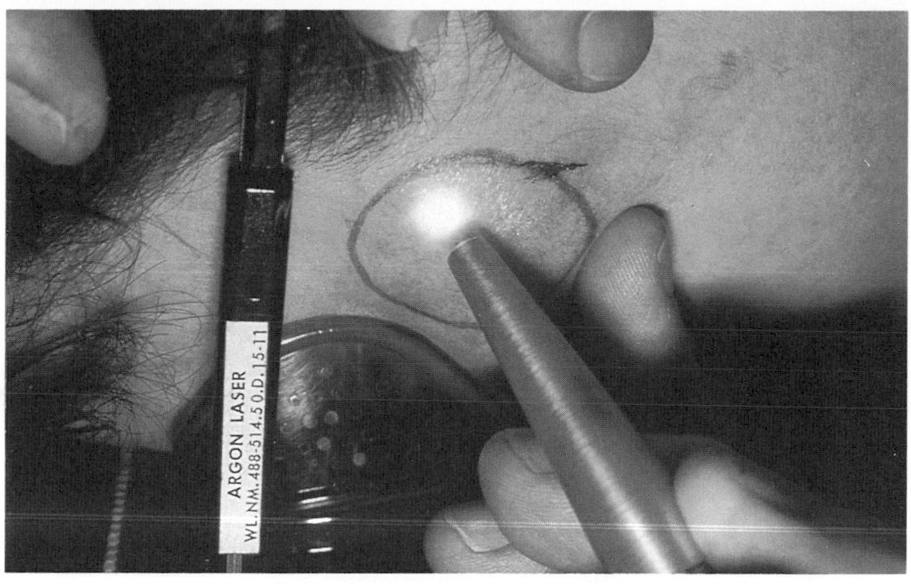

Fig 7.1 An argon laser removing a birthmark.

Light can easily be lost from the surface of the fibre if it is in contact with a substance of equal or higher refractive index. This could be grease or an adjacent fibre. To avoid such leakage the fibres are clad with glass of a lower refractive index than the fibre core. The selection of the core and cladding materials affects the performance of the fibre. In particular, the maximum

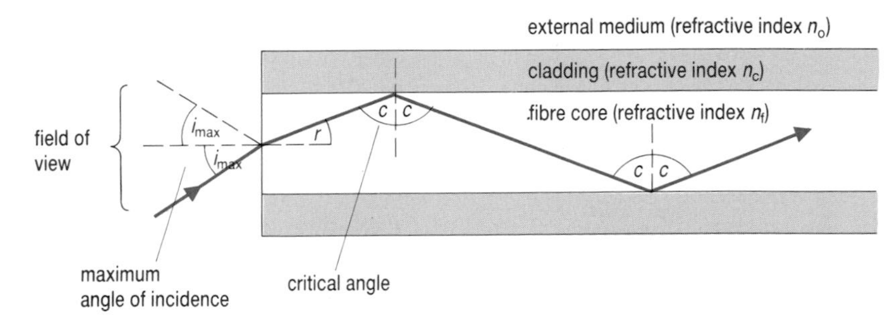

Fig 7.2 Internal reflection in a glass fibre.

The refractive index n_m of a medium is given by

$$n_m = \frac{\sin i}{\sin r}$$

where i is the angle of incidence of a ray of light in air or a vacuum, on the surface of the medium, and r is the angle of refraction within the medium.

angle of incidence from which the light can be admitted, i_{max}, is affected. i_{max} is called the half angle of the fibre since it represents half the field of view that can be transmitted. With reference to Fig 7.2 it can be shown that

$$n_0 \sin i_{max} = \sqrt{(n_f^2 - n_c^2)}$$

The term $n_0 \sin i_{max}$ is called the **numerical aperture**; a typical value is 0.55, giving a half angle i_{max} of about 33° in air. It can be seen that the numerical aperture increases with an increase in the difference between the refractive indices of core and cladding. There is a limit to the difference that can be produced however, since the absorption of light increases if the refractive index of the core is above a certain value.

This analysis is strictly true only when the fibre is straight, any curve will of course change the angle at which light strikes the surface. In practice curves as sharp as about twenty times the fibre diameter result in only minor losses of light.

QUESTION

7.1 An optical fibre has a refractive index of 1.55 and is clad in glass of refractive index 1.42. Calculate:

(a) the numerical aperture of the fibre,

(b) the half angle of the fibre in air.

Intensity losses are usually quoted in dB m^{-1} (see Chapter 3 for details of decibel scale). Typical values are: ordinary window glass: 0.1 dB m^{-1} best optical fibre: 2×10^{-4} dB m^{-1}.

Image formation

If the bundle is to be used to transmit an image of the inside of the body, then attention has to be paid to a number of factors. There will be a loss of intensity due to absorption and scattering by the glass, which will depend on the length of the fibre. Production methods are now such that there is very little loss over the length required for observing parts of the body (maximum length about 1 m).

There will also be losses at the entry to and exit from the fibre, and at the reflections along the length of the fibre.

Each fibre will transmit the light incident on its face, independent of all the other fibres. To transmit an image the fibres have to be kept in the same relative positions, however the bundle is bent (Fig 7.3), with a smooth square surface so that each fibre's element makes the correct contribution to the whole image. This is called a cohèrent bundle. The resolution of the image, the amount of detail that can be seen, depends on the fineness of an individual fibre, and how closely they are packed together. There is a limit to this, if the fibre diameter is very small diffraction effects interfere with the linear transmission. A typical diameter is 10 μm with about 60 per cent of the total area of the cross section of the bundle occupied by core (this is called the packing fraction). The size of the bundle depends on its use, some examples are given in the next section. Diameters range between 0.5 mm and 3 mm, containing between 5000 and 40 000 fibres.

In a bundle designed to transmit light rather than an image, the fibres can be larger and incoherently packed. The fibres are typically 30 μm in diameter, which makes them more efficient at transmitting light. It also makes the light guide much cheaper to produce than the coherent image guide.

The manufacture of the fibre bundle is an automated process. An individual glass rod is coated with the low refractive index glass and bundled together with thousands of others in a cover sheath of glass which can be dissolved in acid. This large bundle is heated and drawn to reduce the size. It takes many passes until the rods are reduced to the fibre diameter of about 10 μm (Fig 7.4). The glass cover has fused during this process and kept the fibres coherently arranged, it is then dissolved away to leave the fibres free to be flexible, but fixed in position by an external sheath.

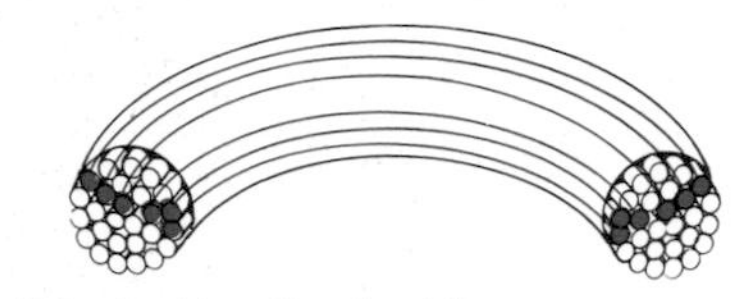

Fig 7.3 A coherent bundle of fibres.

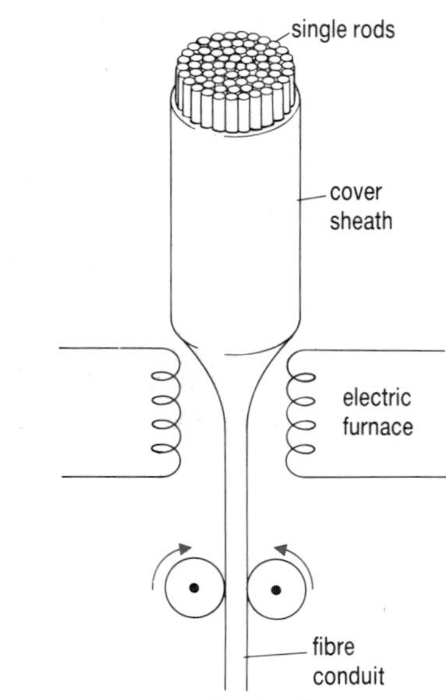

Fig 7.4 Production of a fibre bundle.

QUESTION

7.2 (a) What factors affect the following features of an image produced by an optical fibre:
(i) brightness, (ii) detail, (iii) coherence?

(b) A particular fibre loses 20 per cent if the incident intensity over a length of 0.75 m. If it is used in a bundle with a packing fraction of 60 per cent, what is the percentage of the incident intensity that will be received at the far end?

7.2 ENDOSCOPY

The endoscope is basically a tube for looking into the body. The person credited with its invention is a German, Kassmaul, in the late nineteenth century. His device was hardly practical, but the potential was realised and the technique of endoscopy was born. Most eminent of those who developed the instrument in the twentieth century was Rudolf Schindler. He produced a semi-flexible tube using arrangements of prisms and lenses to bend the light into an arc. The device probably still felt very inflexible to the patient; a key element in its initial use was Schindler's wife Gabrielle, who carefully reassured and manipulated the patient to enable the tube to be inserted.

The invention of the optical fibre in the 1960s came as a real breakthrough for the science of endoscopy, and a great relief to patients! (Fig 7.5)

The original endoscope was a straight, rigid metal tube, illuminated by an oil lamp. The first subjects for experimentation were sword swallowers. Legend has it that one of the first recruits exclaimed 'I'll swallow a sword anytime, but I'm damned if I'll swallow a trumpet!'.

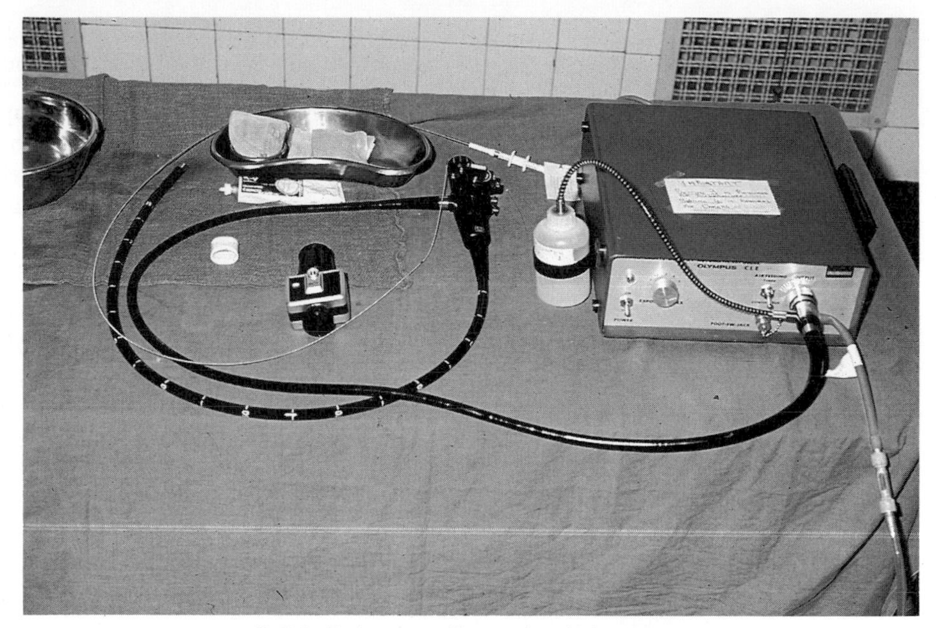

Fig 7.5 Equipment used in gastrointestinal endoscopy.

The fibre optic endoscope

The instrument is shown in Fig 7.6. The long flexible shaft containing the fibres is usually made of a helical steel band inside steel mesh, to prevent the glass fibres from damage whilst permitting the wide range of movements required. This is enclosed in a plastic sheath to provide waterproofing, chemical protection, and ease its passage into the body. This shaft

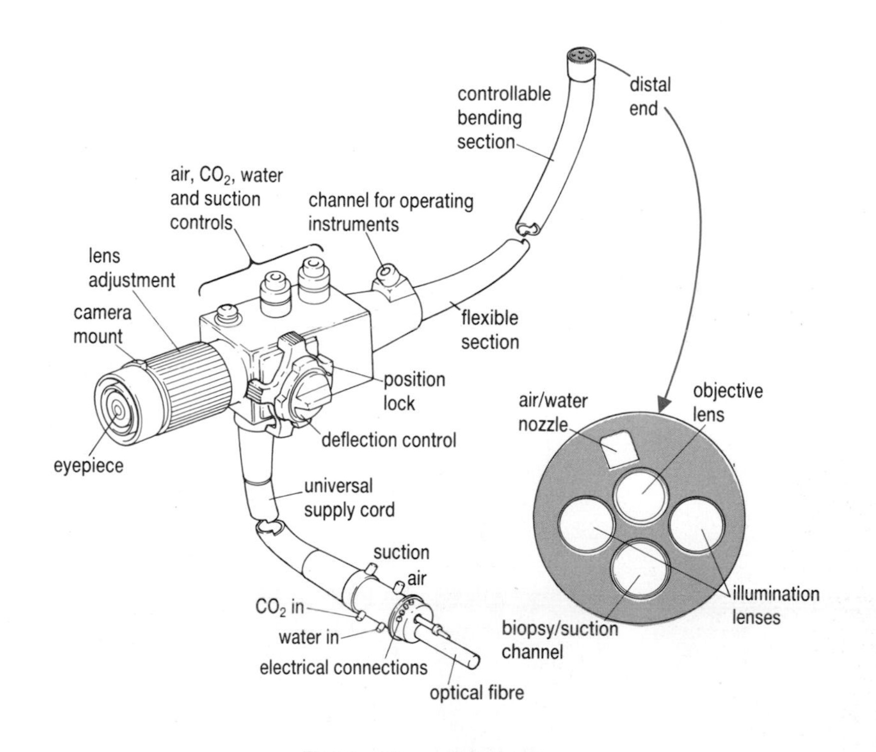

Fig 7.6 A fibre optic endoscope.

is about 10 mm in diameter and up to 2 m long depending on the application. At the far or distal end is a bending section which is fully controllable from the operator's end, to carry out manoeuvres and minor operations.

At the distal tip the fibres bringing the light and carrying away the image are fitted with lenses. The endoscope shaft contains the following:

- usually two non-coherent fibre optic bundles – the light guides,
- a coherent fibre optic bundle – the image guide,
- a water pipe to wash the distal face of the optical system,
- an operations channel used to insert surgical instruments,
- control cables to operate the end which is adjustable,
- in some cases a channel for suction and one for pumping in air or carbon dioxide gas.

The viewing, or proximal end of the endoscope contains the controls for all of these functions, an adjustable eyepiece and a connection to the light source and camera. The light source is a high intensity xenon lamp, with a lens coupling arrangement to permit wide angle illumination. The camera can be attached to the eyepiece of the image guide. A photocell monitors the brightness and informs the light source so it produces a flash for a suitable exposure.

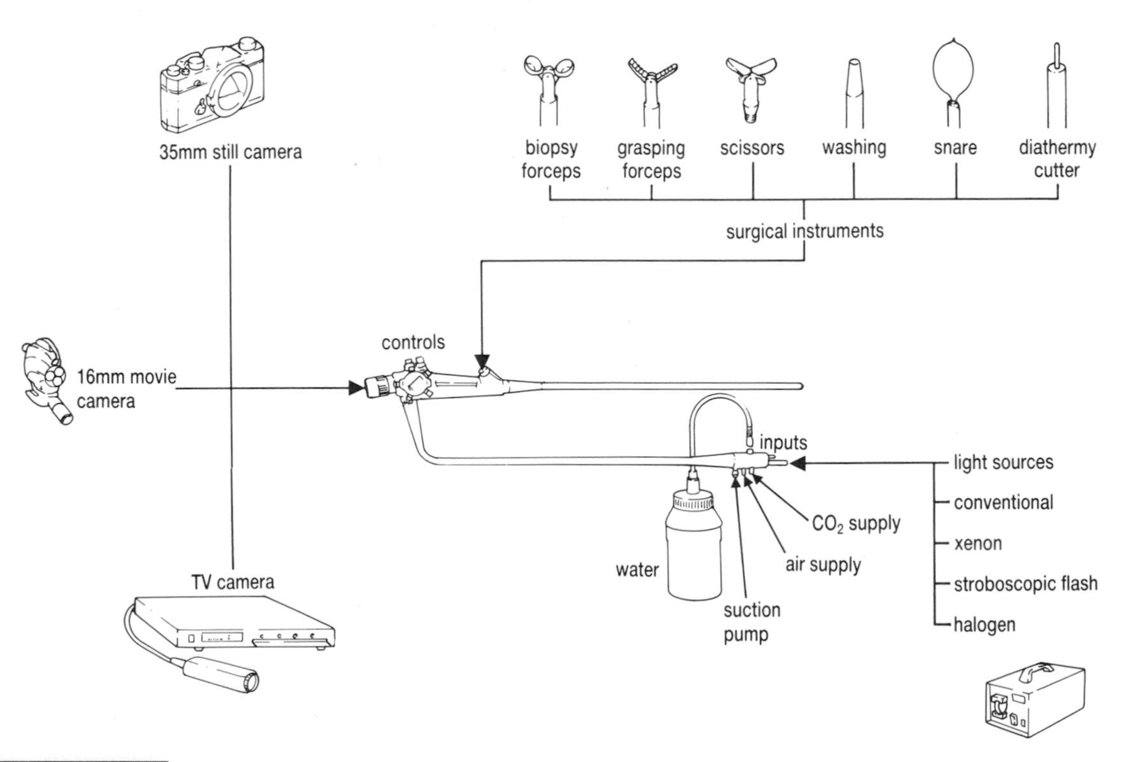

Fig 7.7 Some accessories for an endoscope.

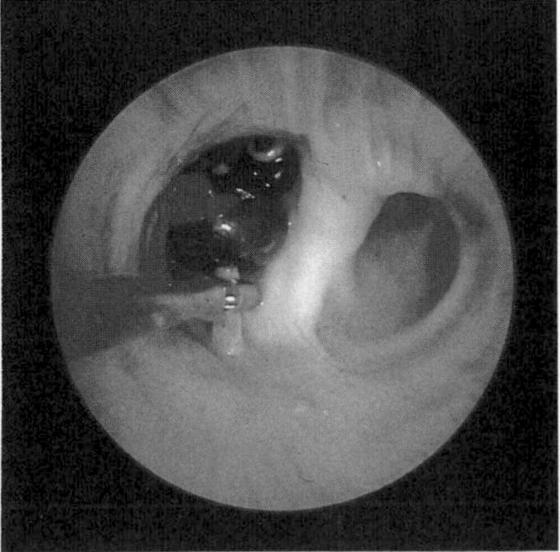

Fig 7.8 Endoscope image showing the removal of the top of a martini stirrer from the windpipe of a young child who had accidentally swallowed it.

Uses

A large variety of accessories and ancillary equipment is available to support the numerous applications of endoscopy, a selection of which is shown in Fig 7.7. Endoscopes have probed much of the body's plumbing, including the gastrointestinal tract (oesophagus, stomach and duodenum), the pancreas and biliary tract, and the colon. In the technique of laparoscopy the viewing tube is inserted through the wall of the abdomen to study the liver, spleen and other organs. The diagnostic information obtained from these observations is invaluable, providing direct and often very clear evidence of, for example, bleeding, ulcers, constrictions, benign and malignant tumours, and degenerative conditions such as cirrhosis of the liver. The endoscope also allows a range of minor surgical treatments, examples include: using forceps for taking samples of tissues (called biopsies), electrodes to apply heat to stop bleeding and various snares or extractor devices to remove obstructions such as foreign bodies (Fig 7.8).

QUESTION

7.3 Write a brief description of some of the uses of endoscopes for diagnosis and treatment, with reference to this passage and illustrations.

The video endoscope

In America in 1983 a radically different endoscope was shown in action for the first time. It makes use of a device which converts the received light into an electrical signal at the distal end of the scope. This is the **charge coupled device (CCD)**, originally developed to measure low levels of light detected in astronomy. The principle is like measuring the distribution of rainfall in a field by setting out an array of buckets before the rain and afterwards moving them on conveyor belts to a metering station where the amount in each bucket is recorded. The buckets are equivalent to small zones of low potential below an array of electrodes formed on the surface of a thin wafer of silicon. When the photons strike the silicon they produce charge in each potential well in proportion to the number of photons, (Fig 7.9). Each well becomes a small electrical image or pixel. The images are read off by sequential changes in the voltage on the electrodes placed across the grid. This is the charge coupling, and it enables the information from individual small areas to be recorded and used to build up the com-

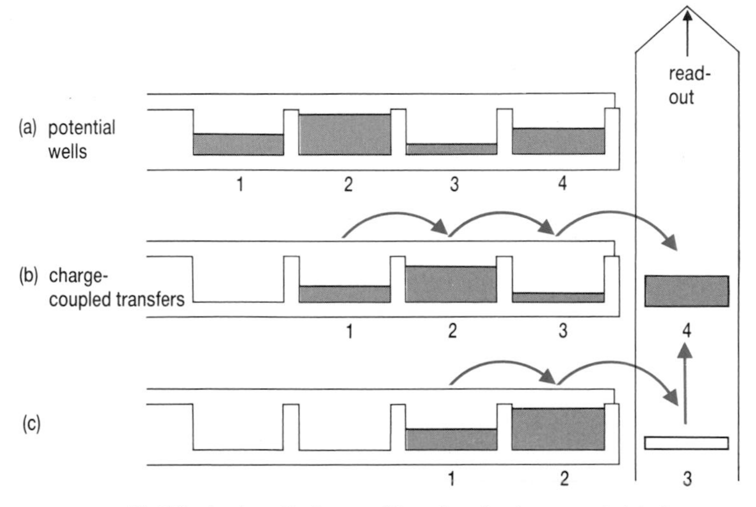

Fig 7.9 A schematic diagram of the action of a charge coupled device.

plete image. This electrical signal is passed by wire to a video processor which digitises the information for display on the TV monitor and for video or computer tape storage. The endoscope tube is similar to the fibre optic tube, except for the absence of the image guide bundle. The current designs of CCD have between 32 000 and 56 000 pixels, which is twice the number of fibres in a typical fibre endoscope. Use can also be made of the wavelength sensitivity of silicon, to identify particular types of tissue by the use of special photoactive compounds. The main advantage of the system is that of a permanent display and storage of everything illuminated. The major disadvantage is, as always with new technology, the price, but in time this is certain to reduce.

QUESTION

7.4 Explain in your own words, the advantages and disadvantages of a video endoscope system. How do you think this will change in say the next ten years – do you think it will have more or less advantages?

7.3 LASERS

The laser offers the opportunity to deliver energy to the body at a high intensity, and a very precise location, usually with an optical fibre, and internally by use of an endoscope. The energy can be used to cut, coagulate or destroy tissue, so that the laser can be used as a surgical tool. A further advantage is that lasers produce light of a particular wavelength. The interaction between tissue and light is very dependent on the wavelength of the light, the total energy absorbed, the rate of energy supplied, and the type of tissue. Over the past thirty years, this has resulted in the development of a range of lasers for different medical applications.

Laser – the coherent substitute for a scalpel?

Laser is an acronym for Light Amplification by the Stimulated Emission of Radiation. Stimulated emission depends on changes in the energy of electrons orbiting the nuclei of the atoms of the lasing medium.

By absorbing a photon, an electron can move to a more energetic orbit around the atom. This 'excited' state is less stable and tends to lose energy by spontaneous emission of a photon of light. The energy of the photon emitted, and hence its wave-length, depends exactly on the difference in energy between the electron's orbits. If a photon with this amount of energy hits an excited atom, it is absorbed and re-emitted, together with a second photon of exactly the same wavelength and phase as the first. This phenomenon is stimulated emission. The emitted photons travel in the same direction as and in phase with the incoming photon. (Fig 7.10)

In a laser, the aim is to ensure that more atoms in the lasing medium are in an excited state than in the lower-energy state. To this end, light or electrical energy is pumped into the lasing medium. When excited atoms are in the majority, a population inversion is said to exist. A stray photon of the correct wavelength, produced by spontaneous emission, is enough to set off a chain of stimulated emissions (hence light amplication). Because the lasing medium lies between two mirrors, one totally reflecting and one partially reflecting, photons can bounce back and forth, stimulating more and more atoms to emit photons, thus rapidly increasing the intensity, as they leave through the partially reflecting mirror. If the pumping energy is applied continuously, population inversion is maintained and gives rise to a continuous-wave laser. If the pumping energy is applied intermittently, as in a pulsed laser, the stimulated emissions die down as the atoms that are in an excited state are depleted, destroying the population inversion.

It is possible to reduce the laser cavity's ability to bounce photons back and forth during pumping by inserting a shutter between the mirrors. This technique termed spoiling the quality (Q) of the oscillation of photons, allows a very high population inversion to build up. When the shutter is open, all the energy leaves in a short pulse of very high power that lasts about 10 nanoseconds. Such a device is called a 'Q-switched laser'; it is possible to Q-switch most pulsed lasers by fitting a shutter inside the laser cavity.

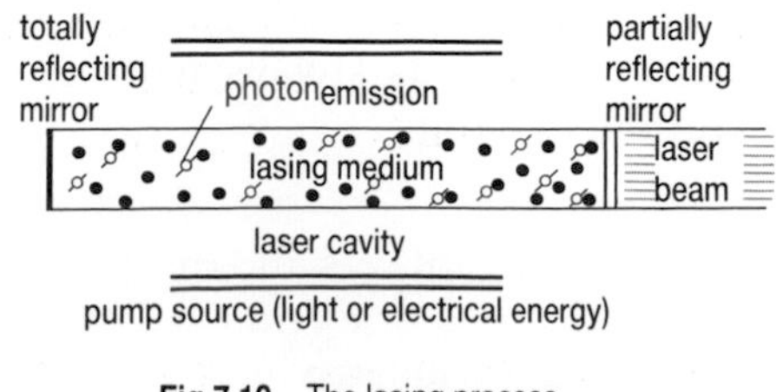

Fig 7.10 The lasing process.

The average power output of a continuous-wave laser and of a pulsed laser is about the same, but the power of the pulsed laser at the peak of its pulse is much greater than that of the continuous model. (Turning a laser on and off very quickly does not produce the same effect as a pulsed laser.)

The arrangement of the mirrors at either end of a laser cavity ensures that the light produced by the laser is collimated (parallel) and coherent (the waves are in phase). Photons emitted in other directions are lost through the walls of the laser cavity. The light produced is monochromatic (of one wave-length), as it is composed of many photons that have exactly the same energy. The colour depends on the nature of the lasing medium.

These properties suit the laser to medical purposes very well. Because the beam is collimated, it can pass easily along optical fibres within an endoscope, a narrow tube that a doctor can pass into the lungs or the alimentary canal, for example. As the beam is monochromatic, it will be absorbed by some kinds of tissues and not others. This property is useful in surgery on the eye, for example, when the blue-green beam of the argon-ion laser passes without effect through the cornea and lens in order to treat the retina.

The name of laser refers to the lasing medium used. Lasers that rely on the changes in the energy levels of electrons within the lasing medium include the argon-ion laser (pumped directly be electrical energy) which can work only as a continuous wave laser, and the neodymium-YAG laser. In the latter, the excited neodymium atoms are held in a crystal lattice of yttrium, aluminium and garnet. Light energy, either continuous or pulsed, "pumps" the neodymium atoms.

The carbon dioxide laser works by taking advantage of the changes in the vibrational and rotational energy levels in the molecule that electromagnetic radiation of radio frequencies induces. The helium-neon laser is a gas-discharge laser, similar to the argon-ion laser, but its maximum output is around 5 milliwatts, so it is used as an aiming beam for the infrared lasers. A recent introduction is the excimer laser, which releases photons in the ultraviolet region of the electromagnetic spectrum.

Dye lasers are an exception to the single-wavelength devices. Here, the lasing medium is a complex organic dye. Stimulated emission can occur from complex molecules of this kind as a result of changes in their vibrational and rotational states, as well as from electrons moving between energy levels. These lasers have a band-width of about 100 nanometres. Individual wavelengths can be selected with a tuning device inside the cavity. Changing the dye will alter the wavelength more markedly.

The lasers used in medicine together with their operating characteristics and wavelengths, are listed in the Table.

This article is taken from 'Lasers take a shine to medicine', by Frank Cross, *New Scientist* p. 20.2.86.

Table 7.1 Laser – the coherent substitute for a scalpel

Continuous-wave lasers

Laser	wavelength / *nm*	Power / *W*	Fibre transmission
CO_2	10 600	0.1-50	No
Nd-YAG	1064	0.5-100	Yes
Argon	488 514	1-10	Yes
Dye	Tunable	0.05-5	Yes

Pulsed lasers

Laser	wavelength / *nm*	Pulse duration	Energy	Fibre
Nd-YAG(QS)	1064	nanoseconds	0.1-1J	No
Nd-YAG	1064	microseconds	0.1-1J	Yes
Dye	tunable, visible	microseconds	0.01-0.1J	Yes
Excimer	ultraviolet	nanoseconds	0.01-0.1J	Yes

Energy and power parameters are those used specifically in medical applications rather than the whole range obtainable from a particular laser.

QUESTION

7.5 Read the boxed article on the operation of medical lasers and answer the following questions:

(a) Explain, with the help of a labelled diagram, the way that laser light is produced, and why the process is called Light Amplification by the Stimulated Emission of Radiation.

(b) In the article, which 'properties suit the laser to medical purposes very well' and why?

(c) How does Q-switching produce a pulsed output?

(d) What is the difference of output between a pulsed and continuous wave laser?

(e) Which type(s) of laser are gas discharge? How is the energy supplied to these?

Absorption of laser light

Tissue such as skin absorbs the far infrared light of the carbon dioxide laser in about 0.1 mm depth. Near infrared radiation from the neodymium-YAG laser is absorbed over distances of a few centimetres. The presence of the black pigment melanin in the skin, increases the rate of energy absorption. So too does any carbon which may be produced if the tissue is burnt. The haemoglobin in blood is a good absorber of the blue-green light of an argon laser.

The extent of the damage to tissue is reduced by a pulsed laser, compared to a continuous wave. There are difficulties in passing the higher power pulsed lasers down an optical fibre, since it can be shattered, by thermal shock. The long wavelength of the carbon dioxide laser is absorbed by the fibre, so this cannot be used in a fibre optic endoscope.

The damage process in tissue is essentially due to the heating effect. The effect at high energy levels is to boil the water in cells, causing local damage to the cell and more widespread physical disruption as a result of the pressures produced by the steam generated. This is called photovaporisation. At lower energy levels proteins photocoagulate and change their chemical and physical nature, in a way similar to what happens in cooking egg-white.

Uses of lasers

The laser has found a place in medicine for specific applications, its more general use being limited by the high cost of the equipment (scalpels are simple and cheap if nasty!). Some examples of present uses are given here;

in general they are operations that would be very difficult to carry out in any other way.

The carbon dioxide laser is used for cutting surface skin, and for delicate surgery on the brain and the fallopian tubes. The neodymium-YAG laser is used for internal removal of blockages such as that shown in Fig 7.11. The laser beam is passed down an additional fibre in the endoscope, whilst the site is viewed with the image guide fibre. This laser can also be used to seal bleeding ulcers, by cauterising them and to remove fatty deposits *(atheroma)*, from the blood vessels, as these can lead to heart failure.

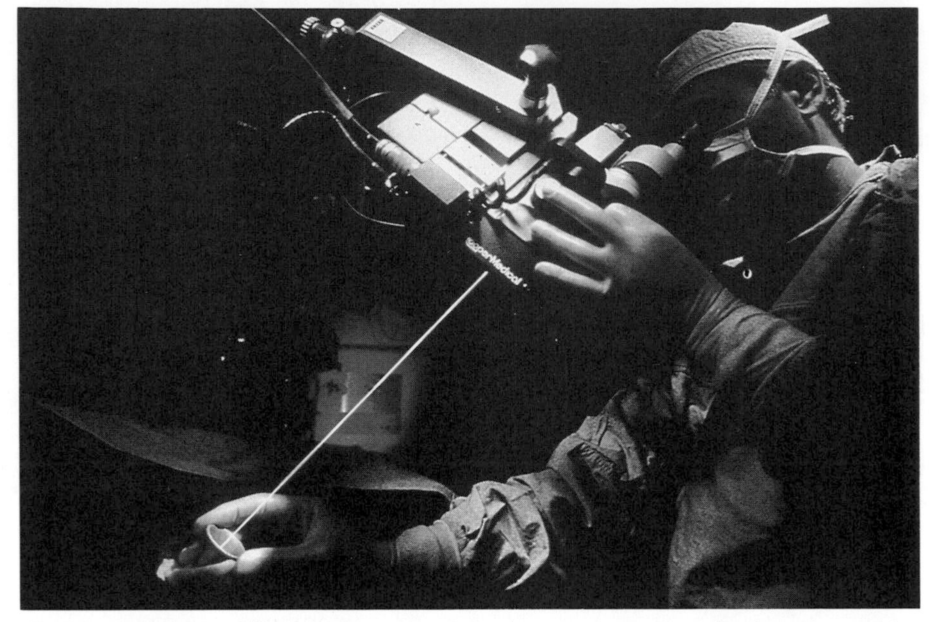

Fig 7.11 An argon laser drilling inside the ear; it can be used to remove a tumour or to free the movement of the small bones.

The argon laser has been particularly successful in treating eye disease which results from diabetes. The laser is able to stop the bleeding which results, and prevent blindness. It is also able to repair small tears in the retina by photocoagulation. In a process which is referred to as 'spot-welding' a pulse of about 0.1 J of energy is delivered in less than 1 ms over an area of about 10^{-3} mm^2. Thus only a few cells are targeted and the operation is virtually painless. This laser is also used to remove unsightly birthmarks such as the port wine stain which results from having abnormal blood vessels at the surface of the skin. Present development work is concentrating on the possible use of lasers in the treatment of cancers. One approach is to inject into tumours a chemical that is particularly sensitive to light – a photoactive substance, which the laser would then react with,

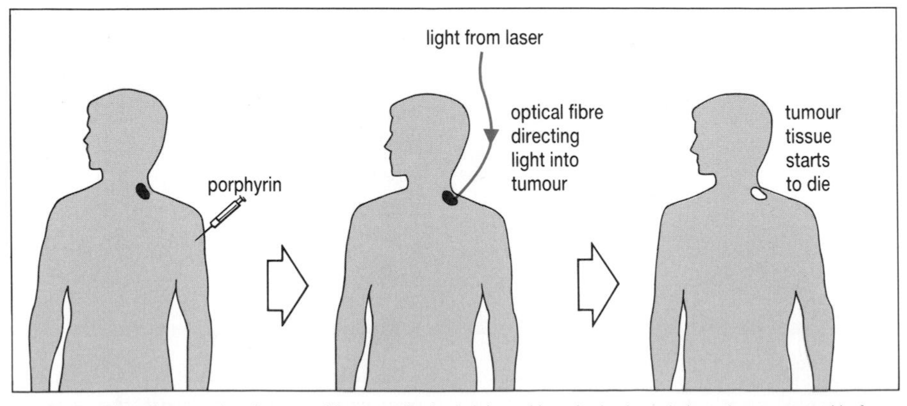

Fig 7.12 Phototherapy: the photosensitiser, porphyrin, is injected into the body and given time to spread before light is passed down an optical fibre into the tumour. A few hours later, the cancer cells start to die.

making the chemical toxic to tissue. This technique called phototherapy is illustrated in Fig 7.12; it is beginning to emerge from the development stage into hospital use. More speculative is the possibility of producing a spot of laser light small enough to carry out surgery on a chromosome – destroying genes responsible for inherited abnormalities such as haemophilia and cystic fibrosis.

SUMMARY ASSIGNMENTS

7.6 Relate the uses of lasers that are described here to the mechanisms of their interaction with tissue, for example the use of argon lasers to destroy blood cells because this light is preferentially absorbed by red haemoglobin. (See Fig 7.1)

7.7 This question compares the power of various lasers.

(a) Calculate the intensity of the light used in 'spot-welding' the retina, in the example given above (in W mm^{-2}).

(b) The laser you may have used in physics demonstrations is a helium-neon continuous wave type with a maximum output of 1 mW and a minimum diameter of 1 mm.

(i) What is the maximum intensity of light from this laser?

(ii) If this intensity falls on the outer surface of the eye and the area of the spot is then reduced by 10^{-3} by the focusing of the cornea and lens, what is the maximum intensity on the surface of the retina?

(c) Research has shown that the retina can be damaged by intensities as low as 10^{-2} W mm^{-2}. How does this compare with your answers to parts **(a)** and **(b)**?

(d) Suppose that you accidentally allowed light from the He–Ne laser into your eye.

(i) What should you do?

(ii) What would you observe as a result of the exposure?

7.8 **(a)** (i) Explain what is meant, in fibre optics, by an *incoherent bundle* and by a *coherent bundle*.

(ii) Explain what is meant by an *endoscope*.

(iii) Describe, with the aid of a diagram, a fibre-optic flexible *endoscope* and explain the purposes of its chief parts.

(b) (i) Fig 7.13, which is not to scale, shows a cylindrical optical fibre, diameter 0.20 mm, of refractive index 1.5, surrounded by air. A ray of light travelling along the axis of the straight portion reaches the region where the axis of the fibre is curved into an arc of radius R. The ray will be reflected at the surface and remain in the fibre provided the angle of incidence exceeds the critical angle i_c where $\sin i_c$ is the reciprocal of the refractive index. Calculate the smallest value of R which allows the ray to remain within the fibre.

(ii) Describe how a fibre with cladding differs from that described in **(b)**(i) and give **one** reason why, in making an optical fibre bundle, fibres with cladding are preferred.

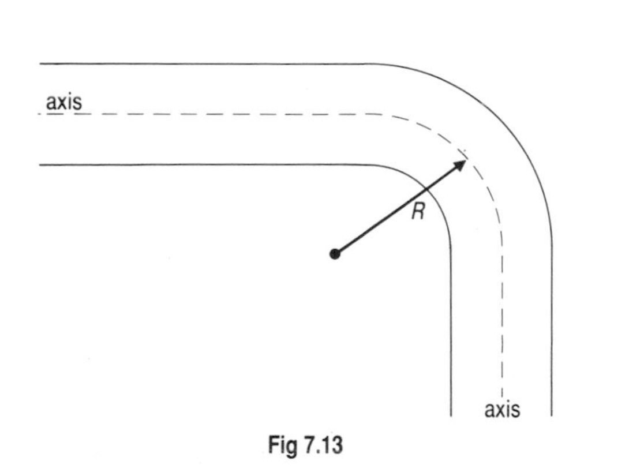

Fig 7.13

(JMB 1982)

Further reading

Bonnet, R. A death ray for cancer? *New Scientist*, 28.1.89

A detailed account of the physical and chemical basis of phototherapy, and its current state of development.

Cross, F. Lasers take a shine to medicine, *New Scientist*, 20.2 86

A full description of how laser light is generated and supplied, to perform a wide range of medical applications, including endoscopy and surgery.

Hill, P.D. Lasers in Medicine, *Pyhsics Education* vol 24, no 4 July 1989 p 211.

Chapter 8

ULTRASONICS

The limit of human hearing is a frequency of about 20 000 vibrations per second (20 kHz), though animals make use of sound at considerably higher frequencies. Bats and dolphins, for example, emit pulses of sound in the 30–100 kHz range, and navigate by listening to the echoes. During the Second World War a way was found for humans to use these ultrahigh frequencies. The emissions are called ultrasonic waves, or ultrasound and their field of use, ultrasonics, has developed greatly from the original aim of locating underwater objects such as submarines. Ultrasonics offers a method of investigating objects internally without causing any damage. It is widely used in industry; in this chapter we will study its growing use in medicine.

LEARNING OBJECTIVES

After studying this chapter you should be able to:

1. use the following scientific terms correctly: piezoelectric, sonar, acoustic absorption, acoustic reflection, coupling medium;
2. describe the production and detection of ultrasound by piezoelectric devices;
3. describe the process of ultrasound reflection by the body, and account for the formation of signals by tissue boundaries;
4. explain the principles of operation of the different types of ultrasound scanners: 'A','B','M', and 'real-time', and give examples of their use in medical diagnosis;
5. explain how the Doppler effect is used in the measurement of blood flow and foetal heart beat;
6. describe the physiological effects of ultrasound on the body, and explain how these are used in therapy.

8.1 GENERATION AND DETECTION OF ULTRASOUND

The ultrasonic **transducer** is the device which can both generate and detect ultrasound vibrations. In principle a number of methods are possible, such as the use of magnetic effects. In practice those used in medicine invariably convert electrical energy into ultrasound, and the opposite, by means of the **piezoelectric effect**.

The piezoelectric transducer

In a piezoelectric material, when an electric field is applied there is a change in the physical dimensions. This is caused by an interaction between the charges bound in the crystal lattice and the applied electric field. Conversely, if a mechanical stress is applied, a potential difference is produced across the ends of the crystal (Fig 8.1). If the applied voltage is alternating, a vibration of the crystal will result. The maximum transfer of energy will result at a natural frequency of vibration of the crystal, this is **resonance**. Resonance

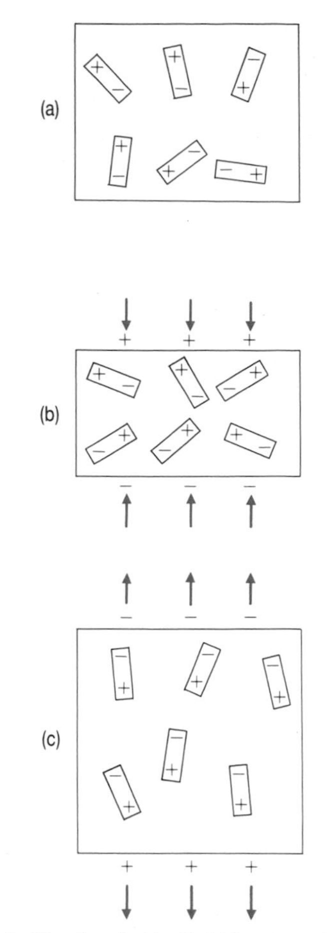

FIG 8.1 The piezoelectric effect **(a)** unstressed **(b)** compressed **(c)** extended.

also ensures that the ultrasound energy is in a certain frequency range, which produces better images. To facilitate this the crystal has a thickness of one half of a wavelength of the ultrasound, producing resonance at the fundamental or lowest frequency.

The piezoelectric effect is present in a number of crystalline substances including quartz, lithium sulphate and barium titanate. The most commonly used substance is the artificial ceramic, lead zirconate titanate (PZT), which is available in a range of forms for different uses.

The transducer is made from a thin slice of the crystal, protected by a cover which usually incorporates a lens to converge the beam slightly (Fig 8.2). The beam consists of short pulses so the crystal is backed by epoxy resin which damps the vibrations quickly. The two faces of the crystal are covered with thin layers of silver as electrodes. One of these is connected to the earthed case which provides both mechanical and electrical protection. The other electrode is connected by a coaxial cable to (i) the power supply when acting as a transmitter, and (ii) an amplifier and cathode-ray tube when acting as a receiver.

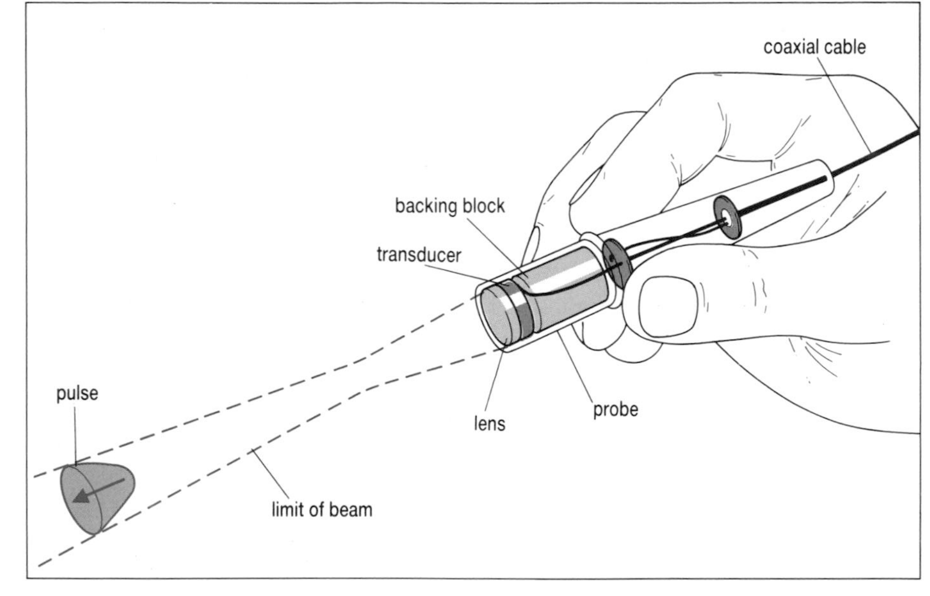

Fig 8.2 The piezoelectric probe.

8.2 ULTRASOUND IN THE BODY

Information about the internal structure of the body can be obtained by echolocation techniques using frequencies of several megahertz (MHz). The ultrasound is transmitted into the body where it is reflected at the boundaries between different types of tissue. It is also absorbed by tissue. So from the time lapse and intensity loss of the returning signal, the nature and position of the organs and tissues can be deduced. This technique is sometimes called **sonar, so**und **na**vigation and **r**anging.

Properties of ultrasound

These sound waves behave in the same way as audible sound, which is described in Chapter 3. In particular the wave equation

$$c = f\lambda$$

applies. The speed of sound in soft tissues is about 1500 ms^{-1} (Table 8.1), so for frequencies in the MHz range the wavelength is typically a few millimetres. The basic echolocation method involves measuring the time interval between the emission of the pulse and its reception.

Table 8.1 Ultrasound properties of materials

Medium	Density (ρ) /kg m^{-3}	Ultrasound velocity (c) /m s^{-1}	Specific acoustic impedance Z (=ρc) /kg m^{-2} s^{-1}
Biological			
Air	1.3	330	429
Water	1000	1500	1.50×10^6
Blood	1060	1570	1.59×10^6
Brain	1025	1540	1.58×10^6
Fat	925	1450	1.38×10^6
Eye			
(aqueous humour)	1000	1500	1.50×10^6
(vitreous humour)	1000	1520	1.52×10^6
Soft tissue (average)	1060	1540	1.63×10^6
Muscle (average)	1075	1590	1.70×10^6
Bone (varies)	1400		5.6×10^6
	1908	4080	7.78×10^6
Transducers			
Barium titanate	5600	5500	30.8×10^6
Lead metaniobate	5800	2759	16.0×10^6
Lead zirconate			
titanate (typical)	7650	3791	29.0×10^6
Lithium sulphate	2060	5437	11.2×10^6
Quartz	2650	5736	15.2×10^6

QUESTION

8.1 (a) With reference to Table 8.1, calculate for an ultrasound wave of 1.5 MHz:
(i) the wavelength in bone, fat and the piezoelectric crystal PZT,
(ii) the thickness of a PZT crystal in a transducer, producing its fundamental resonance.
(b) The time delay for an echo from ultrasound in soft tissue was 0.133 milliseconds. At what depth was it reflected?

Reflection

This occurs at a boundary because of a difference in the characteristic **acoustic impedance** Z, of each substance. This is simply the product of the density ρ and the velocity of sound c through it. As Table 8.1 shows, the impedance is similar for most soft tissues, so the amount of energy reflected at their boundaries is small, typically about 1 per cent. It can be shown that the ratio of the reflected intensity I_r to the incident intensity I_i is:

$$\frac{I_r}{I_i} = \frac{(Z_2 - Z_1)^2}{(Z_2 + Z_1)^2}$$

This is called the **intensity reflection coefficient** α. Clearly there will be no reflection if the impedance of the two media is the same, $Z_1 = Z_2$.

QUESTION

8.2 (a) What is the value of α going from bone to soft tissue?
(b) What will happen at the interface between the lung (full of air) and the surrounding tissue?

The answer to **8.2(b)** shows that it is important to ensure there is no air between the transducer and the skin. This is done by interposing a **coupling medium** such as a gel or oil. In a similar way the presence of gas anywhere in the field of view of the ultrasound will cause a problem for imaging.

A reflection of 1 per cent of the incident energy at a soft tissue boundary is sufficient for a sensitive detector, and this has the advantage that the remainder of the beam can penetrate more deeply into the body, so that reflections from a series of tissue boundaries can be detected (Fig 8.3). A strong reflection such as that from bone, can cause a problem of a duplicate or multiple reflection. The beam reflected from the vertebra, is reflected back into the body and can then be re-reflected from the vertebra. Such multiple images can usually be recognised by the regularity of their spacings. They are an example of an artifact, something which is not part of the subject but a product of the imaging system.

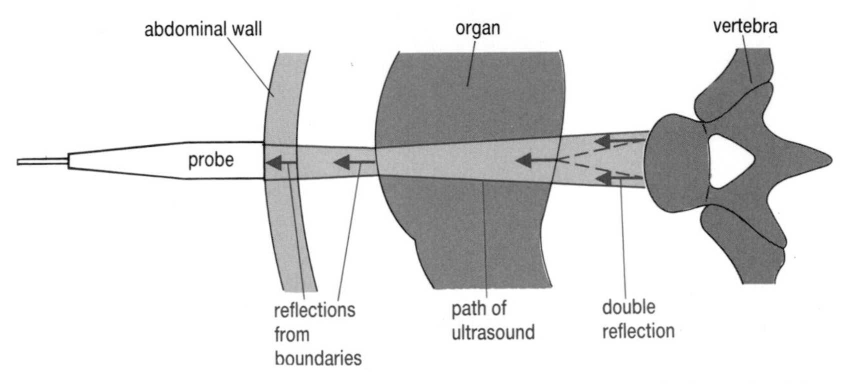

Fig 8.3 Reflection and double reflection of ultrasound pulses, cross section laterally through the abdomen.

Resolution and attenuation

Resolution is a measure of the smallest detail that can be detected, and so is an important feature of all imaging systems (see Section 2.2 and 2.3 for example). It is limited by diffraction effects so depends on the wavelength of the ultrasound. The smaller this is the better the resolution; rather like using a finer point to a brush to paint smaller features. This means that higher frequency ultrasound gives more detailed pictures.

The ultrasound beam spreads out as it passes through tissue, and is scattered and absorbed by the molecules it meets. Unfortunately this attenuation also increases as the frequency increases. In practice there has therefore to be a compromise between resolution and penetration, in the choice of frequency for any particular application. It turns out that in most cases the optimum frequency is one for which the organ concerned is at a depth of about 200 wavelengths. For example, the optimum frequencies for examination of the brain and the abdomen are 1–3 MHz.

QUESTION

8.3 What range of frequencies would be used to examine the eye? (Estimate its depth.)

INVESTIGATION

The properties of ultrasound

Ultrasound transmitters and receivers are produced for laboratory use, for investigating their wave properties, for example, Ultrasonic kit from Unilab. If this equipment is available to you, follow the suggested methods of investigating the reflection, refraction and interference of ultrasound waves. Try to devise an arrangement to simulate the passage of ultrasound through different media (e.g. water and bone) to investigate acoustic coupling.

Alternative sources of ultrasound are the motion sensor from Educational Electronics, or you could try your local hospital for redundant equipment.

8.3 SCANNING AND IMAGING

The pulse echo received from the body is converted by the transducer into an electrical signal. It is usually displayed on the screen of a cathode-ray tube, and can then also be stored, directly on a storage CRT, on videotape, or converted to digital information and stored by computer. There are several different modes of display available for different purposes.

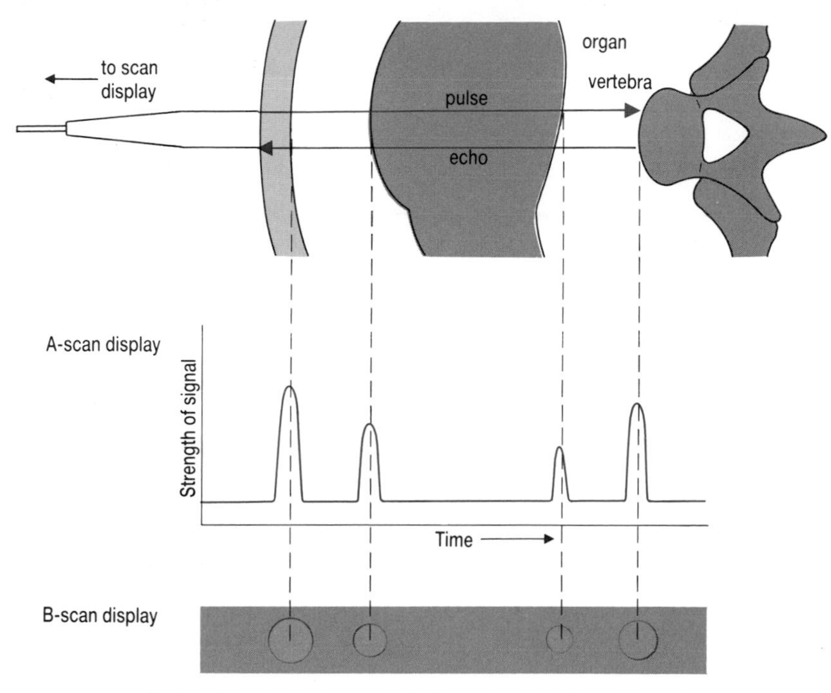

Fig 8.4 Pulse display by amplitude (A-scan) and brightness (B-scan).

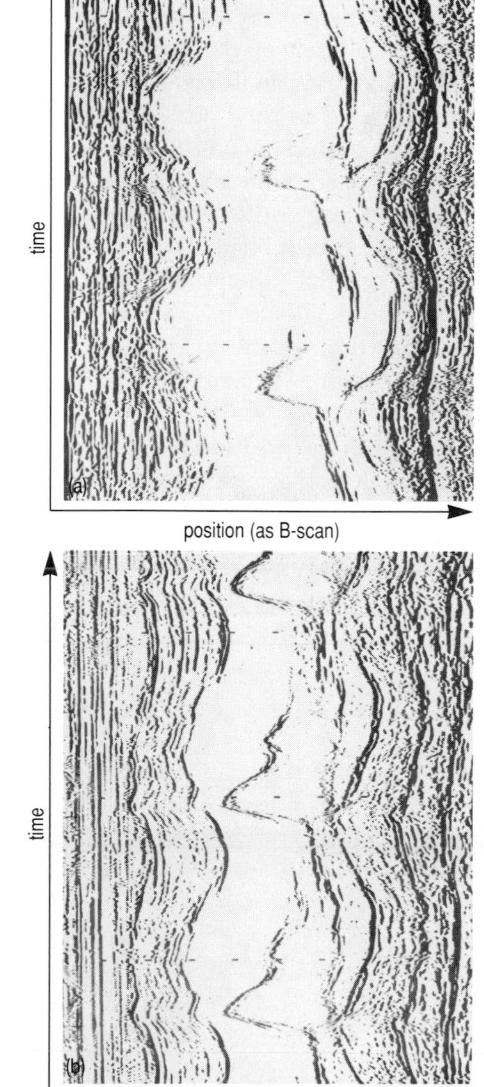

Fig 8.5 **(a)** M-scan showing abnormal mitral valve and left ventricle movements **(b)** M-scan of same patient after corrective treatment.

A-scan

This is the range-finding display, in which the horizontal axis represents time, corresponding to the distance into the reflecting tissue, and the vertical axis represents the amplitude of the echo. Fig 8.4 shows how this works. The first pulse received is from the abdominal wall. The time for the pulse to echo is a function of the thickness of the wall, and so the A-scan could be used to measure it – assuming that the clinician knew the anatomy of the body section being scanned! Likewise the two boundaries of the organ give pulses of decreasing size because of attenuation, whereas the bony vertebra at the back gives a higher intensity because it reflects a greater proportion of the sound.

A-scanning is useful in those situations where the anatomy of the section is well-known, and a precise depth measurement is required. One such example is echoencephalography, the determination of the position of the midline of the brain. This is the gap between the two hemispheres of the brain, which is normally symmetrically placed between the two sides of the skull, both of which can be accurately positioned. Any delay or advance of the echo from the midline could indicate the presence a space occupying mass such as a tumour, blood from a haemorrhage or fluid as a result of hydrocephalus. A-scans are also used in some detailed measurements of the eye in ophthalmology. For most purposes however, the **a**mplitude-modulated **A**-scan information can be more usefully presented as a **b**rightness-modulated display called a **B-scan** (Fig 8.4). This is the basis of several types of display.

Time-position scan (M-scan)

This is a useful technique for monitoring the regular movement of an organ such as the heart. It is a combination of the A-scan and the B-scan. The dots are produced in positions along the x-axis, according to the depth of the reflection. The Y-plates are connected to a time-base so that the dots move down the screen at a low speed. A change in the position of the interfaces, such as the heart wall, will therefore result in horizontal displacement of the dots and a non-straight line. A regular oscillation will produce the wavy trace shown in Fig 8.5. The viewing of the heart is limited by surrounding bone and air, so care has to be taken to position the probe to avoid the ribs and lungs. In addition the heart valves are not plane surfaces so do not produce strong reflections. The technique of ultrasound cardiography (UCG) is therefore one which requires considerable skill.

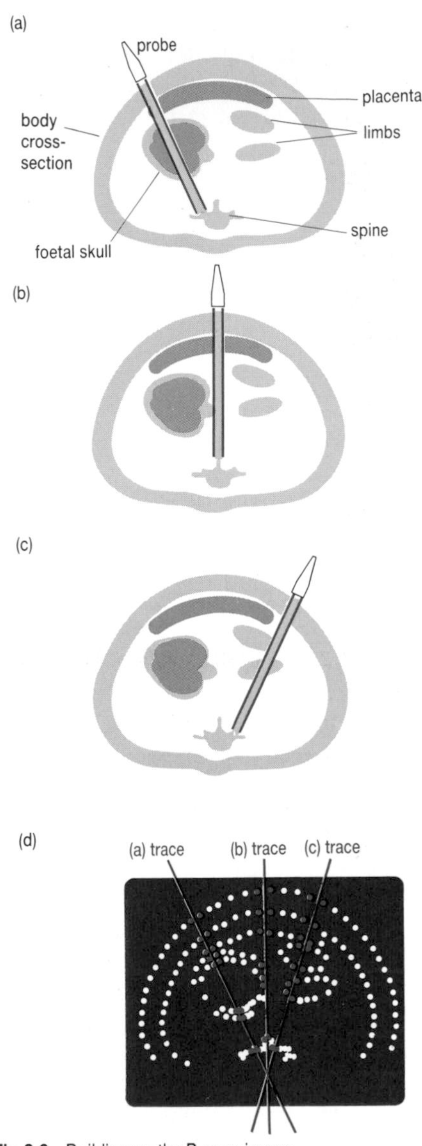

Fig 8.6 Building up the B-scan image.

Two-dimensional B-scan

This is particularly important in medical diagnosis as it results in the production of a pictorial display of the organs, at a particular cross section. In order to produce these the transducer probe is moved so that the body is viewed from a range of angles. These different views give rise to a set of brightness points. These are automatically correlated with information about the position and orientation of the probe to build up the two-dimensional picture (Figs 8.6, 8.7). The movement can be along the surface of the body (linear scan), as shown in Fig 8.7, or rotation of the transducer at a fixed position (sector scan). Or the two movements can be combined to produce a compound scan (Fig 8.8). This gives the maximum amount of information but the method requires some manipulative skill to keep the probe coupled with the skin and also some detailed knowledge of anatomy. One way of ensuring that the moving probe remains coupled to the body, without air intruding, is to immerse the patient in a water bath. This has its limitations. The method is unsuitable for soft and sensitive structures such as the breast and the eye that would be distorted by the pressure of the probe.

The B-scan technique has proved most valuable in obstetrics where it is now a routine procedure for monitoring the health of the pregnant woman and the growth of the foetus. This latter is usually done by accurately measuring the widest diameter of the skull, the bi-parietal diameter (see Fig 8.6). Various complications of pregnancy such as faults in the amniotic fluid and the placenta, can be diagnosed early; later scans can show any developmental problems in the foetus and check the position before birth.

Two-dimensional images are also used for detecting cysts, abscesses and tumours in organs such as the liver, kidney and ovary. They are invaluable

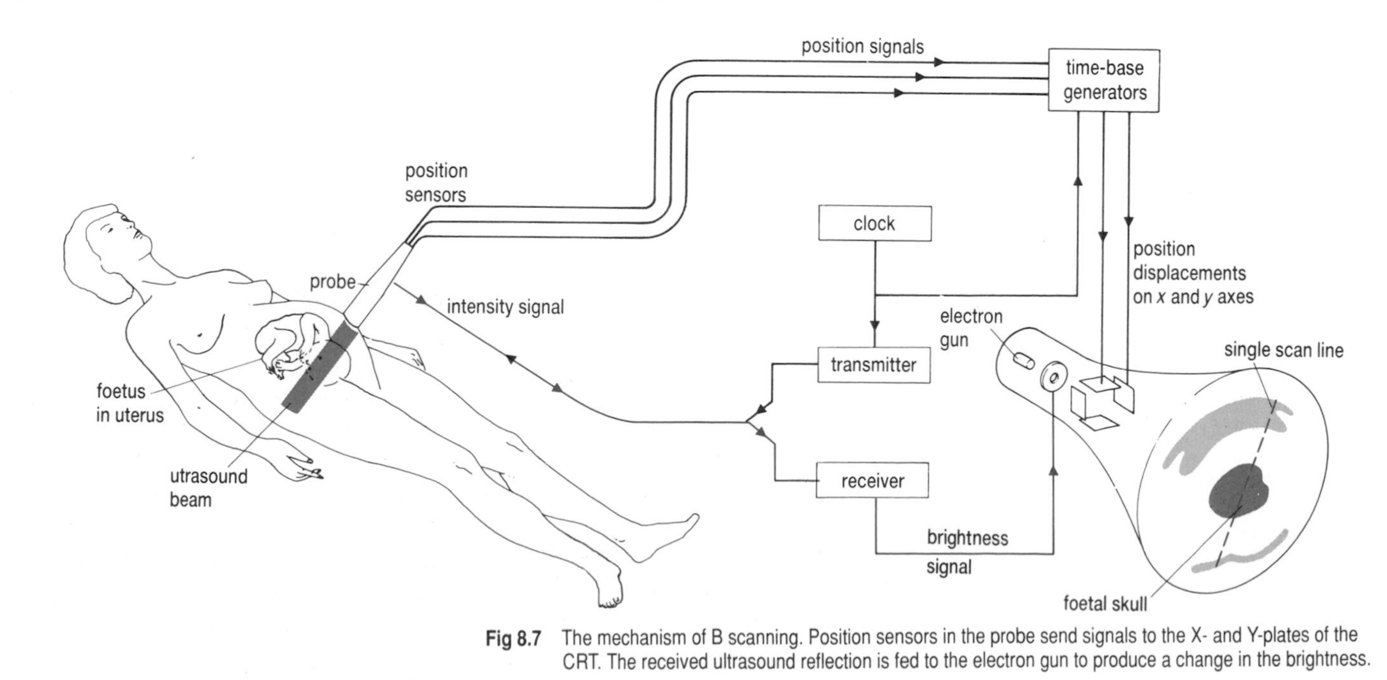

Fig 8.7 The mechanism of B scanning. Position sensors in the probe send signals to the X- and Y-plates of the CRT. The received ultrasound reflection is fed to the electron gun to produce a change in the brightness.

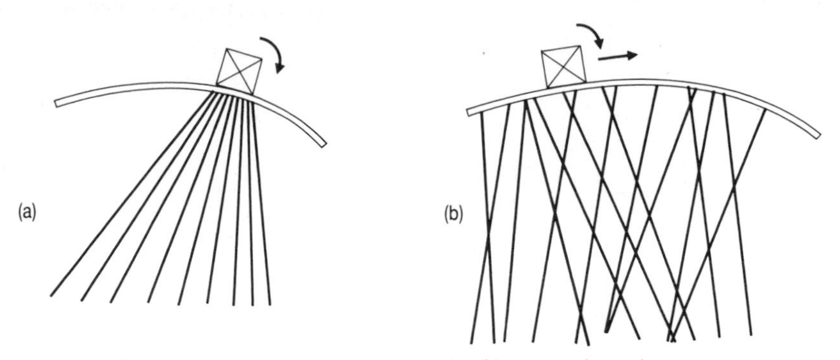

Fig 8.8 B-scan methods **(a)** sector scanning **(b)** compound scanning.

in the examination of aneurysms, that is weak, ballooned-out parts of arteries. Follow up to these investigations may be complementary X-ray or radioisotope imaging, radiotherapy or surgery. In addition there is a growing use of this type of imaging to assist the surgeon with minor operations through the skin with very small incisions. Cysts can be drained and minor blood vessels repaired quickly and without major trauma to the patient.

QUESTION

8.4 With reference to the description and illustrations of this section, write an account of how a two-dimensional image of varying brightness is produced. Include in your notes a simple block diagram to represent the required instrumentation, and discuss the merits of the different scanning movements.

Real-time B-scans

The scanner described above takes several seconds to build up the complete image of the body cross section. This is not useful when the body structures are moving (which is quite often, and almost always in the case of obstetric examinations!). Real-time systems take a series of pictures in very rapid sequence, rather like cine filming. Although the detail of any one picture is reduced by the speed at which it was taken, seeing the sequence reveals more than the best still picture. Movement brings features to life so that diagnosis with a real-time scanner is easier,requiring less manipulative skill and detailed anatomical knowledge. There are two types of scanner. One is simply a speeded-up version of the two-dimensional system, so that each image is a 'freeze-frame' of the movement. A single image looks like a poor version of the 'still' picture.

The second type has an **array** of transducers, which gives it its name. There are many variations of these, the latest developments use computer control to operate a linear array in rapid sequence, and to process the received pulses to give frame rates of up to 150 per second. These systems are able to combine a recently stored image with the current real-time image to overcome the problems of flicker and low resolution of the non-array type.

8.4 DOPPLER METHODS

Measurement of a continuous movement such as the flow of blood, can be made using the effect of the moving reflector on the frequency of the ultrasound. This change is called the Doppler effect. You will have noticed this even though you may not have realised what it was. When siren on a vehicle moves past you hear a sudden fall in the pitch. The frequency of the siren note was higher when it approached you than when it receded from you. The effect is used in police 'radar traps' to measure whether vehicles are travelling faster than the speed limit. These use electromagnetic radiation, for the effect is found in all types of wave. We will see how this occurs.

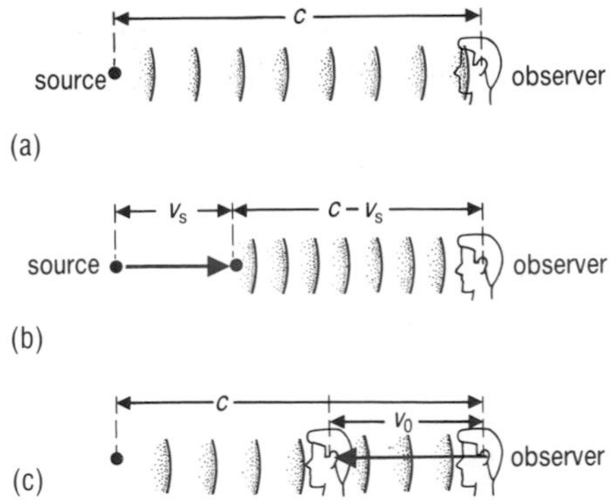

The Doppler effect

Moving source Suppose the source of sound has a frequency f_s in a medium in which the wave speed is c. In one second f_s waves will be emitted and will occupy a distance c (Fig 8.9(a)). If however, the source is moving at a speed of v_s towards the observer, then the f_s waves will be compressed into a distance $c-v$ (Fig 8.9(b)). Let us see what effect this has on the frequency. The observed wavelength λ_o is the distance divided by the number of waves:

$$\lambda_o = \frac{c - v_s}{f_s}$$

so the observed frequency

$$f_o = \frac{c}{\lambda_o} = \frac{cf_s}{c - v_s}$$

which is higher than the source frequency.

This change from the original frequency is called the Doppler shift, in this case:

$$\Delta f_1 = f_o - f_s = \frac{cf_s - f_s}{c - v_s}$$

simplifying,

$$\Delta f_1 = \frac{f_s v_s}{c - v_s}$$

In the case of ultrasound and blood flow, $c \gg v_s$ so this further simplifies to

$$\Delta f_1 = \frac{f_s v_s}{c}$$

Moving observer Suppose the observer moves towards the stationary source with speed v_o (Fig 8.9(c)), then the wavelength remains λ_s but the relative speed has increased to $v_o + c$, so that the observed frequency becomes

$$f_o = \frac{c + v_o}{\lambda_s} = \frac{(c + v_o)\ f_s}{c}$$

In this situation the Doppler shift is

$$\Delta f_2 = f_o - f_s = \frac{(c + v_o)}{c} f_s - f_s$$

$$\Delta f_2 = \frac{f_s v_o}{c} \qquad (1)$$

Fig 8.9 Doppler effect (a) stationary source and observer (b) moving source (c) moving observer.

Ultrasound reflector Let us now apply this to the ultrasound wave frequency f, reflecting off a surface which is moving towards the detector with speed v (Fig 8.10). The total Doppler shift is made up of two relative motions:

(i) the surface as observer is moving towards the transmitter source so receives a Doppler shifted frequency f' from equation 1 above,

$$\Delta f_2 = f' - f = \frac{fv}{c}$$

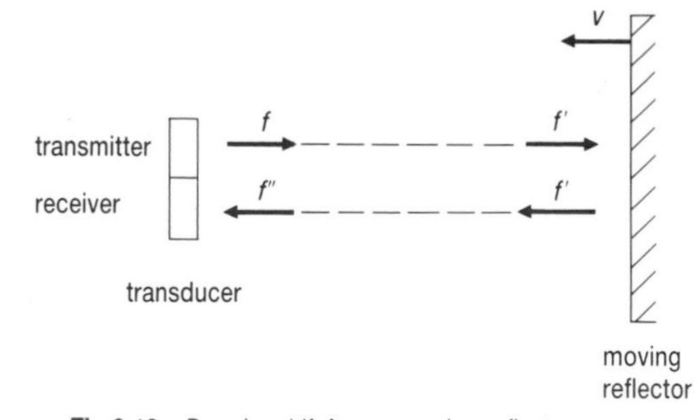

Fig 8.10 Doppler shift from a moving reflector.

(ii) the surface as source is moving towards the receiver as observer, which results in the Doppler shifted frequency f'', from equation 1 above,

$$\Delta f_2 = f'' - f' = \frac{fv}{c}$$

Therefore the sound has a total Doppler shift which is double the one-way shift:

$$\text{Total Doppler shift } \Delta f = \frac{2fv}{c}$$

For example, a 2 MHz ultrasound beam travelling at 1.5 km s^{-1} will have a Doppler shift of about 270 Hz from blood flowing at a speed of 0.1 m s^{-1}.

Doppler system

The frequency of the ultrasound is chosen to give the optimum balance between resolution and penetration. For measurements near the surface the range used is 5–10 MHz, for the centre of the body it falls to 2 MHz. The beam for ultrasound is a continuous wave, unlike the pulse used in previously described methods. This means that the system cannot be used for depth measurement, and that it requires a separate *transmitter* and *receiver*. The signals from both are *electronically mixed* and the output is *filtered* so that only the Doppler-shift frequency is passed on to be *amplified*. These frequencies are usually in the audio range, so the operator wears *earphones* to 'hear' the moving structure. The tone will vary with the shift frequency, and so with the speed of the movement. If the structure is the beating heart then the tone will vary with the heartbeats. The operator may work completely on this 'real-time' effect, recognising characteristic pitches or patterns, but there is a *recording* facility available.

QUESTION

8.5 Draw a block diagram to show the processes in a Doppler system (these are in *italic* in the previous paragraph).

Uses

Foetal heart monitoring This is the most common use of Doppler-shift ultrasound. The instrument is actually being used as a highly directional and sensitive stethoscope, though the sound that is heard is like galloping horses, which might alarm the expectant mother! It is normally possible to hear the foetal heart after the twelfth week; checks after that can determine not only whether the foetus is alive, but whether the foetus is in any difficulties, particularly during labour. This monitoring also includes the motion of the placenta and umbilical cord.

Blood flow measurement In most cases the blood is travelling at an angle θ to the direction of the beam, so the frequency shift is given by:

$$\Delta f = \frac{2fv}{c} \cos\theta$$

as shown in Fig 8.11. In this case it is necessary to measure θ, (which will depend on the inclination of the probe relative to the blood vessel), as well as Δf. A blockage due to a clot *(thrombosis)*, or a constriction due to plaque on the vessel walls *(atheroma)*, is immediately apparent as a sudden change in the shift frequency. More detailed investigations can analyse the range of shifts produced by the reflection. This arises from the variation of the blood flow speed across the diameter of the vessel, usually parabolic in distribu-

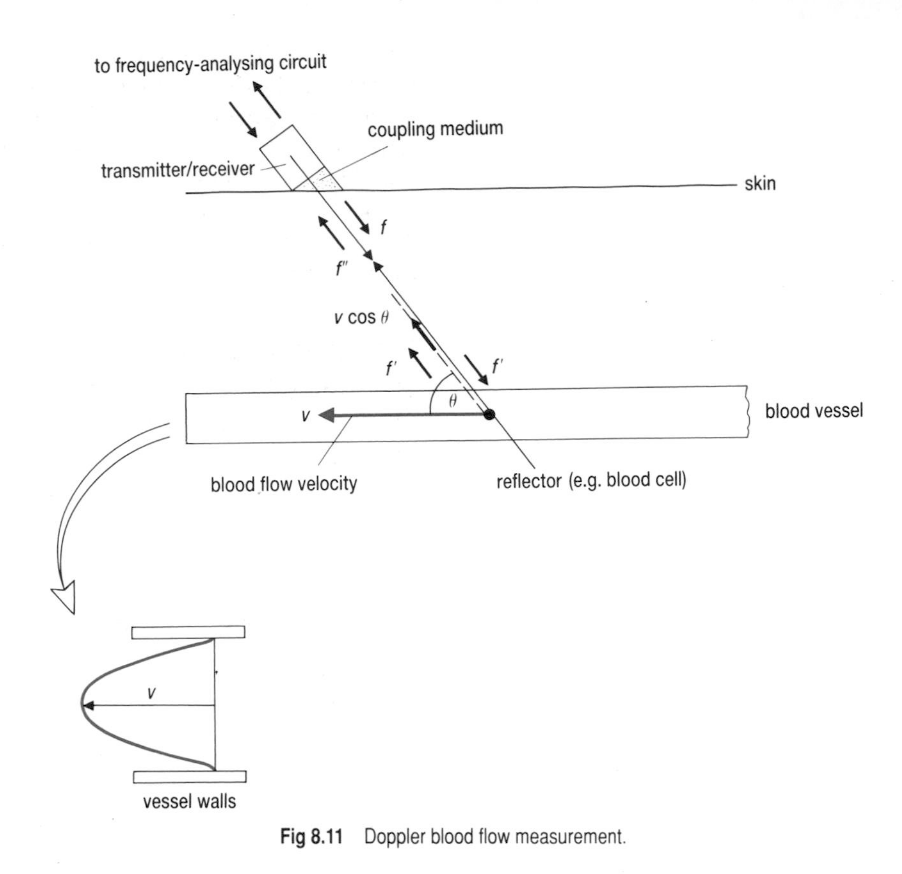

Fig 8.11 Doppler blood flow measurement.

tion, as shown in Fig 8.11 inset. Small obstructions will alter this pattern, so affecting the reflection. One problem is that arteries and veins often lie close together, so the two signals can interfere with each other.

QUESTIONS

8.6 (a) Why does the fact that blood flows in opposite directions in arteries and veins, not enable them to be distinguished?

(b) In a blood flow investigation, a source of 5 MHz ultrasound directed at an angle of 60°, produces a Doppler shift of 150 Hz. Assuming that the speed of the wave is 1.5 km s^{-1}, what is the speed of the blood?

More sophisticated ultrasound instruments are now incorporating electronic techniques which enable both direction and position to be determined.

8.5 PHYSIOLOGICAL EFFECTS OF ULTRASOUND

Ultrasound is a very safe way of investigating the interior of the body, hence the rapid growth of its use in recent years to replace X-rays. As a high frequency vibration of the body tissues it can produce physical and chemical changes. These depend on the intensity of the beam; in diagnostic use this is very low (usually about 0.01 W cm^{-2}), much higher intensities (up to 10^3 W cm^{-2}) are used for therapy.

Damage mechanisms

Ultrasound waves interact with the thermal motion of the molecules of body tissue. At energy levels of 1 W cm^{-1} and above, this results in measurable **temperature** increase, which can cause chemical changes (like cooking!), or the destruction of tissue at levels of around 1000 W cm^{-1}.

The **pressure** changes between the compressions and rarefactions are the second cause of damage. The rapid physical movement can rupture tissues, just as the eardrum can be torn by an intense external source of sound. At an intensity of 35 W cm^{-1} a wave of frequency 1 MHz can produce a change of 10 atmospheres in pressure over a distance of less than a millimetre (half a wavelength). This can cause dissolved gases, e.g. O_2 and CO_2 in the blood to come out of solution and form bubbles. These bubbles collapse violently as the pressure changes. This release of energy can cause both **physical damage** and **chemical change.** For example water can be changed into H_2 and H_2O_2 and DNA molecules can be broken. Free radicals can result in oxidation reactions which disrupt the metabolism of that part of the body. These chemical effects are similar to those resulting from ionising radiations described in Theme 3; the cause is very different however.

Ultrasound therapy

Diathermy Ultrasound is used to cause local heating of tissue deep inside the body in a similar way to that produced by microwave and electrical methods. The aim is to produce relief from pain for example in the joints of an arthritis sufferer. Typical treatments use intensity levels of several W cm^{-2} for periods of a few minutes several times a week. The transducer is moved over the surface of the joint during treatment to avoid 'hot spots' developing. Ultrasound is particularly useful as a 'deep heat' source as it does not deposit much of its energy in the soft tissues near the surface of the body. It is not used in delicate organs such as the eye or the gonads.

Tissue destruction. At intensity levels of 1000 W cm^{-2} it is possible to selectively destroy tissue by focusing the beam on the required place. **Meniere's disease,** in which dizziness and hearing loss is caused by blockage in the middle ear, is treated very successfully by this method. In this case it is easy to locate the site. In treatment of Parkinson's disease in which there is a lack of muscle control due to the absence of dopamine, a neurotransmitting substance, it has proved more difficult to focus on the correct region of the brain and this method has been discontinued. Investigations have shown that ultrasound can destroy cancer cells, but that it may also stimulate growth. It has also been successful in breaking up gallstones which avoids the need for surgery.

Developments in ultrasound

The past ten years have seen ultrasound expand rapidly in range and extent of its applications in medicine. The main reasons for this are that it is inexpensive, non-invasive and, at the intensity levels required for diagnosis, no harmful effects have been detected. It is particularly good at imaging soft tissue, compared with radiological methods. These advantages have resulted in it being the only imaging technique used in obstetrics. It is also very important in gynaecology, cardiology and the studying the vascular system. Research is continuing to explore the use of ultrasound in other organs such as the prostate, breast and muscles.

Developments in ultrasound equipment have made possible much of this increased use. Improved resolution of the image is enabling foetal hearts to be examined for cogenital defects earlier in the pregnancy. The electronic processing of the image has made it possible to compare changes in an image of a tissue over time to study, for example the degeneration of a muscle in muscular dystrophy, so that prognosis and treatment can be determined. New electronic techniques are making the display (on monitors), recording (on printed 'hard copy'), and storage (on videotape and computer tape) easier and cheaper. Electronic control of the transducer arrays allow more rapid scanning and the availability of relatively inexpensive, portable, real-time scanners is now very general (Fig 8.12). You will have an opportunity to find out the effects this is having on hospital practices, in the final chapter of this book.

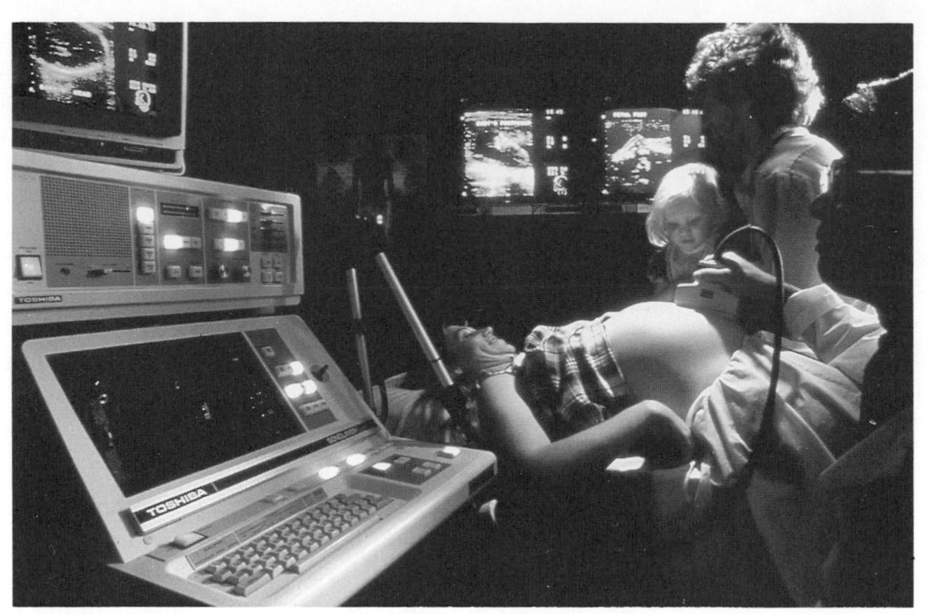

Fig 8.12 Routine ultrasound foetal monitoring; a family views its newest member, a healthy boy. The screen shows his foot, bent arm and fist.

QUESTIONS

8.7 (a) Over what distance in soft tissue, does the maximum pressure change occur, for a 1 MHz wave?

(b) Why would you expect that Meniere's disease would be easier to treat than Parkinson's disease?

(c) Make a table of intensity levels of the various uses of ultrasound, and of the levels at which types of physiological effects occur.

8.8 Collect any information you can on new uses of, and equipment developments in ultrasound, these may be useful for Chapter 13. (See further resources.)

SUMMARY ASSIGNMENTS

8.9 (a) When sound is transmitted from one medium (medium 1) to another (medium 2), the ratio, R, of the reflected intensity to the incident intensity is given by

$$R = \frac{(Z_2 - Z_1)^2}{(Z_2 + Z_1)}$$

where Z_2 is the acoustic impedance of medium 2 and Z_1 is that of medium 1. In the use of ultrasound for medical diagnosis, a coupling medium such as a water-based cellulose jelly is used between the ultrasonic transducer and the patient's skin. Explain why this is so.

Acoustic impedance of air $= 0.430 \times 10^3$ kg m^{-2} s^{-1}
acoustic impedance of water-based jelly $= 1500 \times 10^3$ kg m^{-2} s^{-1}
acoustic impedance of tissue $= 1630 \times 10^3$ kg m^{-2} s^{-1}

(b) Explain the basic principles behind the ultrasound method of obtaining diagnostic information about the depths of structures within a patient's body. Illustrate your answer by reference to a block diagram of a simple (A-scan) scanning system.
What factors limit the quality of the information obtained?

(ULSEB 1988, part)

8.10 (a) A transducer emits ultrasonic pulses towards a moving surface inside a patient's body. When the transducer detects the signals reflected from this surface they are found to have undergone a 'double' Doppler shift in frequency. Explain why this is so. (Detailed derivations of the expressions for the Doppler shift are not required.)

(b) Ultrasound of frequency 5.0 MHz reflected from red blood cells flowing in an artery was found to be Doppler shifted in frequency by 1.5 kHz when the blood flow was at 30° to the direction of propagation of the sound waves. If the speed of the ultrasound is 1.5 km s^{-1}, calculate the speed of blood flow in the artery.
(The expression $\Delta f \sim 2fv \cos \theta / c$, where the symbols have their usual meaning, may be assumed without proof.)

(c) Why, in practice, is a range of Doppler shifts detected?

(d) Give one advantage of this method of measuring the speed of blood flow.

(ULSEB 1988, part)

8.11 **(a)** With the help of a sketch of a typical transducer describe the generation of ultrasonic waves for medical diagnosis purposes. Explain
(i) why it is essential to use short pulses of ultrasound,
(ii) how it is ensured that the sound energy enters the body,
(iii) the means employed for detection of the received signals.

(b) For pulses of a given power emitted by the transducer mention **two** factors which affect the intensity of the received ultrasound signals.

(c) If the transducer were to be part of a B-scanner, what other fact would you need to know about the received ultrasonic signals? Apart from such data about the received signals, state the additional information needed by the B-scanner if an image is to be produced.

(JMB 1986)

Further reading

Gosling, R., Medical imaging with ultrasound: Some basic physics, *Physics Education*, vol 24, no 4 July 1969 p 215.

Theme **3**

IONISING RADIATIONS

Roentgen's discovery of X-rays in 1895 immediately captured popular imagination. His first paper included a radiograph of his wife's hand. Within a few months, the diagnostic use of X-rays was widespread. Many of the other claims for the new rays, including their potential for abolishing vivisection, smoking and alcoholism, were not sustained. Nor was their potential for damage to the body recognised for some time, despite the impression given by a contemporary cartoon.

This theme considers the effects of X- and other ionising radiations on the body. Examples of diagnostic and therapeutic use are presented, and the controversial issue of suitable safety precautions is reviewed.

Prerequisites

Before you study this theme you should have some familiarity with the following:

- The nature and properties of X-, alpha, beta and gamma radiations.
- The notation used in writing nuclear reactions.
- The decay and half-life of radioisotopes.

Cartoon from Life 17 February 1896

Chapter 9

X-RAYS

You will probably have had at least one X-ray examination, by the time you read this book, as X-rays are used routinely for diagnosis in dentistry and many branches of medicine (Fig 9.1). In this chapter you will study the nature and uses of X-rays in some detail. The subject is a large and complex one; for ease of reference it has been organised as follows:

Section 9.1 : the nature of X-rays, and how this depends on their production.
Section 9.2 : how X-rays interact with tissue.
Section 9.3 : the equipment which generates, controls and detects the X-ray beam.
Section 9.4 : the use of X-rays in diagnosis and therapy.

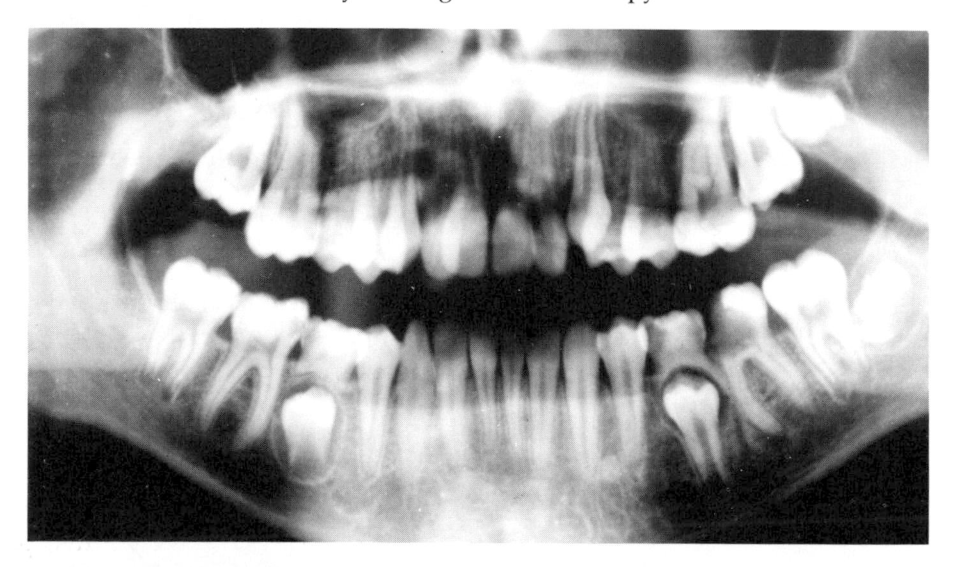

Fig 9.1 X-radiograph of the teeth of a child, showing some second teeth below the first or milk teeth.

LEARNING OBJECTIVES

When you have studied this chapter you should be able to:

1. use the following terms correctly: photon energy, hardness, total linear attenuation coefficient (μ), positron, fluoroscopy, tomography;
2. recall the nature of X-radiation and the intensity–wavelength spectra produced by a tungsten target and their dependence on tube voltage and current and on filtration;
3. describe the mechanism of interation of X-rays with the human body, and list some of the resulting types of damage;
4. use the relationships:
   (a) Intensity in vacuum $I = I_0/r^2$ (inverse square law)
   (b) Intensity in a medium $I = I_0 e^{\mu x}$
   (c) Half value thickness $x_{1/2} = \dfrac{\log_e 2}{\mu}$
5. describe the principle of production of X-rays by a rotating anode tube and the formation of a radiograph of body tissue;
6. Give examples of the use of X-rays in diagnosis and therapy.

9.1 THE NATURE OF X-RADIATION

Origin

X-radiation is electromagnetic radiation of high photon energy. It is produced when charged particles, usually electrons, moving at high speed are slowed down rapidly by striking a target.

Only about one per cent of the energy of the electrons is converted to X-rays, the remainder being dissipated as substantial heating of the target. X-radiation is always emitted with a range of wavelengths, corresponding to the range of photon energies produced. There are two mechanisms by which this occurs: these are illustrated in Fig 9.2. In (a) an electron is slowed which results in Bremsstrahlung radiation, which is German for braking radiation. In (b) an inner electron is removed causing ionisation, and subsequently an X-ray is emitted.

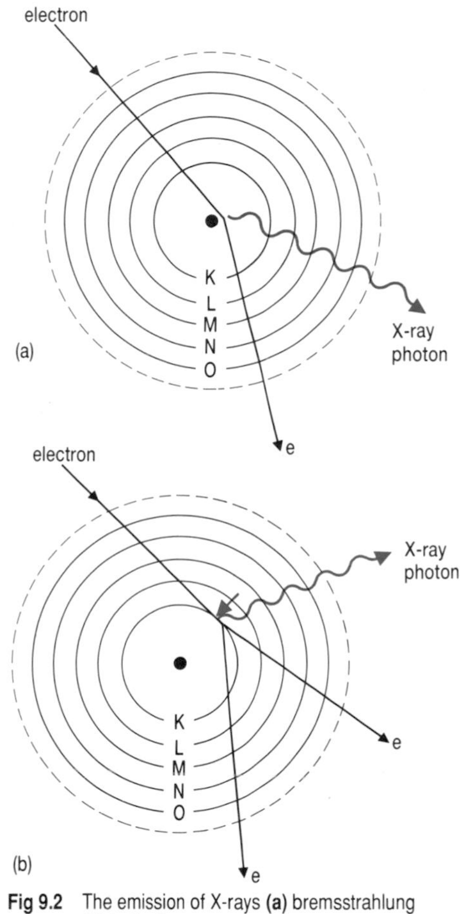

Fig 9.2 The emission of X-rays **(a)** bremsstrahlung **(b)** ionisation.

Spectra of X-radiation

Bremsstrahlung radiation results in a continuous spectrum. There is a short wavelength limit to the spectrum, which is when all the energy of the incident electron is converted to a single X-ray photon. This energy

$$E_{max} = eV_0$$

where e = the charge on an electron

V_0 = the peak value of the applied voltage

From the Quantum theory

$$E_{max} = hc/\lambda_{min}$$

where h = Planck's constant c = the velocity of light
and λ_{min} = the minimum wavelength

Therefore the short wavelength limit is given by;

$$\lambda_{min} = hc/eV_0 = 1.24 \times 10^{-6} \text{ m/peak voltage (in volts)}. \quad (1)$$

This important relationship is sometimes known as the **Duane–Hunt Law.** The long wavelength limit of the spectrum is not so clearly defined. Being of low energy, these X-rays are easily absorbed by the X-ray generator.

Ionisation results in the **characteristic** or **line** spectrum. Each line is the result of the change in energy of a particular electron in an atom of the target. The lines are named according to the shell in which the electron terminates. Those with the shortest wavelength which result from a transition to the innermost K-shell, are K-lines. The applied voltage must be above a critical value for this to occur. For tungsten, the usual target material, this value is about 70 kV, for K-lines. A typical spectrum is shown in Fig 9.3.

The intensity and quality of X-rays

The effects of X-radiation depend both on the intensity (energy per unit area), and on the spectral spread. It is the latter which determines how penetrating the radiation is, and is called the **quality.** To describe this fully requires a graph such as Fig 9.4. In practice however, it is usually sufficient to specify the quality in simpler terms, for example, the thickness of material to reduce the intensity to a half. This is called the **half-value thickness** and is explained in the next section.

We will now consider the factors affecting the intensity and quality of the X-ray beam, resulting from changes in the Bremsstrahlung and characteristic radiations.

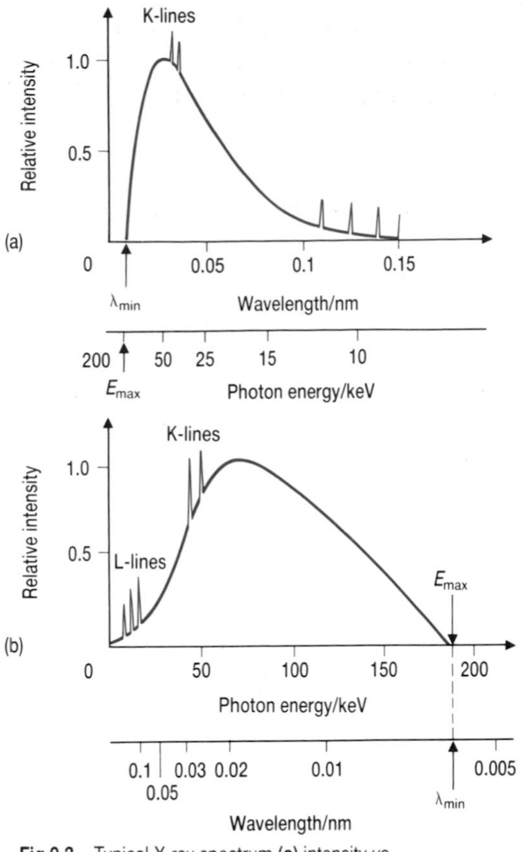

Fig 9.3 Typical X-ray spectrum **(a)** intensity vs wavelength **(b)** intensity vs photon energy.

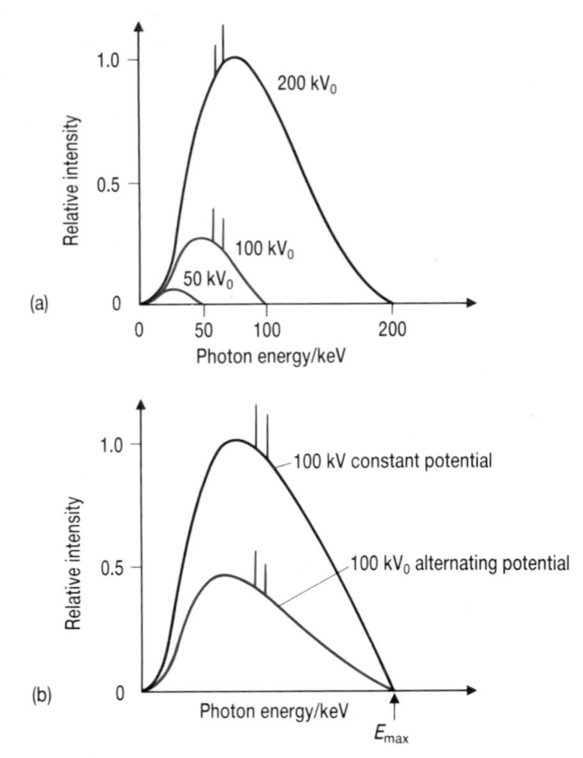

Fig 9.4 Effect of tube voltage on spectrum **(a)** change of peak voltage **(b)** alternating power supply.

Tube voltage As the tube voltage is increased the following changes occur, as shown in Fig 9.4(a):

- E_{max} increases and λ_{min} decreases;
- the peak intensity of the continuous spectrum moves to higher energies;
- the total intensity, which is given by the area under the curve, increases rapidly; it is proportional to V_0^2;
- more characteristic lines may appear in the spectrum.

In addition the beam will be affected by variations in the supply of the tube voltage over time. Some X-ray generators have an alternating power supply which results in a greater proportion of low energy radiation (Fig 9.4(b)).

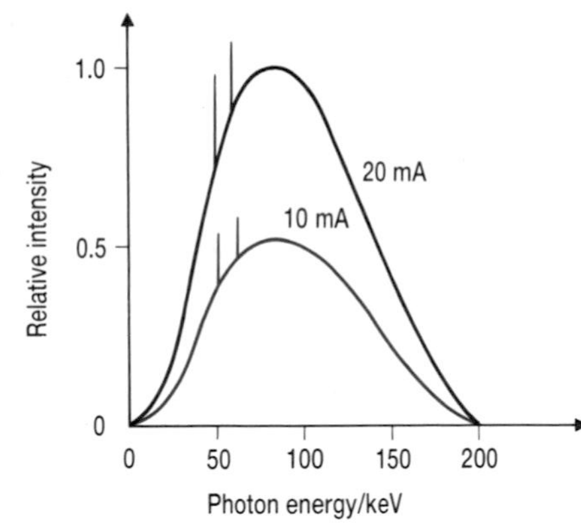

Fig 9.5 Effect of tube current on X-ray spectrum.

Tube Current As the tube current is increased, more electrons are emitted from the cathode, which results in the following changes (Fig 9.5):

- E_{max} remains unchanged, as V_0 is constant;
- the shape of the spectrum remains unchanged;
- the total intensity increases; it is proportional to the tube current.

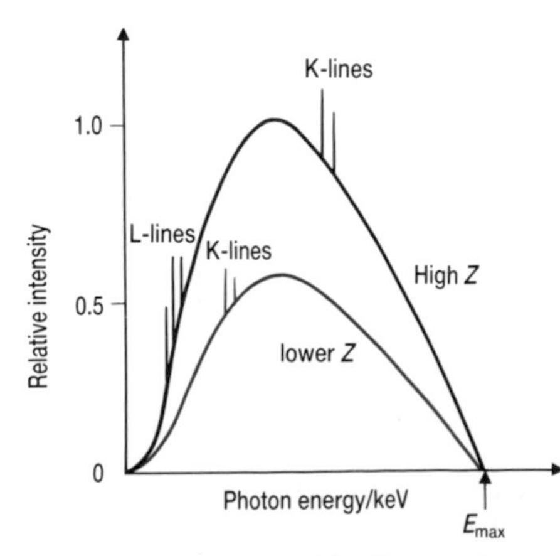

Fig 9.6 Effect of target material on X-ray spectrum.

Target material As the proton number Z of the target material increases, the following effects are observed (Fig 9.6):

- E_{max} remains constant;
- the total intensity increases because there is a greater probability of collision between the bombarding electrons and the larger, more positively charged target nuclei;
- the characteristic line spectra are shifted to higher photon energies.

The target material needs to have a high proton number Z, but also a high melting point because of the heat generated. Tungsten is the ideal material, as it has $Z = 74$ and a melting point of 3650 K.

Filters A thin sheet of material placed in the beam, will selectively absorb more low energy photons than high energy photons, resulting in a change of the kind shown in Fig. 9.7. The filtered beam is more penetrating, and said to be **harder**.

QUESTIONS

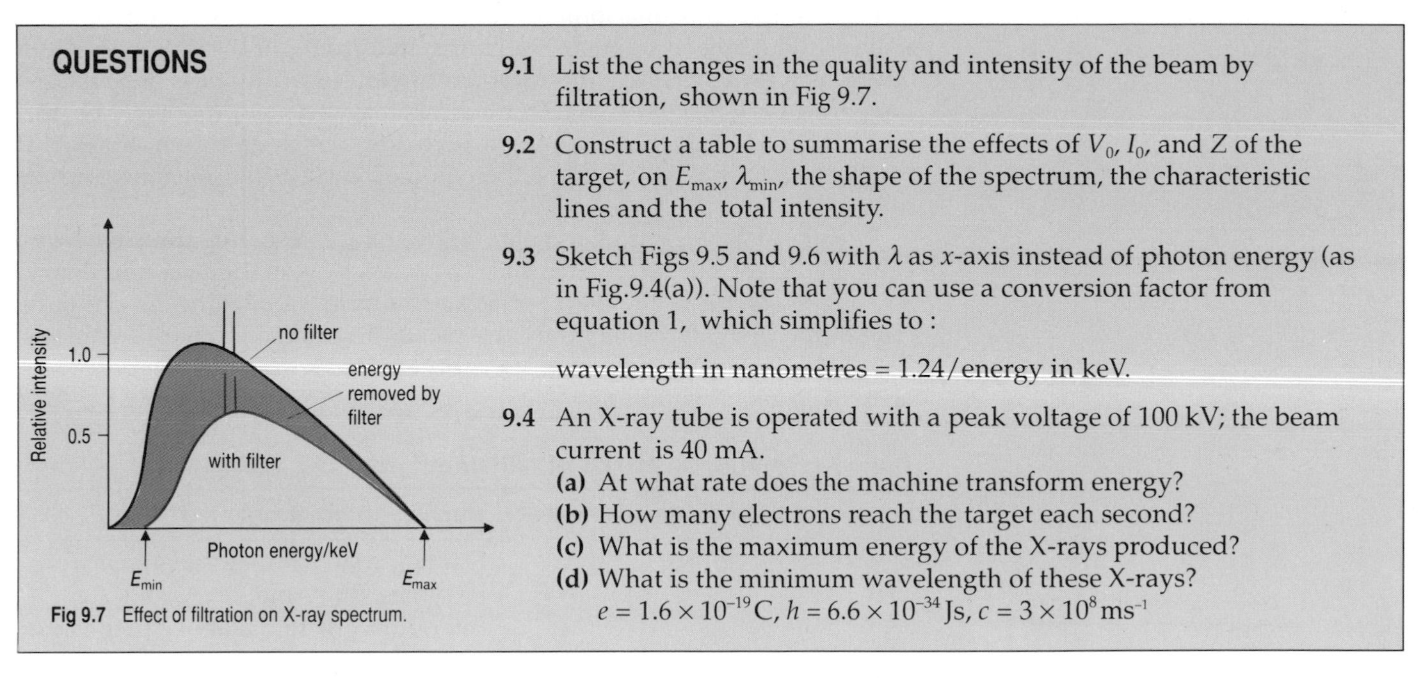

Fig 9.7 Effect of filtration on X-ray spectrum.

9.1 List the changes in the quality and intensity of the beam by filtration, shown in Fig 9.7.

9.2 Construct a table to summarise the effects of V_0, I_0, and Z of the target, on E_{max}, λ_{min}, the shape of the spectrum, the characteristic lines and the total intensity.

9.3 Sketch Figs 9.5 and 9.6 with λ as x-axis instead of photon energy (as in Fig.9.4(a)). Note that you can use a conversion factor from equation 1, which simplifies to :

wavelength in nanometres = 1.24/energy in keV.

9.4 An X-ray tube is operated with a peak voltage of 100 kV; the beam current is 40 mA.

(a) At what rate does the machine transform energy?

(b) How many electrons reach the target each second?

(c) What is the maximum energy of the X-rays produced?

(d) What is the minimum wavelength of these X-rays?

$e = 1.6 \times 10^{-19}$ C, $h = 6.6 \times 10^{-34}$ Js, $c = 3 \times 10^{8}$ ms^{-1}

9.2 THE INTERACTION OF X-RAYS WITH MATTER

The sequence of events

This is shown in summary in Fig 9.8.

When a beam of X-rays passes through matter, such as body tissue, energy is given to the tissue and the intensity of the beam is reduced, or **attenuated**. In the interaction, some of the energy is **scattered,** and some is **absorbed**. Absorption can cause **excitation**, in which an electron is raised to a higher energy level within the atom or molecule, or **ionisation**, in which the

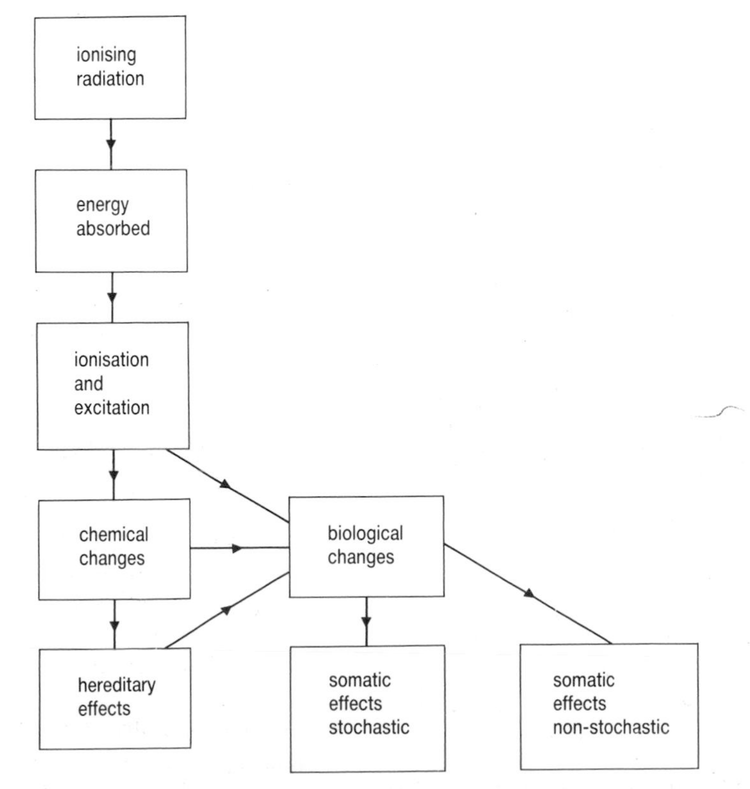

Fig 9.8 Absorption of radiation by body tissue.

electron is removed. This ionisation may damage or destroy biologically important molecules directly. Alternatively, this may occur indirectly through chemical changes in the surrounding medium, which is mainly water.

For example, the ionisation of water molecules produces O and OH free radicals, which have unpaired electrons and are highly reactive. (H^+ and OH^- ions which are also produced, are present in ordinary water and are not particularly reactive). Two OH radicals may react to produce H_2O_2, the powerful oxidising agent hydrogen peroxide, which can then attack the complex molecules which form the chromosomes in the nucleus of each cell.

A range of biological effects can follow this. These are summarised in Table 9.1. They can be divided into two classes, **somatic** and **hereditary.** Somatic (meaning of the body) effects arise from damage to the ordinary cells of the body and affect only the person irradiated. Hereditary effects arise from damage to the reproductive organs and so damage can be passed on to the person's children.

Table 9.1 Biological effects of radiation

Level of biological organisation	Important radiation effects
Molecular	Damage to macromolecules such as enzymes, RNA and DNA, and interference with metabolic pathways.
Subcellular	Damage to cell membranes, nucleus, chromosomes, mitochondria and lysosomes.
Cellular	Inhibition of cell division; cell death; transformation to a malignant state.
Tissue; organ	Disruption of such systems as the central nervous system, the bone marrow and intestinal tract may lead to the death of animals; induction of cancer.
Whole animal	Death; 'radiation lifeshortening'.
Populations of animals	Changes in genetic characteristics due to gene and chromosomal mutations in individual members of the species.

The presence and severity of these effects depends of course on the amount of energy absorbed by the irradiated tissue. This will be considered in Chapter 11. A distinction is made between **stochastic** and **non-stochastic**. For stochastic effects are those which there is always some possibility of the effect occurring; the probability increases with the dose of radiation received. Non-stochastic effects occur only above a certain threshold, then the severity of the effect increases with the dose of radiation received. Hereditary effects are stochastic and can be seen in the statistics for large populations (because the probability is small). Somatic effects can be either stochastic, e.g. leukaemia, or non-stochastic, e.g. burns of the skin.

Attenuation mechanisms

Four processes can result in loss of energy from an X-ray beam to soft tissue. They are shown in order of increasing photon energy in Fig 9.9.

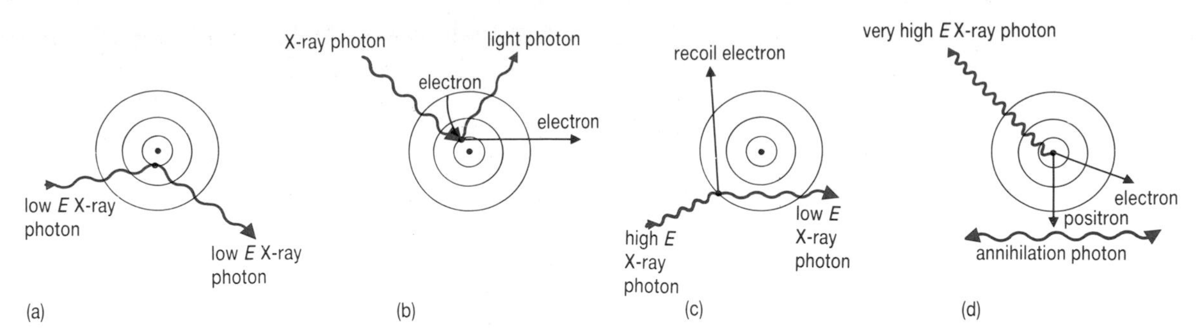

Fig 9.9 The four processes of attenuation **(a)** simple scatter **(b)** photoelectric effect **(c)** Compton scatter **(d)** pair production.

Scatter The energy of the X-ray photon, E is smaller than the energy required to remove an electron from its atom (this is the binding energy E_b).

Photoelectric effect The photon, with energy slightly greater than E_b, ejects an electron from its atom. Another electron in the atom then 'drops into' the vacancy with the emission of characteristic photons.

Compton scatter The photon has much more energy than the binding energy, so that a photon of reduced energy is scattered from the interaction with the ejected electron, which is known as a recoil electron.

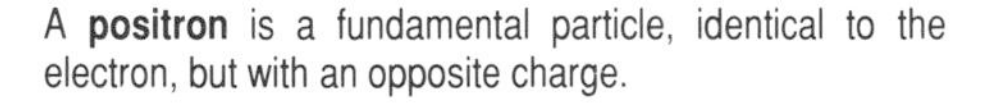

A **positron** is a fundamental particle, identical to the electron, but with an opposite charge.

Pair production At very high photon energies, the photon can interact with the nucleus of an atom. The photon disappears and an electron and a positron emerge. These two particles lose their energy by ionisation, until a positron is annihilated by an electron, with the generation of two identical photons.

Attenuation and distance

If the energy of the X-rays radiate from the source in all directions, the intensity will fall in proportion to the square of the distance from the source. This arises simply from the **geometry** of the situation. The energy is spread out over the surface of a sphere; as the radius r increases from R to 2R, the area increases from $4\pi R^2$ to $4\pi(2R)^2$. Thus the intensity falls to one quarter.

$$I \propto 1/r^2$$

This is the inverse square law for radiation. It describes the attenuation of any radiation in a vacuum, and is a reasonable approximation for the attenuation of X, γ and β radiations in air.

In a medium, where absorption processes are occurring, the intensity of a beam I, falls by a constant fraction dI/I, through each unit distance travelled dx. That is

$$-dI/I = \mu dx$$

or

$$\frac{-dI}{I\,dx} = \mu$$

where μ is the **total linear attenuation coefficient**, a constant which depends on the medium and the photon energy of the X-rays. Integrating this differential equation gives

$$\log_e I - \log_e I_0 = -\mu x$$

where I_0 is the incident intensity and I is the intensity at distance x.

$$I = I_0 e^{-\mu x} \qquad (2)$$

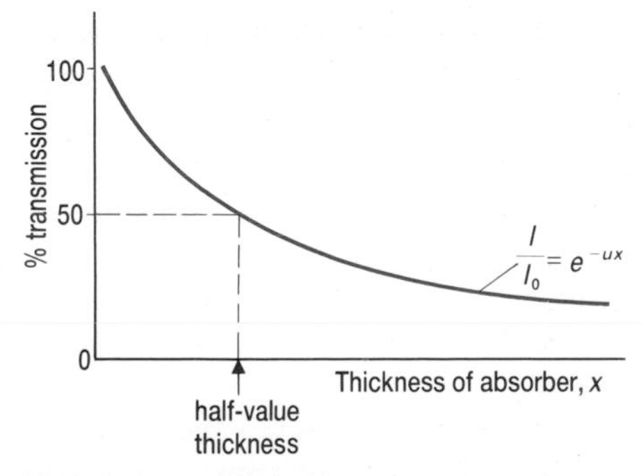

Fig 9.10 Exponential absorption of radiation by a medium.

This results in an exponential fall in the intensity with distance, as shown in Fig 9.10. It is useful in radiology to define the **mass attenuation coefficient**

μ_m, which refers to the attenuation per unit mass of material traversed.

$$\mu_m = \mu/\rho$$

where ρ is the density.

The penetrating power, or quality, of a radiation can conveniently be described in terms of the thickness of material needed to reduce the intensity to half the original value. This is called the **half-value thickness**, $x_{1/2}$ (HVT). If we put $I = I_0/2$ and $x = x_{1/2}$ in equation 2 , we obtain

$$x_{1/2} = \log_e 2/\mu = 0.693/\mu$$

A beam of 80 keV X-rays has a HVT of 1 mm in copper, and a cobalt-60 gamma source (1 MeV), has a HVT of 10 mm in lead. You may be able to check the latter in the laboratory, by placing various thicknesses of lead between the cobalt-60 source and a Gieger tube and counter.

INVESTIGATION

Half value thickness

Measure the half-value thickness of cobalt-60 gamma radiation, for lead and for wood. Estimate the percentage reduction in gamma radiation produced by the box in which the source is supplied.

Attenuation and photon energy

The mass attenuation coefficient of a medium is a function of its proton number Z (because of the probability of collision between photon and electron). Penetration is also dependent on photon energy of the X-rays, in different ways, depending on the mechanism of the attenuation. This is summarised in Table 9.2.

Table 9.2 Attenuation mechanisms

Mechanism	Variation of μ_m with E (photon energy)	Variation of μ_m with Z	Energy range in which it is the dominant mechanism in soft tissue
simple scatter	$\propto 1/E$	$\propto Z^2$	1 to 30 keV
photoelectric effect	$\propto 1/E^3$	$\propto Z^3$	1 to 100 keV
Compton scatter	falls very gradually with E	independent	0.5 to 5 MeV
pair production	rises slowly with E	$\propto Z^2$	above 5 MeV

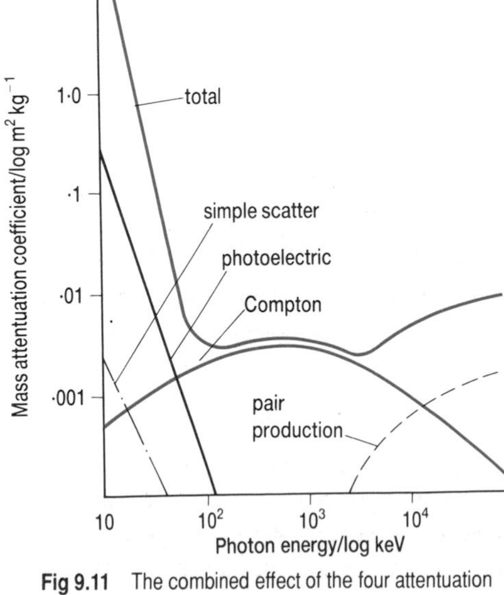

Fig 9.11 The combined effect of the four attentuation mechanisms.

The contributions of these mechanisms to the total attenuation in the body, is shown in Fig 9.11, where the envelope curve is the sum of the attenuation of the four individual mechanisms. You will notice that as the incident photon energy is increased (by increasing the tube voltage) the radiation becomes more penetrating. The optimal photon energy for radiography is around 30 keV, when the photoelectric effect predominates, as this gives maximum contrast between different tissues, because of its dependence on Z^3. For example bone, with a high atomic density, attenuates the beam about eleven times more than the surrounding tissue.

For therapeutic uses, higher energies are preferred (between 0.5 and 5 MeV), where Compton scatter predominates, as this avoids the preferential absorption of the beam by the bones, because this mechanism is independent of Z.

Filtration

The above relationships hold for radiation of a particular energy. Most X-ray beams are, as we have seen, heterogeneous, containing a range of energies. It is possible to make the beam more homogeneous by the use of selected high Z materials which filter out the lower energy photons, by the mechanism of photoelectric emission. During radiotherapy, these would only be absorbed by the patient's skin, adding unnecessarily to the dose. For the low tube voltage used in diagnosis and superficial therapy, filters usually consist of a few millimetres of aluminium. For therapeutic radiation deep in the body, higher voltages are used. Aluminium absorbs too much energy by Compton scatter, so filters with higher Z are used. Composite filters of copper, tin, lead and gold may be used, to select a narrow range of photon energies. The resulting hardened beam has a HVT which is greater than the original beam.

QUESTIONS

9.5 Explain the origin of the photoelectric and Compton effects in absorbing energy. Which of these is preferred for radiography and which for therapy? Give reasons.

9.6 Explain what kind of filter would be used with X-rays used for
(a) radiography,
(b) radiotherapy? Explain why.

9.7 (a) An X-ray tube operates at a steady tube voltage of 70 kV and tube current of 120 mA. It produces a beam of cross-sectional area 4 mm^2, with an efficiency of 1 per cent. Assuming there is no inherent filtration calculate the intensity of the beam as it emerges from the tube.
(b) Assuming that the intensity in (a) is at a distance of 0.1 m from the point source of the X-rays, calculate the intensity at a distance of 1 m from the tube.
(c) Calculate the intensity of the same source of X-rays after they have travelled through 4.5 mm aluminium, placed 1 m from the tube. The half-value thickness for the beam in aluminium is 1.5 mm.

9.3 X-RAY EQUIPMENT

We can now take a brief look at the stages in the creation of a radiographic image. An X-ray generator produces the radiation; apertures and grids control the geometry of the beam and so define the image; film or fluorescent screens detect the image, and various methods are used to increase its brightness; the irradiated tissue can be modified by the addition of a contrast medium to improve the image.

The X-ray generator

Fig 9.12 shows the structure of the rotating anode X-ray tube. Tubes are evacuated glass envelopes, immersed in oil to provide cooling, and shielded by lead. The filament is a coil of tungsten wire, heated to incandescence by a low voltage supply. A metal focusing cup converges the emerging thermionic electrons to a focus on the target. The target is a solid tungsten disc, with a bevelled edge. It is rotated rapidly during operation, normally at about 3600 revolutions per minute. In making a radiograph the energy supplied can be as high as 10 kJ in 0.2 s but only about 1 per cent of this is converted into X-rays, the rest has to be dissipated. Tubes used for therapy need to supply more total energy, but over a longer period; typically 2.25 MJ over 600 s. For these the tungsten target is fixed and mounted on a copper anode block through which oil is pumped.

The X-ray tubes operate at potential differences of up to 40 MV, and at tube currents in the range of tens of mA. These are supplied by high voltage transformers, with a low voltage transformer for the heater current to the filament.

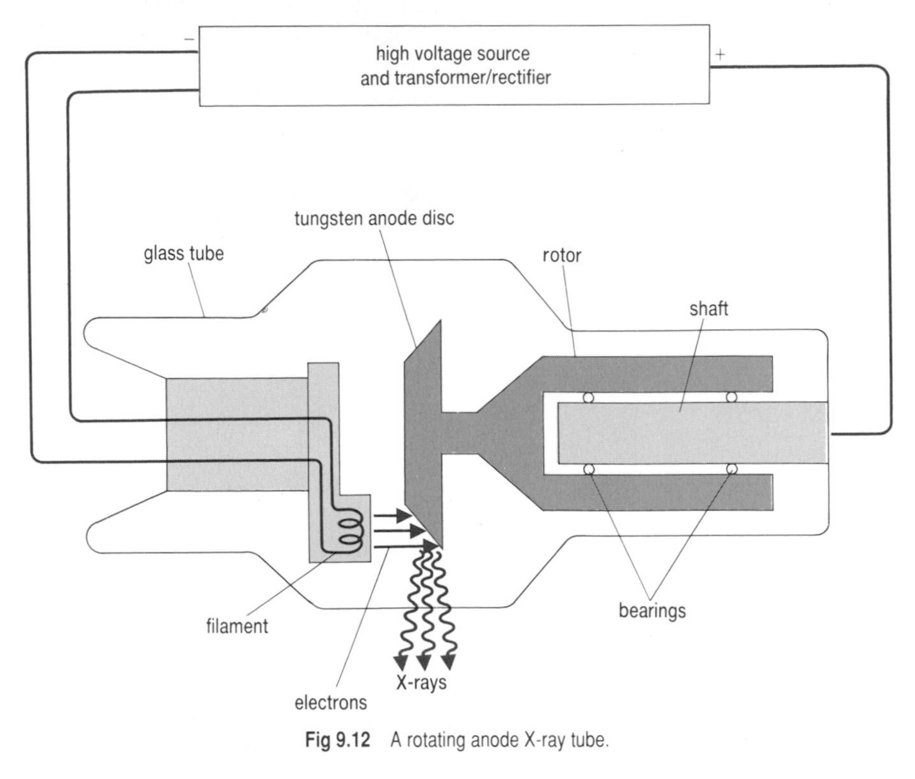

Fig 9.12 A rotating anode X-ray tube.

QUESTION

9.8 How is the energy dissipated in the two types of tube described above? Why do you think different methods of cooling are used in the two types of X-ray tubes?

Controlling the X-ray beam

The ideal beam is small and sharp edged, so that it will produce clear images in radiography and can deliver all its energy to the selected tissue in radiotherapy. The beam of electrons is focused onto the target in the X-ray tube, and the X-rays then radiate from this spot. Since they cannot be focused, X-ray images are basically shadows of objects placed in the beam. You can investigate the process of improving the sharpness of a shadow by placing your hand in the beam of a lamp.

QUESTION

9.9 Why can X-rays not be focused like electrons?

INVESTIGATION

Shadows

If you take an ordinary reading lamp and the bulb is large the shadow of your hand will be blurred.

1. What happens to the shadow when the lamp is small or a pinhole is interposed as in Fig 9.13(a)?
2. The sharpness can also be improved by changing the relative positions of hand, screen and lamp. Try this. The region of blur or partial shadow is called the **penumbra**. Its dependence on separation can be seen from the geometry of similar triangles – Fig 9.13(b).

3. The material surrounding the object will also affect its image. If you place your hand in cloudy water which is in the beam, some of the light is scattered and some absorbed, so the shadow loses sharpness and brightness – Fig 9.13(c).

4. It will be clear to you without a demonstration, that any movement of your hand will also reduce the sharpness of the image.

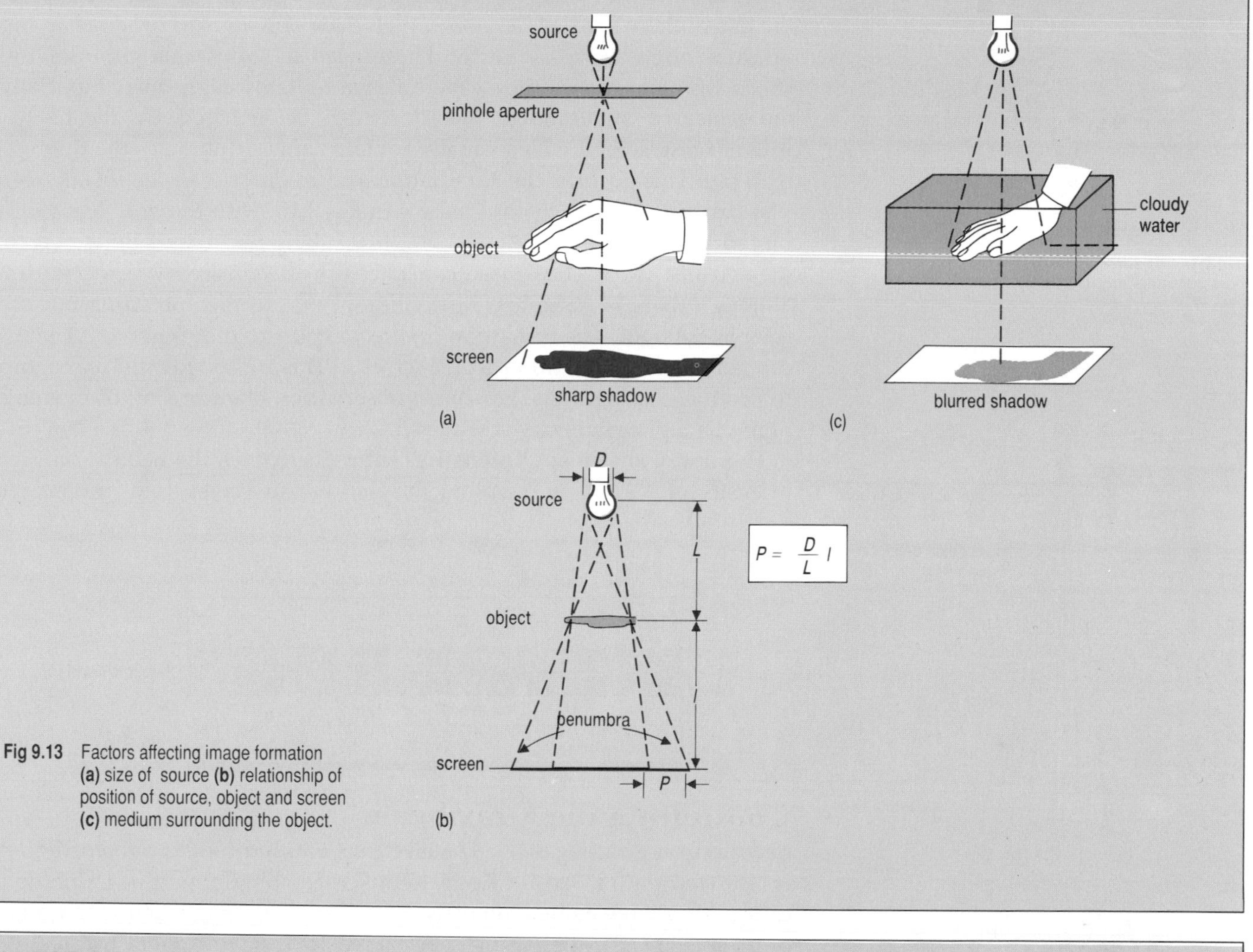

Fig 9.13 Factors affecting image formation **(a)** size of source **(b)** relationship of position of source, object and screen **(c)** medium surrounding the object.

QUESTION

9.10 Relate these observations to the formation of an X-ray image by the lungs of a patient.

We will now take a look at the ways in which the beam has to be limited physically. Orienting the target at an acute angle (typically 17°), to the direction of the electron beam, spreads the area over which the X-rays are generated. This minimises the temperature rise while reducing the apparent area of the X-ray beam. This **line-focus** principle, illustrated in Fig 9.14, is simply a foreshortening of the line (actually an area in three dimensions), by the angle of view.

The beam can then be limited or defined, by the use of apertures. Two types are shown in Fig 9.15, the adjustable diaphragm and the simple cone. They are made of lead to absorb scattered radiation. Scattering will occur in the tissue being examined, this can cause blurring of the image. It can be reduced by the use of a grid, directly in front of the detector. It consists of strips of lead, 0.05 mm thick and 5 mm long, sandwiched between material which is transparent to X-rays. This arrangement allows direct radiation to fall on the screen, and absorbs scattered radiation.

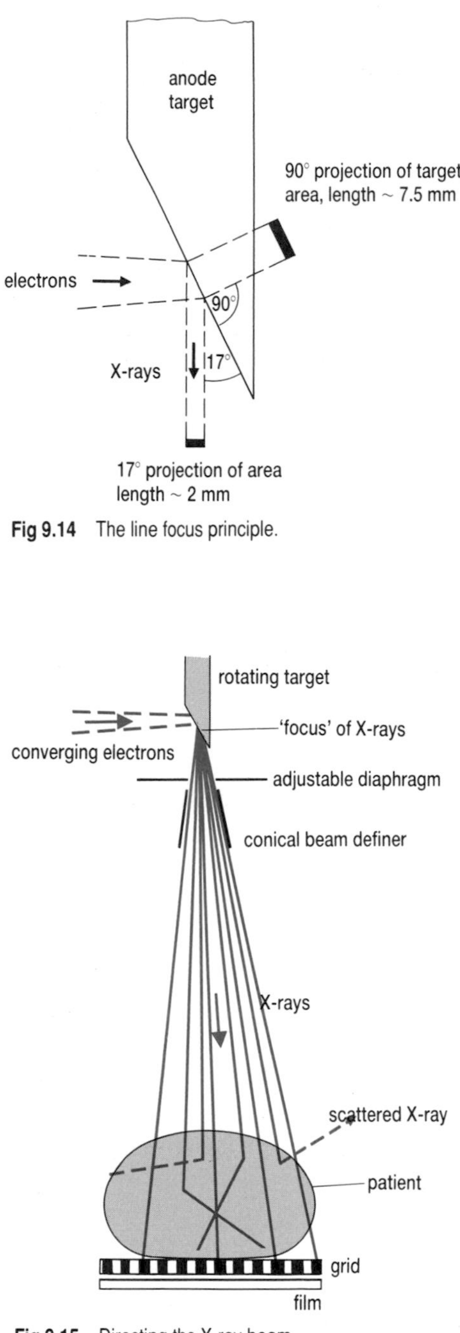

Fig 9.14 The line focus principle.

Fig 9.15 Directing the X-ray beam.

Creating the image

Since the time of Roentgen, the photographic film and the fluorescent screen have been used to display X-ray images. The film is used when a record of the image is required. The use of screens, called **fluoroscopy**, enables real-time investigations to be carried out. If for example a radiologist is examining the bowel, she can move this around whilst viewing (with her hand suitably protected with a lead glove!), to see all the details. In practice the use of these screens requires a high dose of X-rays to produce an image bright enough to view directly. Those used in shops as a gimmick to sell shoes were banned as they gave an unacceptably high dose, especially to the testes of small boys. The dose can be reduced by means of an **intensifying screen** which usually contains zinc sulphide as the fluorescent material. This absorbs the X-radiation and re-emits it in the visible region. The structure of the screen is shown in Fig 9.16, with the double sided film placed to receive X-radiation direct from the patient, and light from both the screens. In practice these can intensify the image by up to 40 times, although there is some loss of definition, due to the spreading out of the fluorescent light. The resolution of direct exposure film is about 0.1 mm this can fall to 1 mm with fluoroscopic screens. This is used for still photography, to produce radiographs. For direct observation or recording of movement, light of a higher intensity is required.

The use of the **image intensifier tube** overcomes the need for a greater intensity of X-rays by coupling the fluorescent screen to a photocathode

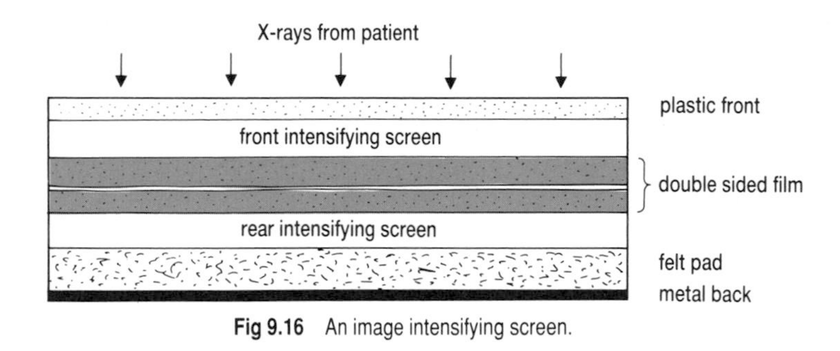

Fig 9.16 An image intensifying screen.

and electron gun (Fig 9.17). The electrons emitted by the photocathode are accelerated onto a second fluorescent screen containing zinc sulphide, the energy they have gained intensifying the original image, usually by a factor of up to 1000. The signal can then be fed to a video camera for 'real-time' viewing using a television monitor or video recording.

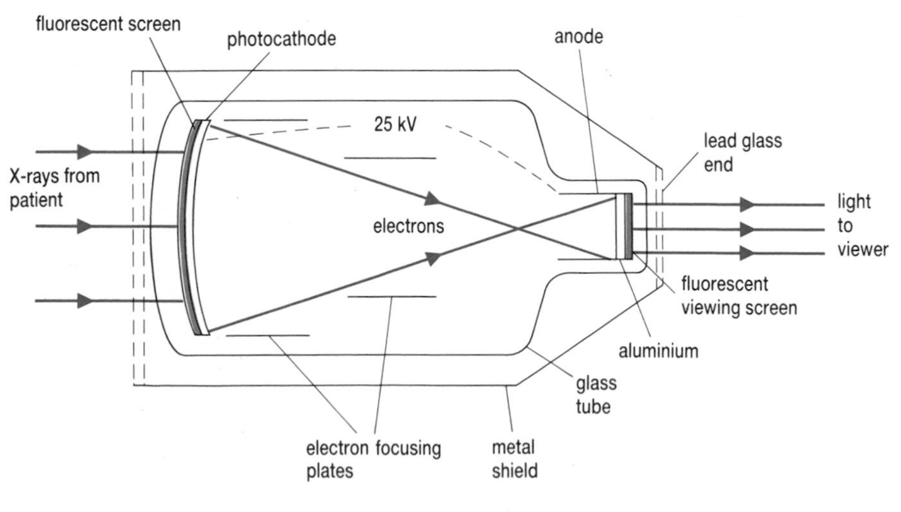

Fig 9.17 An image intensifier tube.

Godfrey Hounsfield was born in England in 1919 and educated in Newark and London as an electrical engineer. He joined the medical systems section of the firm EMI in 1951 and his long career in medical research and engineering culminated in his invention of the computed axial tomography scanner, often called the EMI scanner. This earned him the Nobel prize for medicine in 1978.

Tomography

This is a technique for producing an image of a slice of the object, taking its name from the Greek *tomos*, a section. At its simplest it is a way of arranging to have one plane produce a clear image, regions above and below being blurred. Since X-rays cannot be focused by lenses, this is arranged by moving the X-ray tube and the film cassette. The two are connected mechanically, so that as they move the image of the chosen plane stays in the same place on the film. In Fig 9.18 the image of A stays at A'. Objects not in this plane, such as B are imaged in different positions as the unit moves, from B' to B'' in Fig 9.18. This simple translation movement can introduce artifacts (mistakes) in the images, so more complex cycloidal or elliptical paths are often used.

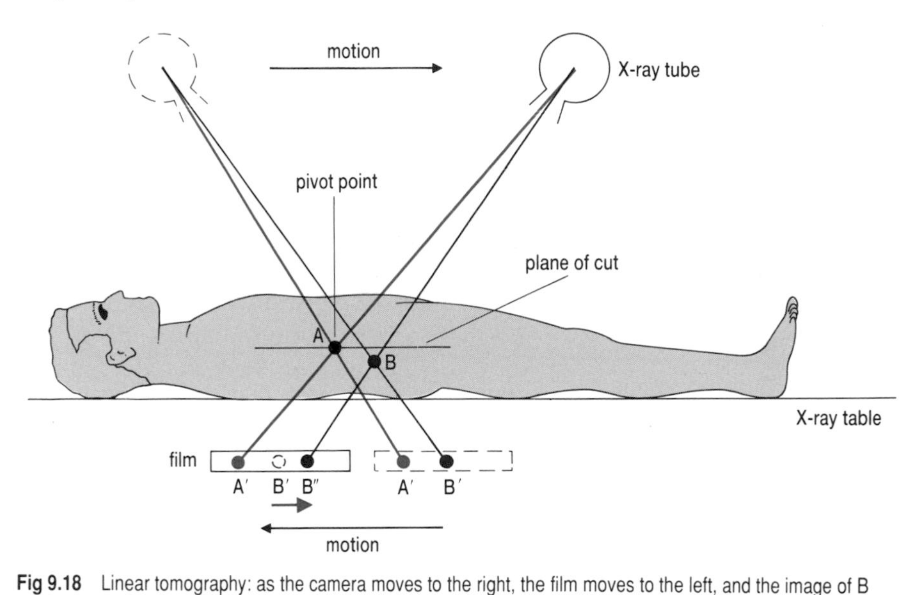

Fig 9.18 Linear tomography: as the camera moves to the right, the film moves to the left, and the image of B moves from B' to B'', while the image of A stays at A'.

The plane imaged is a longitudinal one, in **axial tomography** the film and tube must rotate around the body to image a cross section. These techniques were greatly enhanced in 1971 when Hounsfield developed a computer method of analysing and displaying the images. This **computed tomography** is described in Chapters 12 and 13.

Contrast media

The X-ray shadows are clearest when there is the greatest difference in the photoelectric effect, for example between bone (high Z), soft tissue (moderate Z), and air (low Z). If the natural contrast produced in the tissue is insufficiently high, it can be improved by the introduction of an artificially introduced contrast agent. For example, in investigations of the gastrointestinal tract, which is all soft tissue, the patient is given a 'barium meal' or barium enema which places a high Z, radio-opaque barium compound in the tract. (Fig 9.19).

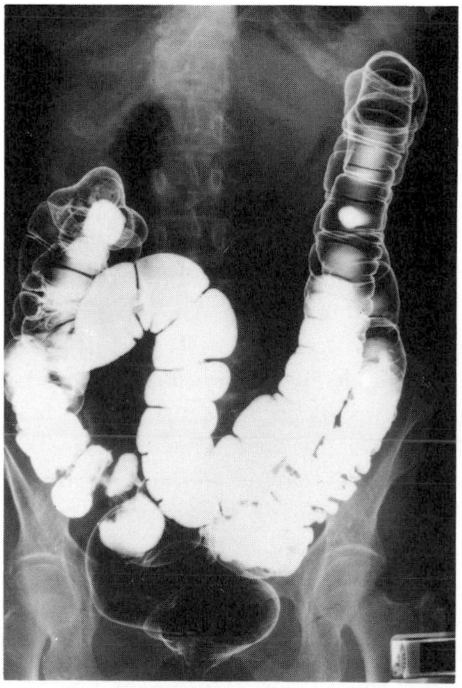

Fig 9.19 Barium enema in the large bowel, showing as white contrast. The darker parts of the bowel are filled with air.

QUESTION

9.11 The barium contrast medium is allowed to drain out of the bowel and replaced with air. Explain how this might give a particularly good image of the inner surface of the gastrointestinal tract (Fig 9.19).

9.4 RADIOTHERAPY

The purpose of radiotherapy is to destroy malignant cancer cells in the body, without damaging normal cells or interfering with the unaffected part of the body. It may seem paradoxical that radiations can cure the very diseases that they can cause. It turns out that all cells are susceptible to radiation damage but those which are malignant are, under certain circumstances, more susceptible than normal cells. In Chapter 11 you will learn more of this. At this stage we will concentrate on descriptions of the techniques and equipment used.

Energy delivery

The most important feature of therapy is delivering the energy in the correct amount, to the affected tissue and not to other parts of the body. This is increasingly difficult the further into the body the malignant tumour is situated. There is a greater likelihood of damaging the intervening and surrounding tissue. Using X- and γ radiations the beams are aimed at the body from several directions, intersecting only at the tumour, so that the dose elsewhere is much lower. Alternatively the beam can be rotated around the patient, remaining directed at the tumour rather like the arrangement for tomography. The actual dose received by the tumour is critical. Typical results indicate that if a particular dose can cure, then 10 per cent less can leave the tumour unaffected and 10 per cent more can cause other problems.

Radioisotope tracers offer a completely different approach; substances are taken into the body so that they accumulate only at the sites required, to deliver the radiation energy locally. These methods are described in the next chapter. Here we are concerned with the longer established, and still more widespread, use of external sources.

X-ray therapy

The development of X-ray machines for therapeutic use has been concerned with producing higher energies accurately and reproducibly. Although the therapeutic use has a long history, until the 1940s available energies had been similar to those for diagnostic use. In this kilovoltage range only fairly superficial (i.e. near the surface of the body) therapy is possible. The highest energy levels possible with conventional X-ray tubes are of the order of 0.1 MeV. This is called the orthovoltage range. The possibilities were greatly enlarged by the invention of the **Betatron**, first used in 1948. This uses a different method of accelerating electrons to voltages of up to 40 MeV producing very penetrating photons of energies up to 10 MeV. These machines are very large and a further advance was made in the 1970s with the development of the medical linear accelerator, or **linac**, which is much more compact (Fig 9.20). This can also generate potentials of several million volts, so these two machines are said to offer megavoltage therapy.

Teletherapy

A parallel development in the years after the Second World War was the production of powerful artificially radioactive substances in nuclear reactors. One of the easiest to produce is an isotope of cobalt called cobalt-60 (details of the production of isotopes and their naming are given in the next chapter). This emits penetrating gamma rays of an energy of about 1.25 MeV. This is equivalent to X-rays generated at 3 MV, from a very small source – though when suitable shielding and other equipment is added the unit is similar in size to the linac (Fig 9.21). The equipment requirements are much simpler for teletherapy, and the lack of high voltage hazards has to be traded off against the constant radioactivity of the cobalt source. This cannot be turned off like X-rays; cobalt has a half-life of 5.3 years. In use it is surrounded by thick lead shielding, with a shutter which opens during the therapy period.

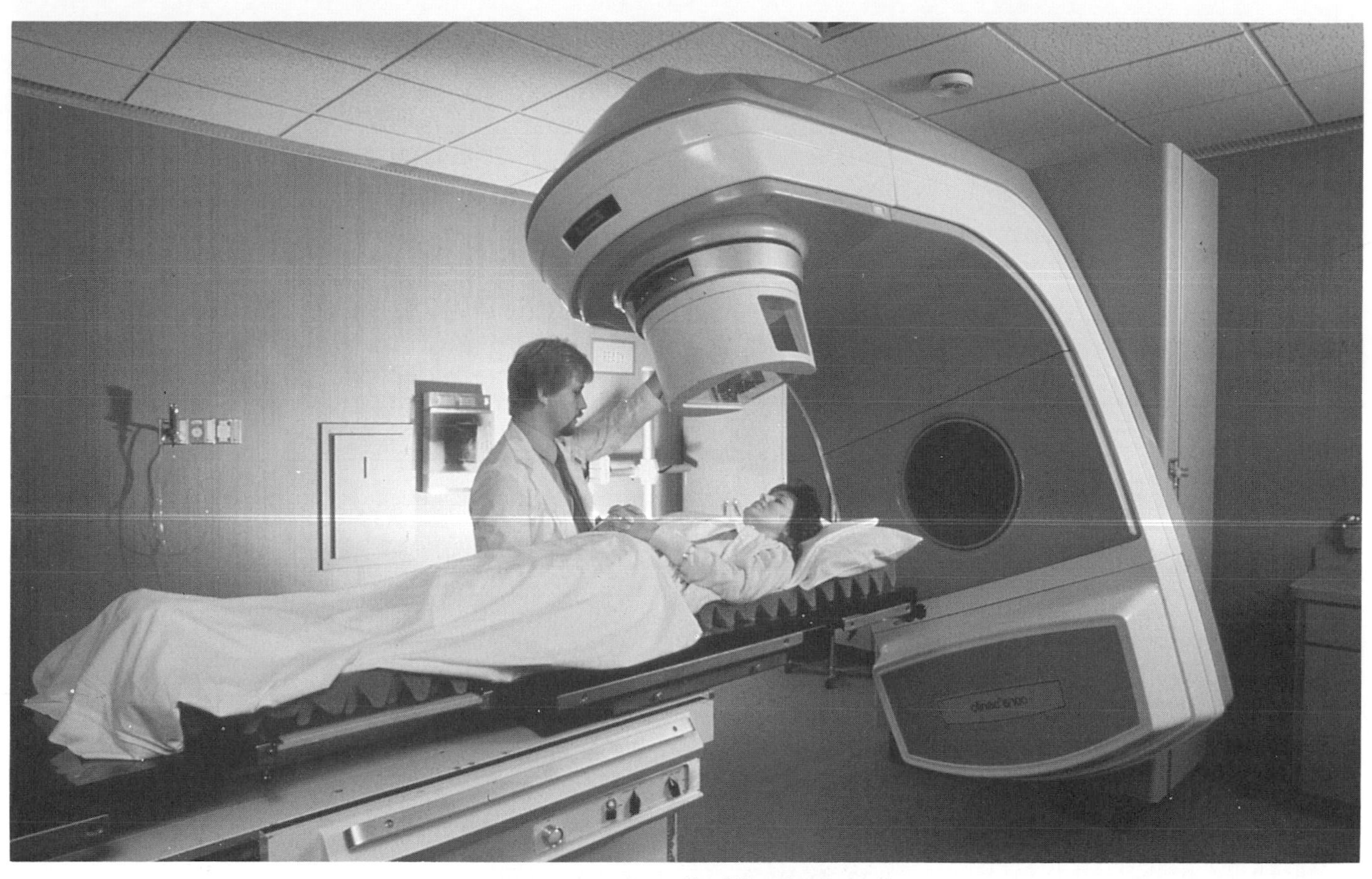

Fig 9.20 Linear accelerator (Linac) X-ray therapy machine.

Latest developments in radiation therapy make use of high-energy protons. Produced in specially designed accelerators at energies of up to 250 MeV, they are very penetrating, but more importantly, they deliver most of their energy to the targeted region of the body. Research at Harvard in the USA, Uppsala in Sweden and Zurich in Switzerland over some 40 years has now led to the setting up of equipment in many countries. The first in Britain, at Clatterbridge Hospital, Merseyside, was established in 1988.

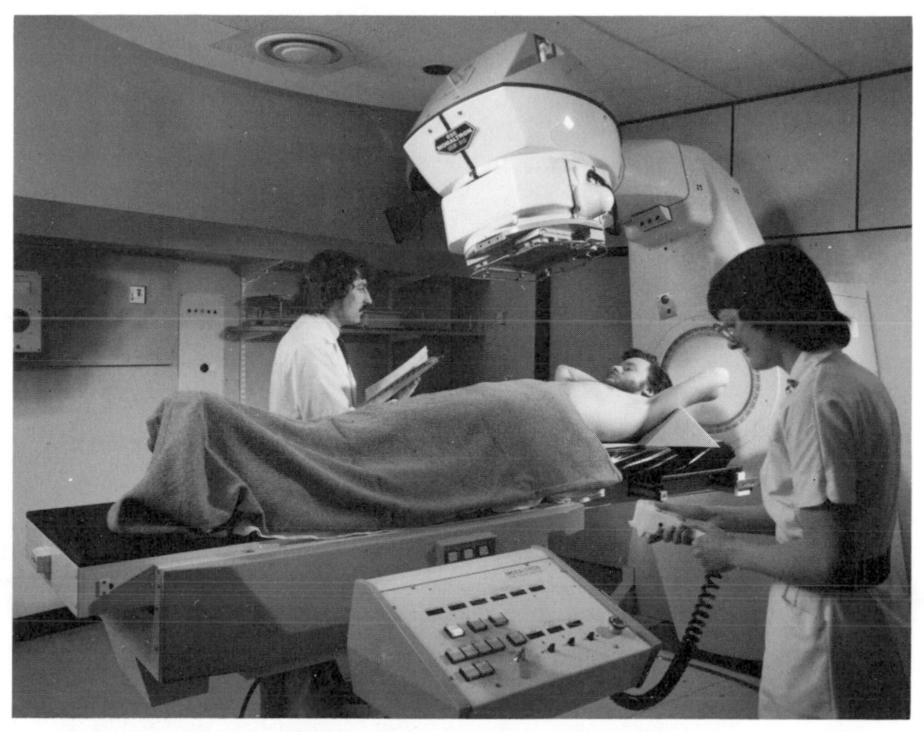

Fig 9.21 A cobalt-60 teletherapy machine.

A summary of therapeutic radiations is given in Table 9.3. The energy of the cobalt-60 and other gamma therapy sources is in the mega- or supervoltage range, the only difference is in the method of production. Other similarities between X-rays and radioisotopes will be considered in the following chapters.

Table 9.3 Therapeutic radiations

Treatment	Voltage range	Typical photon energy MeV
Superficial therapy	20–150 kV	0.03
Orthovoltage therapy	140–300 kV	0.1
Mega- or supervoltage therapy		
Linear accelerator	4–8 MV	2
Betatron	9–42 MV	8
γ Teletherapy		
^{137}Cs		0.66
^{60}Co		1.17, 1.33

QUESTION

9.12 The main advantages of megavoltage therapy compared to orthovoltage are:

(i) ability to treat tumours deep inside the body.

(ii) reduction of the dose to the skin, which considerably reduces the pain from the treatment,

(iii) reduction of the dose to the bone which reduces the possibility of damage.

Explain these features, with reference to the interaction of X-rays of different energies with tissue.

SUMMARY ASSIGNMENTS

9.13 Write an illustrated account of how the following factors affect the quality of an X-ray image of body tissue:
tube voltage, tube current, focal spot size, beam definers, filtration, intensifying screens, contrast media and exposure time.

9.14 **(a)** Distinguish between the *intensity* and *quality* of an X-ray beam.

(b) A beam of X-rays is emitted from an X-ray tube with a tungsten target when the tube voltage is 50 kV_p (peak voltage is 50 kV). (i) Calculate the maximum energy of the emitted photons, and the minimum wavelength of the radiation. (ii) Show by means of a graph how the intensity of the radiation varies with photon energy. Explain the main features of your graph. (iii) If the tube voltage were increased to 100 kV_p what changes would you expect in your graph? Explain briefly the origin of any new feature. (iv) Describe the difference in quality between the 50 kV_p X-ray beam and the 100 kV_p X-ray beam.

(ULSEB 1988)

9.15 **(a)** (i) Sketch a diagram of a typical diagnostic X-ray tube, identifying appropriate parts of the sketch. (ii) Draw a typical X-ray spectrum generated by such a tube and explain how the intensity of the X-rays may be varied.

(b) (i) Define the *half-value thickness* of a material used as an X-ray absorber. (ii) Explain why in producing a radiograph, it is usual to filter the X-ray beams, and suggest a suitable filter material. (iii) Identify the process chiefly responsible for the absorption of X-rays by the filter material used in radiography.

(JMB 1982)

9.16 *A question for research.* What developments outside of medicine led to the availability of **(a)** radioactive sources for teletherapy, such as the so-called cobalt-bomb, in the 1950s? **(b)** linear accelerators for megavoltage therapy, in the 1970s?

9.17 **(a)** Describe briefly the physical processes by which a beam of X-rays for medical diagnosis is attenuated in passing through a patient's body.

(b) A diagnostic radiograph is taken of part of a patient's body containing bones, tissue and air spaces. Describe and account for the appearance of these regions on an X-ray film.

(ULSEB 1987, part)

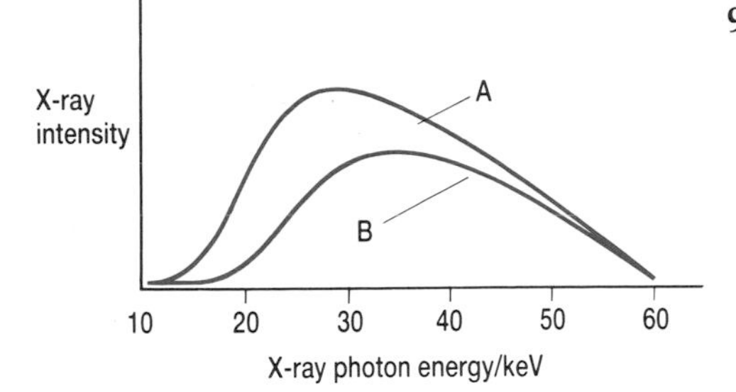

Fig 9.22

9.18 Fig 9.22 shows the relative intensity, *I*, of an X-ray beam as a function of X-ray photon energy, *E*, after the beam has passed through an aluminium filter. For one curve the filter thickness was 3.0 mm, while for the other curve the thickness was 1.0 mm.

(a) Does curve A or curve B correspond to that for the 3.0 mm aluminium filter? Explain your answer by describing the essential differences between curve A and curve B.

(b) Why do both curves have the same maximum value of *E*?

(c) Why is the correct filtration of an X-ray beam of importance in diagnostic radiography?

(d) The intensity of a monochromatic beam of X-rays is reduced to 1/8 of its initial incident value after passing through 3.0 mm of copper. What is the half-value thickness of copper for this radiation?

(ULSEB 1988, part)

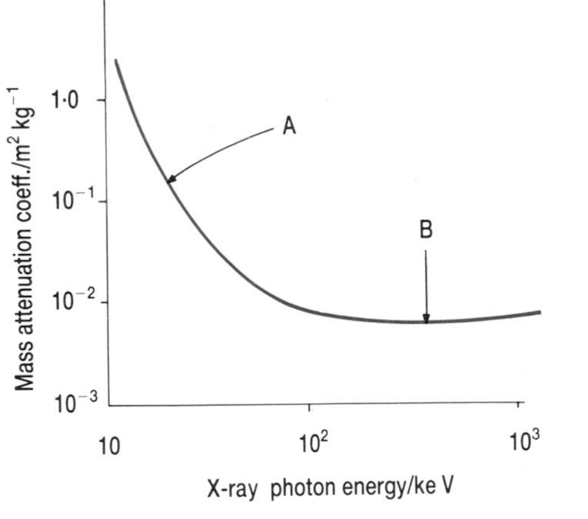

Fig 9.23

9.19 **(a)** Describe briefly the physical processes which take place during (i) *photoelectric absorption* and (ii) *Compton scatter* of an X-ray beam.

(b) The diagram Fig 9.23 shows the variation of the mass attenuation coefficient of water with X-ray photon energy (the scales are logarithmic on both axes).
What are the main attenuation mechanisms in the regions A and B respectively? Give reasons for your answers.

(c) Why is photoelectric absorption more 'useful' than Compton scatter in diagnostic radiography? Hence explain why patients take oral doses of certain barium compounds during X-ray investigations of the upper gastro-intestinal tract.

(ULSEB 1989, part)

Further reading

Hay and Hughes, *First year physics for radiographers*, Balliere Tindall Third edition 1983

A clearly laid out student text giving basic information on phenomena and equipment, with examples of applications.

Hall E.J., *Radiation and life*

Popularly written account of the occurrence and use of radiations, in medicine, biology and power production. Discusses the risks and benefits of their use and draw broadly positive conclusions.

Harrison, R., Diagnostic radiology – the impact of new technology, *Physics Education*, vol 24, no 24 July 1989 p 191.

Chapter **10**

RADIOISOTOPES

Three types of radiation are emitted spontaneously from the nuclei of certain atoms. Gamma radiation is similar to X-rays, but beta and alpha radiations are substantially different in their nature and in the way they interact with matter. In this chapter we will study how this results in a wide range of diagnostic and therapeutic applications of these radiations.

LEARNING OBJECTIVES

After studying this chapter, you should be able to:

1. define and use the following terms: radioactive decay constant, half-life, biological half-life, activity, specific activity, meta-stable radionuclide, tracer;
2. state the properties of alpha, beta and gamma radiations and the ideal radiopharmaceutical;
3. give an example of the production of radionuclides in a nuclear reactor and in a cyclotron;
4. describe the production and use of technetium-99 and iodine-131 as tracers;
5. give examples of other uses of radioisotopes for diagnosis and therapy, explaining their suitability for the function.

10.1 RADIOACTIVE DECAY

An atom of a particular element X, is represented symbolically as ${}^{A}_{Z}X$, where

A is the **nucleon number**, or mass number
Z is the **proton number**, or atomic number.

Examples are ${}^{1}_{1}H$, ${}^{16}_{8}O$, and ${}^{206}_{82}Pb$, for hydrogen, oxygen and lead. Each variety of nucleus is called a **nuclide**, and has a specific value of A and Z. Nuclides with the same proton number (and therefore the same chemical nature) but different nucleon numbers, are called **isotopes**. Most elements have naturally occurring isotopes, but when the ratio of proton to neutrons falls outside a certain range, the nuclide is unstable. It decays to another nuclide, with the emission of radiation. Isotopes of hydrogen are ${}^{2}_{1}H$ (deuterium – stable), and ${}^{3}_{1}H$ (tritium – unstable, β^- emission). Lead has six isotopes, of which three are stable and three radioactive, one of which is ${}^{214}_{82}Pb$. Alternatively this can be referred to as ${}^{214}Pb$, Pb-214, or lead-214. Often there is a series of decays until a stable nuclide is reached. This is natural radioactivity. Radioactivity can also be induced by bombarding stable nuclei with nuclear particles in a nuclear reactor or cyclotron.

Decay law

The process of decay is entirely random, depending only on the nature of the nuclide. The rate of emission of radiation can be described statistically, due to the large number of atoms present, even in a small mass of substance. The number of atoms that decay in unit time is directly proportional to the number of atoms present in the sample, that is:

$$-\mathrm{d}N/\mathrm{d}T \propto N$$

or $$-\mathrm{d}N/\mathrm{d}T = \lambda N$$

where N is the number of atoms remaining and λ is the **radioactive decay constant.** Suppose the number of active nuclei at time $t = 0$ is N_0 and after time t this is N. Then integrating over time t gives:

$$\int_{N_0}^{N} \frac{\mathrm{d}N}{N} = -\lambda \int_0^t \mathrm{d}t$$

$$\log_e \frac{N}{N_0} = -\lambda t \text{ or } N = N_0 e^{-\lambda t} \qquad (1)$$

The number of parent nuclei present decays **exponentially**, as shown in Fig 10.1. (Compare this with the attenuation of X-rays with distance, Fig 9.10 and equation 2 of Chapter 9.)

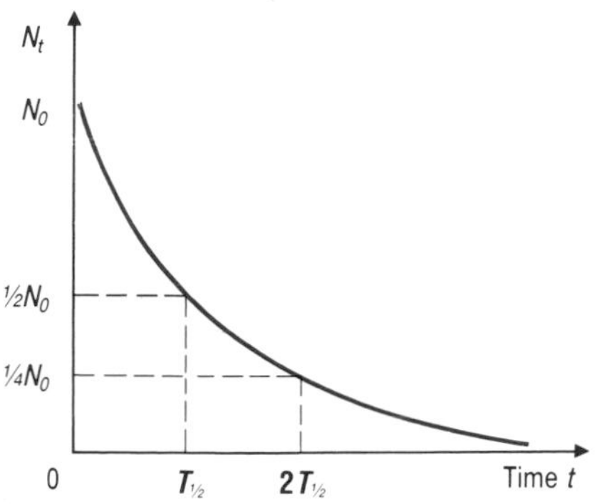

Fig 10.1 Exponential decay of a radioactive nuclide.

Activity

The rate of disintegration of a radioactive nuclide is its activity A, which is defined by;

$$A = \mathrm{d}N/\mathrm{d}t$$

So the unit of activity is second^{-1}, which is called the **becquerel**, symbol Bq. The activity is also given by $A = \lambda N$, and the change in activity from A_0 at time t_0, to A_t at time t can be written by substituting in equation 1, as:

$$A_t = A_0 e^{-\lambda t}$$

showing that the activity also falls exponentially. The term $e^{-\lambda t}$ is called the **decay factor**; it is the fraction of the initial activity present after time t.

The **specific activity** is an important quantity in the handling and preparation of radioisotopes. It is defined as the activity of a sample divided by its total mass, A/m, and is measured in becquerels per kilogram, Bq kg^{-1}. Samples will have varying amounts of active nuclei, depending on their chemical nature, whether there is a mixture of isotopes present and how much decay has occurred. For example, naturally occurring uranium is mainly the isotope U-238. For use in a nuclear reactor it is 'enriched' to increase the proportion of the isotope U-235 with a higher activity. A sample of enriched uranium thus has a higher specific activity.

Half-life

The half-life of a radioisotope $T_{1/2}$, is the time taken for half of the active nuclei to disintegrate, or alternatively the time taken for the activity to fall to half its original value (Fig 10.1).

Since $$N_t = N_0 e^{-\lambda t}$$

$$\tfrac{1}{2} N_0 = N_0 e^{-\lambda T}$$

therefore $\lambda T_{1/2} = \log_e 2 = 0.693$

and $T_{1/2} = 0.693/\lambda$

The half-lives of a number of important radioisotopes are shown in Table 10.1.

When a radioisotope is taken into the body it is subject to various biological processes which will remove it, such as respiration, urination and defecation. This means that its effective half-life T_E in the body is less than the physical half-life T_R from radioactive decay alone. The **biological half-life** T_B is defined as the time taken for biological processes to remove half the original active material. The effective half-life is then given by the equation:

$$\frac{1}{T_E} = \frac{1}{T_B} + \frac{1}{T_R}$$

Table 10.1 The half-lives of some common radioisotopes

Radioisotope	Half-life
Natural	
hydrogen-3 (tritium or triton)	12.3 a (years)
carbon-14	5730 a
potassium-40	1.3×10^9 a
radon-220	55.5 s
radium-226	1622 a
uranium-238	4.51×10^9 a
Artificial	
sodium-24	15 hours
phosphorus-32	14.3 days
cobalt-60	5.26 a
iodine-125	60 days
iodine-131	8 days
caesium-137	30 a
thallium-208	3.1 minutes

For example iodine-131 is used to label the human serum albumen. Iodine-131 has a physical half-life of 8 days, but is removed from the body with a half-life of 21 days. The effective half-life can be calculated as

$$\frac{1}{T_E} = \frac{1}{21} + \frac{1}{8}$$

so $T_E = 5.8$ days.

The biological half-life depends on metabolic rates which vary between different organs of the body and between individuals. It is difficult therefore to give accurate values and so to calculate accurate figures for the effective half-life which would enable the radioactive dose to be known.

QUESTIONS

10.1 The body's ability to metabolise radioisotopes can be used to reduce the dose by substituting the absorbed radioactive nuclide with a non-radioactive one. One of the principal isotopes in the fallout from the Chernobyl nuclear power station accident, was iodine-131. Find out what treatment was given to those affected, to reduce their exposure.

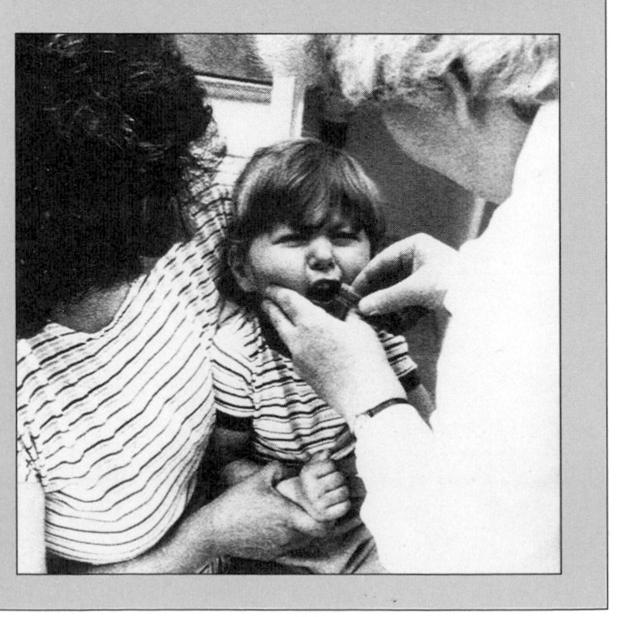

A Polish child being dosed after Chernobyl.

10.2 This question requires a calculation on half-life, activity and specific activity. Naturally occurring uranium contains 99.28 per cent U-238 with a half-life of 4.51×10^9 years, and 0.72 per cent U-238 with a half-life of 7.13×10^8 years. The specific activity can be calculated as follows:

Activity $A = \lambda N = N \log_e 2/T_{1/2}$

One mole of uranium has a mass of 0.238 kg, and contains 6.022×10^{23} atoms of uranium (the Avogadro constant). The activity of the U-238 is therefore

$$\frac{0.9927 \times 6.022 \times 10^{23} \times 0.693 \text{ Bq}}{4.51 \times 10^9 \times 365 \times 24 \times 60^2}$$

which is a specific activity of 2913000/0.238 Bq per kilogram = 12.24 MBq kg^{-1}

The activity of the U-235 is

$$\frac{0.0072 \times 6.022 \times 10^{23} \times 0.693 \text{ Bq}}{7.13 \times 10^8 \times 365 \times 24 \times 60^2}$$

which is a specific activity of 136890/0.235 Bq per kilogram = 0.57 MBq kg^{-1}

So the total specific activity of the natural uranium is 12.81 MBq kg^{-1}

Uranium for use in a nuclear reactor is enriched by replacing some of the U-238 with U-235, so that 'enriched' uranium contains 3.2 per cent U-235. What is the specific activity of this?

10.2 THE NATURE AND PROPERTIES OF NUCLEAR RADIATIONS

The alpha, beta and gamma radiations all originate from the atomic nucleus. Their emission changes the nature of the nuclide in different ways, and the behaviour of each type of radiation is characteristic. This was discovered by Rutherford and co-workers in the early years of this century, and the work is described in the core textbook and the topic book on nuclear physics. Here we will summarise the results and look in detail at the effect of the radiations on the body.

Emission reactions

An α particle is emitted in the decay of many elements with a proton number greater than lead (i.e. $Z > 82$). It is identical to a helium nucleus, consisting of two protons and two neutrons. It therefore has a charge of $+2e$ and a rest mass of about 7×10^{-27} kg. Emission results in a reduction of the proton number, Z, by 2 and in the nucleon number, A, by 4. An example is radium-226, the radioisotope used in a school cloud chamber, which decays to radon:

$$^{226}_{88}\text{Ra} \rightarrow {}^{222}_{86}\text{Rn} + {}^{4}_{2}\alpha$$

QUESTION

10.3 Another isotope of radon, Rn-220 is the product of the decay of thorium. It decays by an emission with a half-life of 55.5 s, so it can be readily observed in cloud chambers. Write down the equation for Rn-220 decay.

The neutrino is a massless, uncharged particle which carries away some of the energy in β^+ decay. The antineutrino is its counterpart in β^- decay. As it has no mass or charge it is difficult to detect. It was only discovered many years after the quantum theory predicted its existence, because of the need to have a particle of this amount of energy.

A β^- particle is emitted by elements having an excess of neutrons over protons. It is simply an electron which is emitted by the decay of a neutron into a proton:

$$^{1}_{0}n \rightarrow {}^{1}_{1}p + \beta^-(e^-) + \upsilon$$

υ represents an antineutrino. This is the antiparticle to the neutrino and it carries away some of the energy of the disintegration. β^- emission results in no change to the *A* of the nuclide, and an increase in the *Z* by 1. An example is the school laboratory source strontium-90, which decays to produce yttrium:

$$^{90}_{38}Sr \rightarrow {}^{90}_{39}Y + e^- + \upsilon$$

A β^+ particle or positron is emitted from elements which have a deficit of neutrons compared to protons. It is simply a positively charged electron which is emitted by the decay of a proton into a neutron:

$$^{1}_{1}p \rightarrow {}^{1}_{0}n + \beta^+(e^+) + \upsilon$$

υ represents the neutrino. β^+ emission results in no change to the *A* of the nuclide, and a decrease in the *Z* by 1. Take for example the isotopes of oxygen, proton number 8. The isotopes with 5, 6, and 7 neutrons decay by β^+, those with 8, 9, and 10 are stable, and those with 11 and 12 neutrons decay by β^-.

QUESTION

10.4 Write down the nuclear equation for the β decays of **(a)** O-14 and **(b)** O-19.

Gamma rays are emitted from the nucleus following β or α emission. The daughter nuclide is left in an excited state by the decay; γ emission enables it to return to a low energy state. Usually this occurs immediately after the main decay, sometimes the daughter can exist in a metastable state for some time, delaying the γ emission. A source of pure γ radiation can be produced by creating such a metastable radioisotope, an example is the physics laboratory source cobalt-60. Gamma emission does not change either the *A* or the *Z* of the nuclide.

Electron capture is a type of nuclear change in which a proton combines with an electron from an inner shell, usually the K-shell, to produce a neutron. This has the same effect on the nucleus as β^+ decay, leaving *A* unchanged and reducing *Z* by 1. An example is the decay of chromium into vanadium:

$$^{51}_{24}Cr + e^- \rightarrow {}^{51}_{23}V + \upsilon$$

Subsequently an electron will move from an outer orbit to fill the space left by the captured *K*- electron, and a characteristic X-ray photon is emitted.

Properties of nuclear radiations

Table 2 summarises many of the important properties of the radiation.

The decay of a particular nuclide usually results in emissions of α, β, and γ with a spectrum of energies; continuous in the case of β and with characteristic energies for α and γ. This will affect the radiations' subsequent ionisation of the material into which it passes and so also its penetration. Alpha particles ionise any medium rapidly and give up their energy over a short path distance; they are said to be high Linear Energy Transfer (LET) radiation. Beta interacts in a similar way but less frequently, and the particle is readily scattered in the interaction. Gamma rays ionise only occasionally, by three mechanisms which are the same as those for X-rays, described in Section 9.2.

Table 10.2 Comparative properties of nuclear radiations

	α	β	γ
Nature	helium nucleus	electron β^- or positron β	high frequency photon or e–m radiation
Charge	+2*e*	–*e* or + *e*	none
$\frac{\text{Mass}}{\text{Mass of proton}}$	4	1/2000	none
Energy range from a particular nuclide	spectrum with characteristic emissions	continuous spectrum	spectrum with characteristic emissions
Ionisation mechanisms	frequent collision with orbital electrons	ejection of electrons, some secondary ionisations	low energy– photoelectric effect medium energy– Compton effect high energy– pair production
Number or pairs produced per mm of air	1000	10	1
Radiation absorbed by	10^{-2} mm Al or a few cm air	1 mm Al	10 cm Pb

QUESTION

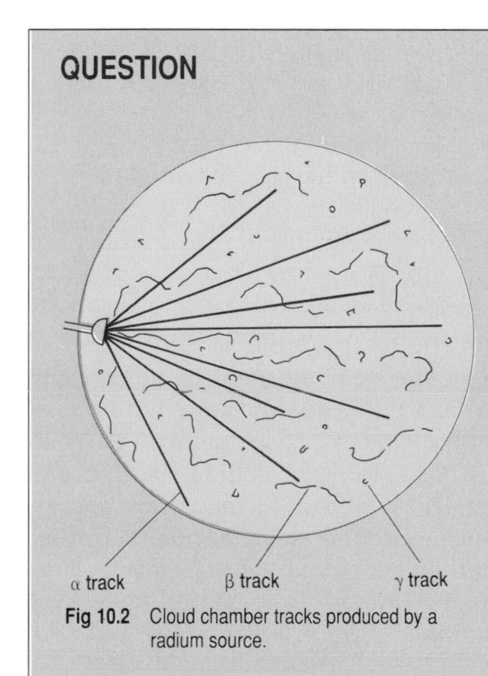

Fig 10.2 Cloud chamber tracks produced by a radium source.

10.5 Use the information in Table 10.2 to answer the following:

(a) How would you expect the deflection of the three radiations, by electrical and magnetic fields, to compare?

(b) Why does γ radiation have a fixed speed? What is changed in γ rays of different energies?

(c) Fig 10.2 shows the tracks that can be observed when radium is placed in the chamber. Explain fully the differences between the three types of track. If you have not made observations like these, you should see if you can set up a cloud chamber now.

Penetration of tissue

The overall process of interaction of nuclear radiations with tissue is very similar to that described in Section 9.2. The absorption of energy causes ionisation and excitation, which results in chemical changes, with a range of biological effects. The absorption of the energy in tissue is quite different, between each type of radiation however, as you would expect from studying Table 10.2. This is summarised in Fig 10.3.

Alpha particles have very little penetration into any type of tissue. An external source will be stopped by the air, unless closer than a few

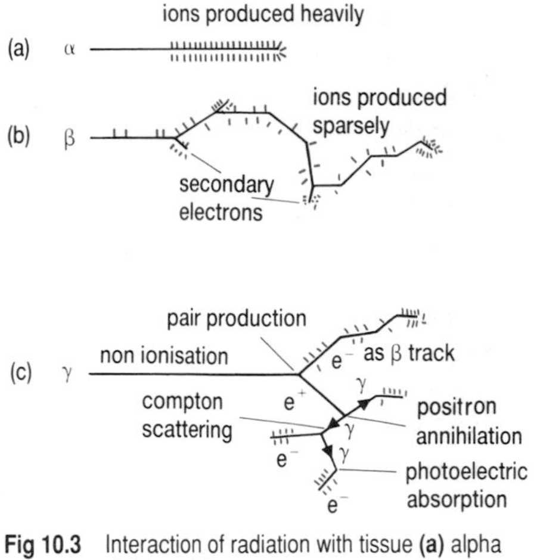

Fig 10.3 Interaction of radiation with tissue **(a)** alpha **(b)** beta **(c)** gamma.

centimetres, when it will be absorbed by a few μm of skin. A source which is ingested however, will be very damaging, because of the density of the internal ionisation.

Beta particles have a wide range of penetrations and because of their irregular paths, their depth in tissue cannot be accurately defined. A maximum depth can be estimated and is considerably greater than α, with the ionisation damage being spread throughout this depth to a decreasing extent as shown in Fig 10.3(b).

Gamma rays do not have a definite penetration depth in tissue, because their interaction causes them, like β to follow irregular paths, and their intensity shows an exponential attenuation like X-rays. The intensity I at thickness x is given by:

$$I = I_0 e^{-\mu x}$$

where I_0 is the incident intensity and μ is the linear attenuation coefficient of the medium, (see Section 9.2). The penetration is described by the **half-value thickness**, for cobalt-60 the HVT with lead is 12.5 mm. It is not possible to screen from γ completely in most situations. However the damage caused is considerably less than α and β because of the low ionisation rate. Remember also that γ rays, like other electromagnetic radiations, attenuate in free space (or air), according to the inverse square law:

$$I = I_0 r_0^2 / r^2$$

(This is strictly true only for a point source in a vacuum)

QUESTION

10.6 A cobalt-60 source gives a dose rate at 1 m from it, of 80 μSv h^{-1}, (Microseivert per hour; you will learn how the seivert is defined in the next chapter.)

(a) At what distance from the source is the dose rate 25 μSv h^{-1}?

(b) What thickness of lead, placed 1 m from the source would give the same protection?

Neutron radiation

Neutrons can be emitted spontaneously, for example in the fission of uranium-235, or emitted as a result of the bombardment of nuclei with energetic particles. These neutrons, because they are uncharged, can penetrate into target atoms and interact with the nuclei. Neutrons with energies of about 1 MeV are called fast neutrons. They can collide with several nuclei and cause considerable ionisation before losing their energy. Neutrons with energies in the eV range, called slow or thermal neutrons, are usually captured by nuclei, which then emit γ radiation. Neutrons have an approximately exponential attenuation through tissue, like γ rays. They are absorbed particularly well by hydrogen atoms, which are in plentiful supply in water in the body.

10.3 THE PRODUCTION OF RADIOISOTOPES

More than 2000 radioisotopes have been made artificially, in addition to the 200 or so naturally occurring ones. The only naturally occurring radioisotope still used is radium-226 (first used in 1901) and this is now generally being replaced by others. Carbon-14 is also used in some investigations, though the source is artificially produced.There are two methods of production, by bombarding stable nuclei with neutrons in a nuclear reactor or with charged particles in an accelerator such as a cyclotron. Not all of these 2000 are potentially useful; there are many considerations to determine an ideal radioisotope. One of the most significant is whether it is to be used for diagnosis or for therapy.

In addition to the particular species of nuclide which is to be used, the specific activity is an important factor, and the pharmaceutical form in which the isotope is prepared. In this section we shall see that different methods are appropriate for certain radioisotopes, and illustrate this with reference to technetium-99 and iodine-131.

Nuclear reactor production methods

Fission reactors operate with natural uranium which absorbs neutrons to form the unstable uranium-235. This splits to form two approximately equal fragments, neutrons and the release of energy:

$$^{235}_{92}U + ^{1}_{0}n \rightarrow ^{236}_{92}U \rightarrow ^{138}_{53}I + ^{95}_{39}Y + 3^{1}_{0}n + \text{energy}$$

This results in a large number of useful isotopes as fission products, for example iodine, strontium and caesium. Although it is quite possible to extract these chemically from the fuel rods of the reactor where they are formed, this is expensive, because of the high radiation. A more common use of the reactor is to irradiate samples or targets with the excess neutrons produced by fission. The neutrons are absorbed by stable nuclei, which then typically become β^- emitters.

QUESTION

10.7 Refer back to Section 10.2 to explain why neutron irradiation produced β^- emitters.

There are three main types of reaction.

(n, γ) reactions or neutron captive reactions are the most frequently used type. The irradiated nuclide captures a neutron and emits a γ ray. For example:

$$^{31}_{15}P + ^{1}_{0}n \rightarrow ^{32}_{15}P + \gamma$$

which is commonly abbreviated to:

$$^{31}P(n,\gamma)^{32}P$$

QUESTION

10.8 Other target materials used include ^{23}Na, ^{41}K, and ^{59}Co, the latter to produce the familiar ^{60}Co source. Write the equations for their production, and hence the general equation for a target of $^{A}_{Z}X$.

These products are chemically identical with their target materials, so separation is not possible. Since only a small fraction of the sample undergoes neutron capture, there will be a high proportion of carriers in radioisotopes produced this way and it will have a low specific activity. In some cases the product undergoes β^- decay to produce a useful radioisotope which can then be chemically extracted. The widely used iodine-131 is produced from tellurium this way, for example:

$$^{130}_{52}Te(n,\gamma)^{131}_{52}Te(\beta^-)^{131}_{53}I$$

(n, p) reactions are less common and require higher neutron energies than the previous reaction. A proton is ejected from the nucleus as a result of the absorption of the neutron. This changes Z so the product can be chemically separated from the target. An important example of this is the production of carbon-14 from nitrogen, for use as a labelling tracer of organic molecules in biological and medical research.

$$^{14}N(n,p)^{14}C$$

(n, α) reactions also require fast neutrons, to remove an α particle when the neutron is captured. The only important medical example is the production of tritium, an isotope of hydrogen, from lithium.

$$^{6}Li(n,\alpha)^{3}H$$

QUESTION

10.9 Write the two previous nuclear reactions in the fuller form, using the $^{A}_{Z}X$ notation.

Cyclotron production methods
In order to penetrate the nucleus with charged particles, energies of several MeV must be supplied, for which a cyclotron is usually used. The bombarding particles are usually deuterons, or α particles, though protons are sometimes used. The product is always chemically different, and its decay is by β^+ or electron capture. Examples of reactions are:

(**d,n**) Oxygen-15 is produced by irradiating nitrogen-14 with deuterons:

$$^{14}_{7}N + ^{2}_{1}H \rightarrow ^{15}_{8}O + 4\,^{1}_{0}n \text{ or } ^{14}(d,n)^{15}O$$

(α,**np**)Fluorine-18 is produced from oxygen irradiated with α particles:

$$^{16}_{8}O + ^{4}_{2}He \rightarrow ^{18}_{9}F + ^{1}_{1}p + ^{1}_{0}n \text{ or } ^{16}(\alpha,pn)^{18}F$$

(α,**2n**) Iodine-123 isproduced by irradiating antimony-121 with α particles. This isotope is preferred to iodine 131 because it emits only γ rays.

$$^{121}Sb(\alpha,2n)^{123}I$$

QUESTION

10.10 Complete Table 10.3 which summarises the production of radioisotopes by the two methods.

Table 10.3 Summary of the production of radioisotopes

	Nuclear reactor	Cyclotron
Mode of operation	.... bombardment and fission	Bombardment by protons, deuterons and ....particles
Major reactions	(n,γ), (n, p) (n, α) U (n, f)	(d, n), (α, d) (α, np) (p, n) and others
Neutron-proton ratio	....	....
Mode of decay	....	..............., EC
If carrier free	(n, y) .... (n, p), (n, α), U (n, f) } Yes	....
Cost	Comparatively low	Comparatively high

Fig 10.4 Hospital cyclotron at the Texas Medical Centre, Houston, USA.

The majority of radioisotopes are produced in reactors. Cyclotrons are a more expensive way of producing radioisotopes because only small quantities can be handled at any one time. Nevertheless they are particularly useful in producing a number of radioisotopes of biologically important elements such as oxygen, carbon and nitrogen. These have short half-lives so some hospitals, such as the Hammersmith in London, have their own small cyclotron dedicated to producing isotopes for medical use (Fig 10.4). The original cyclotron at Hammersmith in London was the first in the world to be built for this purpose. It closed down in 1985 after 30 years service and has recently been replaced.

Generator methods

The generator enables short half-life radioisotopes to be supplied without the need for an on-site reactor or cyclotron. The general requirement is that the radioisotope of interest should be the short-lived daughter of a relatively long-lived radionuclide from which it can be separated. The separation or 'milking' of the generator 'cow' can be done either physically or chemically and either continuously or, more usually, as required.

The ^{99}Mo-^{99}Tc^{m} generator

There are a number of generator systems of importance in nuclear medicine. The following description will concentrate on the ^{99}Mo-^{99}Tc^{m} generator as this produces the most commonly used radioisotope, technetium–99^{m} (the m stands for metastable). ^{99}Tc has a half-life of 2.1×10^{5} years so is no use as a radiopharmaceutical (Table 10.4).

Table 10.4 The ^{99}Mo- ^{99}Tc^{m} generator decay

Parent	^{99}Mo	decays with emission of β^{-}, with half-life of 67 hours
Daughter	^{99}Tc^{m}	is produced which decays by emission of γ, half-life of 6 hours
Product	^{99}Tc	which has a half-life of 2.1×10^{5} a (years)

The parent molybdenum is obtained from a nuclear reactor, either by an (n,γ) reaction or by the fission of uranium. The molybdenum is prepared as ammonium molybdenate and absorbed on alumina in a glass or plastic column. The ^{99}Tc^{m} is formed as the pertechnetate ion. When a saline (NaCl) solution is passed through the column the chloride ions exchange with the pertechnetate ions but not the molybdenate ions. This is called an ion exchange and it is a common separation technique. The saline now contains sodium pertechnetate.

The column is said to be **eluted** by the **eluent** (saline) to give the **eluate** (sodium pertechnetate in saline). It is sterile and compatible with the blood. After the elution the ^{99}Tc^{m} activity builds up and the column can be re-eluted. An example of a commercially available system is shown in Fig 10.5. The volume of eluate is between 5 and 25 ml depending on the manufacturer.

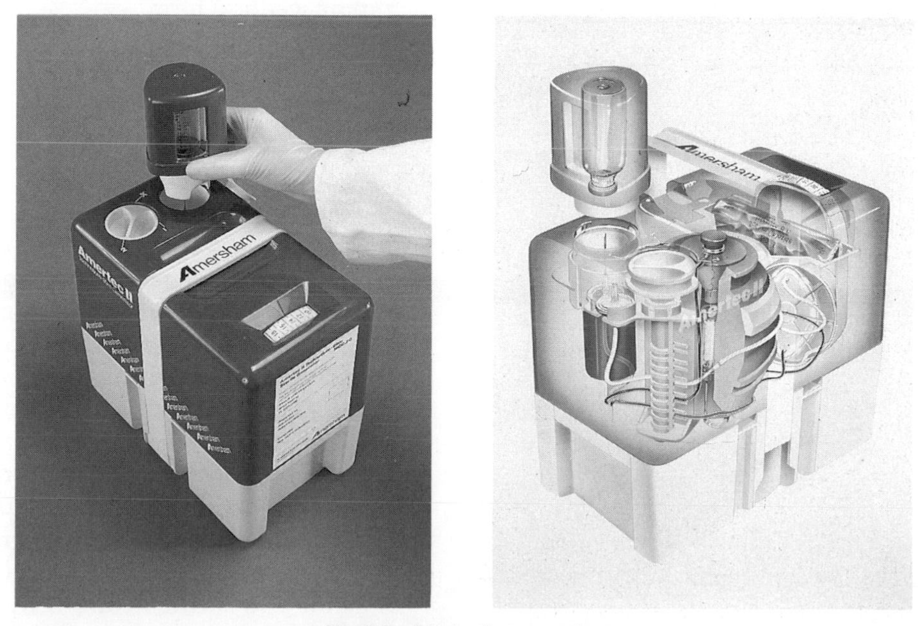

Fig 10.5 A technetium generator.

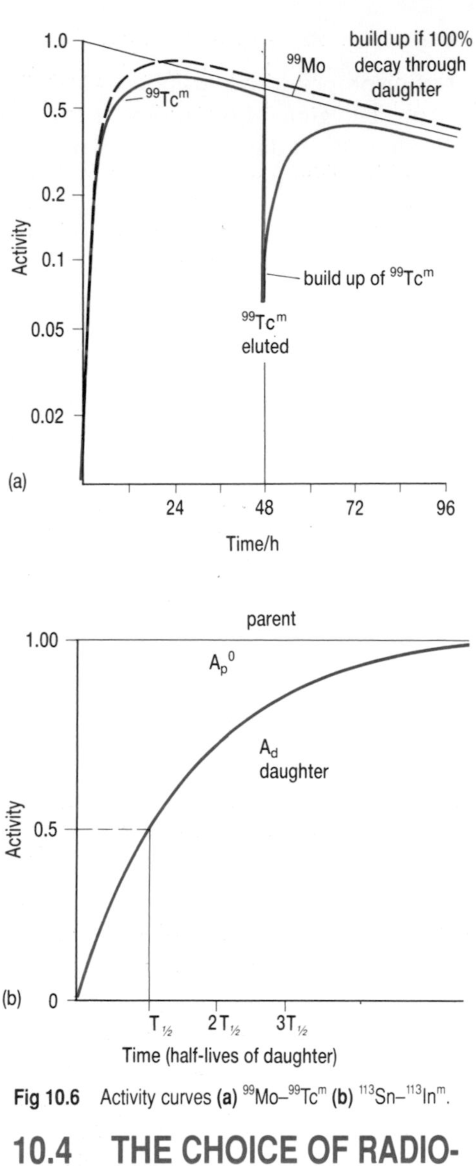

Fig 10.6 Activity curves (a) ^{99}Mo–^{99}Tc^m (b) ^{113}Sn–^{113}In^m.

Yield The activity of the daughter radionuclide at any time depends on:

(a) the activity of the parent,
(b) the percentage conversion of the parent to daughter,
(c) the time since the previous elution.

A typical activity of a ^{99}Mo-^{99}Tc^m generator will be calibrated as 5 GBq of ^{99}Mo at a particular time. This initial value A_p^0 will fall exponentially if the decay constant is λp so that the activity at time t is:

$$A_p^t = A_p^0 e^{-\lambda_p^t}$$

This represents the rate of formation of the daughter. If the decay constant for the daughter is λ_d, then the activity A_d, at time t after an elution is given by:

$$A_d = A_p^0(e^{-\lambda_p t} - e^{-\lambda_d t}) \qquad (2)$$

This is shown for ^{99}Mo-^{99}Tc^m in Fig 10.6(a). In this system $\lambda_d > \lambda_p$ and not all the decay is via the metastable state, about 8.6 per cent decays direct to ^{99}Tc. The dashed line is the curve of equation 2, as if all had decayed to ^{99}Tc^m.

When $\lambda_d >> \lambda_p$, the daughter has a very much shorter half-life than the parent, as in the ^{113}Sn-^{113}In^m system. The decay of the parent can be ignored in comparison with the daughter and equation 2 simplifies to:

$$A_d = A_p^0(1 - e^{-\lambda_d t})$$

which is shown in Fig 10.6(b). It can be seen that after several half-lives of the daughter, its activity approaches that of the parent. This build up is exactly analogous to the production of a radionuclide by neutron bombardment in a reactor. In that case the neutron flux is constant , in this case the rate of decay of the parent is virtually constant. **Continuous elution** In the ^{81}Rb-^{81}Kr^m generator, air passes over the column continuously removing the ^{81}Kr^m. The activity of the daughter is straightforwardly calculated in this case; it is simply the activity of the parent multiplied by the decay constant of the daughter:

$$A_d = \lambda_d A_p$$

10.4 THE CHOICE OF RADIO-PHARMACEUTICALS

From the wide range of preparations available, the following factors will be considered in making the choice:

- The preparation must be **chemically available** in a high purity pharmaceutical form.
- The pharmaceutical must have a suitable **biological behaviour**, distributing itself in the organ or metabolic pathway in an efficient way.
- The radionuclide must have suitable **radiation characteristics** as regards radiation emitted and half-life.

Chemical preparation

Radiopharmaceuticals must be prepared with the same attention to quality control as any other medicinal product, though their radioactive nature and, in many cases, short half-lives pose special problems.

There are two stages in their preparation: the preparation of the primary radionuclide, and the conversion of this into the chemical form required for the radiopharmaceutical use. At the end of the bombardment in a reactor it is usually necessary to separate the radionuclide from the target material and then carry out further purification. Methods of separation used have to be suitable for the small quantities usually involved, and the highly radioactive environment. Various chemical methods are used. Take the preparation of ^{131}I as a typical example. The reaction is given in Section 10.3; at the end of the neutron irradiation, which lasts between one and four weeks, the target is removed and dissolved in 10 per cent sodium hydroxide. The solution is then acidified with sulphuric acid, ferric sulphate is added as oxidising agent, and the ^{131}I is distilled off and collected in dilute sodium hydroxide. Alternatively the TeO_2 may be heated to drive off the ^{131}I.

The formulation of a particular radiopharmaceutical can involve a wide variety of chemical reactions. Their availability is now greatly improved by the development of kits which contain all the ingredients necessary to prepare the radiopharmaceutical to the appropriate standard, without the requirement of extensive on-site facilities.

Biological behaviour

The most important characteristic is that the radiopharmaceutical should localise and concentrate in the organ or system which it is required to image or treat. This maximises the effect required and minimises any side effects, which may be both physiological and radiological.

The normal metabolic activity of the body may result in a satisfactory concentration of the agent. The classic example of this is the concentration of iodine in the thyroid gland. This was in fact one of the earliest nuclear medicine procedures, and can be used both for imaging using scanners, and for therapy in the treatment of *carcinoma* (cancer) of the thyroid. The surprise is that although there are 22 isotopes of iodine to choose from what is now usually used is $^{99}Tc^m$ as sodium pertechnetate. This is taken up in the same way as iodine but is more easily released.

Many other methods of localisation are used. Some are particularly effective in certain parts of the body. Diffusion is used in the brain; small particles in colloids can be ingested by the liver, others can be chosen so that they block the capilliaries and lodge in the lung. One of the most recent developments results from genetic engineering techniques. These techniques make it possible to prepare substances called antibodies which target particular types of cell only. Tagging these with a radioisotope ensures that the radiation is delivered only to those cells. At high doses this can be used to kill tumour cells. Alternatively at lower doses, the radiation can be detected and used to measure the amount and distribution of a particular cell. This technique is called radioimmunoassay (RIA) and one of its uses is to measure growth hormone. Table 10.5 gives examples of commonly available radiopharmaceuticals, and their applications.

Table 10.5 Examples of radionuclides used in imaging

Radionuclide	Radiopharmaceutical	Organ or system	Function
Technetium ($^{99}Tc^m$)	sodium pertechnetate	brain	brain blood flow
Technetium ($^{99}Tc^m$)	coagulated albumen particles	lungs	lung blood flow
Technetium ($^{99}Tc^m$)	colloidal suspension	liver	1. liver blood flow 2. liver function
Technetium ($^{99}Tc^m$)	complex phosphate molecule	bone	1. bone blood flow 2. bone metabolism
Technetium ($^{99}Tc^m$)	red blood cells	heart and circulation	1. blood flow circulation 2. heart contraction
Iodine (^{123}I)	iodide	thyroid	thyroid metabolism
Iodine (^{123}I)	hippuran (sodium iodohippurate)	kidneys	1. renal blood flow 2. renal function 3. urine flow
Xenon (^{133}Xe)	gas	lungs	lung ventilation

Radiation characteristics

For imaging X- or γ radiation of a suitable **energy** is required, particulate radiation is no use as it is absorbed within the patient. A single energy is preferred, not lower than about 100 keV, because of problems with increased scatter, and not higher than about 500 keV, as detectors become less efficient. The preferred type of radionuclide is one which decays by isomeric transition, such as $^{99}Tc^{m}$, as this maximises the amount of energy which is actually available for the imaging process. **Half-life** is an important factor; if the biological half-life is long then the physical half-life is the determining factor. This should be comparable to the time required to carry out the investigation to minimise the dose. When the biological half-life is of a similar period to that of the investigation, and the radionuclide is excreted, a long half-life radionuclide can be used.

For therapy beta emitters are often used, usually because they are absorbed in a small volume and therefore the dose is localised. The energy is not normally critical, though a reasonably high energy is required to give a significant penetration. A half-life of several days is preferred. Beta emission is often accompanied by gamma emission. The advantage of this is that the gamma radiation can be externally measured to monitor the dose. The disadvantage is the increased dose to the patient, and greater protection precautions required.

Table 10.6 Properties of $^{99}Tc^{m}$

Decay mechanism:	isomeric transition
Emission:	140 keV gamma
Half-life:	6 hours
Availability:	Widespread from the ^{99}Mo-$^{99}Tc^{m}$ generator
Chemical behaviour:	can be incorporated into a wide range of pharmaceuticals
Cost:	£10–£30 per dose

QUESTIONS

10.11 Describe the production of a radiopharmaceutical of iodine-131, stating what precautions should be taken to ensure that the products are sterile. What 'special problems' would have to be overcome in preparing a product which is radioactive and which has a short half-life?

10.12 Table 10.6 lists the properties of $^{99}Tc^{m}$, which is the radionuclide most commonly used in nuclear imaging. Explain why it is such an ideal radionuclide.

10.5 NUCLEAR MEDICINE – RADIOISOTOPES IN ACTION

The use of radioisotopes falls into two major categories of therapy and diagnosis. The latter uses are often categorised as *in vivo* or in the body, which covers imaging and other tracer investigations, and *in vitro* or out of the body, which includes sample-taking investigations. In this section we will look at examples of these.

Therapy

Radiotherapy is more commonly carried out using external radiation, usually from X-rays or cobalt-60 γ rays, as described in the previous chapter. The use of unsealed, liquid sources in the treatment of disease, is important in a few, specialised instances. Iodine-131 is taken orally in moderate doses (150–400 MBq), to treat overactive thyroids and in much larger doses (1–3 GBq), in the treatment of some cancers of the thyroid.

Phosphorus-32 and yttrium-90 have been used to treat the blood and the pleura, respectively. The effectiveness of the treatment depends on targeting the affected organ or tissue with high doses of radiation; for this very significant localisation is required. This is the potential of the tagged antibodies mentioned earlier; they are the subject of much current research.

QUESTION

10.13 Find out about the present state of the use of these monoclonal antibodies, or 'magic bullets' as they have been called. Make a note of their potential benefit but also of the difficulties which the early researchers faced, such as the lack of discrimination of the antibodies between cancerous and normal cells. A suitable review article is 'Nuclear medicine homes in on disease' by Christine Sutton in *New Scientist* 15.1.1987.

Thyroid function

This is an example of a dynamic function study, which is a common use for radioisotope tracers within the body, one which cannot be carried out by methods of external irradiation. Iodine-131 is given to the patient orally, usually as sodium iodide solution, it is absorbed from the gastrointestinal tract into the blood stream and collected by the thyroid where it is used to produce hormones. The count rate is measured with a suitably collimated scintillation counter (see Chapter 12 for details) and then compared with a standard or model neck (sometimes called a 'phantom') containing the same amount of iodine-131. The uptake ratio

$$\frac{\text{count rate from patient's thyroid}}{\text{count rate from model or standard}}$$

is recorded at various intervals, usually over a 24-hour period. From the results it can readily be seen if the thyroid is over(hyper-)active, normal or under (hypo-)active.

More recently iodine-123 has come to replace iodine-131 in these tests. It is a gamma emitter with energy of 160 keV and a half-life of thirteen hours. Its use results in a much lower dose than the beta emitting iodine-131 with a half-life of eight days.

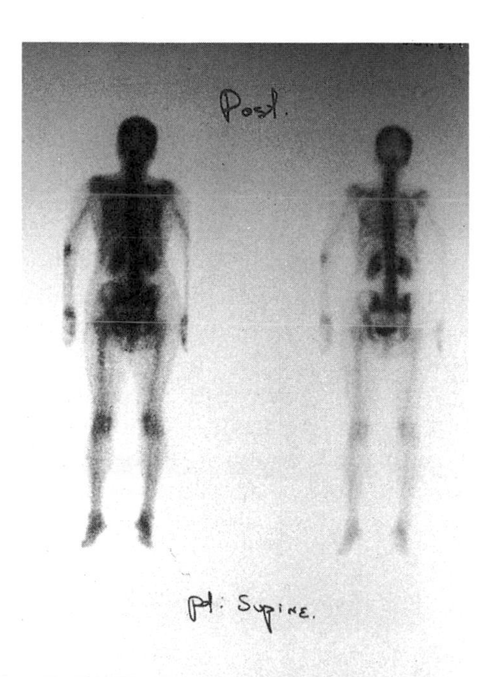

Fig 10.7 A gamma camera scan of a person with secondary cancer in the spine, arm and lower chest, shown by dark spots where the radioisotope has been preferentially absorbed by the cancer cells.

Lung imaging

Albumen, which is part of human blood plasma, can be coagulated (like egg-white during cooking), suspended in saline and labelled with $^{99}Tc^{m}$. This is then injected into the bloodstream where it is trapped in the fine capillaries of the lung. A map is produced of the functioning capillaries and so represents the distribution of blood flow (perfusion) to the lungs. If an *embolus* (blood clot) formed elsewhere in the system reaches the lungs it can block the blood supply with potentially fatal consequences. This area will show on the image as a region without tracer. In order to confirm this another investigation, called a ventilation study, is often carried out. This uses the radioactive gas ^{133}Xe (xenon) which is breathed in with oxygen.

Bone deposits

The turnover of phosphate by the bones can be imaged by $^{99}Tc^{m}$ labelled to a complex phosphate molecule. If a cancer has produced secondary deposits in the skeleton, these show up as increased uptake, or hot spots. Such information can be detected much earlier than with a radiograph, and this can be important in treating the patient (Fig 10.7).

Measurement of body fluids

Following an accident in which there has been significant, but unknown loss of blood, it is desirable to measure what remains in the patient. Since it is not possible to simply drain this out and measure it (!), the technique of **dilution analysis** is used. About 200 kBq ^{131}I-labelled albumen in saline is injected into one arm. After about 15 minutes a blood sample is drawn from the other arm. Since the count rate of the injected material is usually too high to measure directly, on the same instrument for direct comparison, it is usual to dilute an equal amount of the radioactive material with a known volume of water, and count a sample after mixing thoroughly.

Example: A patient is injected with 5 cm^3 of ^{131}I-labelled albumen. 15 minutes later a 5 cm^3 sample is drawn from the patient and found to have an activity of 100 Bq. Another 5 cm^3 sample of the radio-pharmaceutical is mixed with 1000 cm^3 water and the activity of this was measured to be 500 Bq. What is the patient's blood volume?

$$\text{Original activity of sample} = \frac{1000}{5} \times 500\ \text{Bq} = 10^5\ \text{Bq}$$

The patient's blood has diluted the activity by $10^5/100 = 1000$, so the volume of his blood is 1000 times greater than the injected volume, i.e. 5000 cm^3.

A similar technique is used to measure total body water using tritiated water, that is, water containing the radioisotope tritium ^{3}H.

Other examples of *in vivo* investigations are summarised in Table 10.5, along with those described here. There are constant developments to enable new processes and structures to be imaged, to improve those already imaged and to reduce the dose. One way to do this is the *in vitro* test, so called because it is carried out in a test tube. The radioisotope is added to the sample, of blood for example, after it has been taken from the body.

SUMMARY ASSIGNMENTS

10.14 Three of the ways in which nuclear medicine might be considered to complement X-radiography are:

(a) The ability to study how a part of the body functions over a period of time,

(b) The possibility of making quantitative measurements,

(c) The ability to highlight a particular tissue or part of a tissue, by localisation of a radioisotope.

Explain how these processes differ from the conventional X-ray, and give an example of each.

10.15 In a thyroid uptake test, two identical samples of sodium iodide containing radioactive ^{131}I are prepared. One sample is given to the patient by mouth, the other is placed in a model neck at a position corresponding to the thyroid gland.

After 24 hours the radioactivity due to ^{131}I (whose decay produced both β particles and γ radiation) is measured in the patient and in the model by means of a scintillation counter fitted with a collimator. In each case, count rates are observed at the position of the thyroid gland (i) with and (ii) without a lead screen in front of the counter.

$$\text{The ratio } \frac{\text{corrected count rate from the patient's thyroid gland}}{\text{corrected count rate from the model neck}}$$

is compared with that determined for normal healthy subjects

(a) What is meant by the *corrected count-rate* and why is the correction necessary?

(b) What purpose is served by the collimator? Suggest a suitable material for its manufacture.

(c) Why it is unnecessary to know the half-life of ^{131}I for this determination?

(d) Give two reasons why a scintillation counter is chosen rather than a Geiger counter.

(e) Name two mechanisms by which the γ rays (energy 0.36 MeV, 5.8×10^{-14} J) are absorbed in the lead.

(f) Will the β particles contribute significantly to the observed count rate

(i) with the lead in position,

(ii) without the lead?

Give a reason for each of your answers.

(g) Explain why, even though the measurement would be easier if a larger count-rate were achieved by use of samples of greater activity, such samples are not used.

(h) Give the reason why the patient will contain less than half the initial activity of ^{131}I after 8 days, which is the radioactive half-life of the isotope.

(JMB 1984)

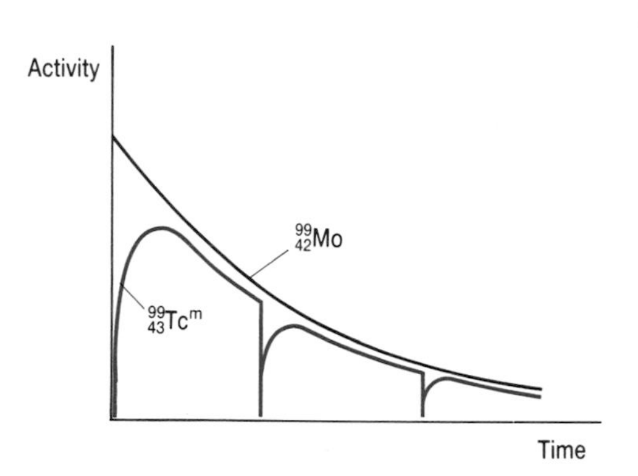

Fig 10.8

10.16 A nuclear medicine investigation requires samples of the metastable radioisotope of technetium $^{99}_{43}Tc^m$, which is produced from the decay of the molybdenum isotope, $^{99}_{42}Mo$, and is eluted from the 'technetium generator' when needed. The activities of these radioisotopes inside the generator before and after various elutions are shown in Fig 10.8.

(a) Explain the shapes of these two curves.

(b) Write down the equation representing the decay of $^{99}_{42}Mo$.

(c) Write down the equation representing the decay of $^{99}_{43}Tc^m$.

(d) Why is $^{99}_{43}Tc^m$ an ideal isotope for diagnostic use?

(ULSEB 1987, part)

10.17 The radionuclide $^{131}_{53}I$ decays with the emission of a β^- particle and a γ photon of energy 0.364 Me V. Its half-life is 8.04 days.

(a) Write down an equation representing the radioactive decay of the nuclide.

(The atomic number Z = 52 for tellurium, Te;
= 53 for iodine, I;
= 54 for xenon, Xe.)

(b) Calculate a value for the decay constant, λ, of $^{131}_{53}I$.

(c) In a certain patient the radionuclide $^{131}_{53}I$ is found to be cleared from the body at such a metabolic rate that its 'biological' half-life is 20.1 days.
Explain why the *effective* half-life of $^{131}_{53}I$ in this patient is less than 8.04 days.

(d) Give a brief explanation of how you would investigate the uptake of $^{131}_{53}I$ by the thyroid gland.

(ULSEB 1989, part)

Further reading

Irradiation and radioactivity, Science support series, Hobson's 1984

A brief, illustrated account of the wide range of applications of ionising radiations.

Sutton, C., Subatomic surgery takes on the tumours, *New Scientist*, p 50. 25.8.88

Gives details of the research and development of a new anti-cancer therapy using accelerated protons.

Sutton, C., Nuclear medicine homes in on disease, *New Scientist* p 48 15.1.87

A review of the range of techniques for diagnosis and therapy using radioisotopes. Topical, readable and well-illustrated with examples of the images produced.

Williams, D., Radionuclides in diagnosis, *Physics Education*, vol 24, no 4 July 1989 p 196.

Chapter 11

MEASURING AND CONTROLLING RADIATION

Fig 11.1 The nuclear genie, an image of potential for good or evil.

The mushroom cloud of a nuclear explosion has come to symbolise the dilemma of our use of nuclear changes (Fig 11.1). The nuclear 'genie' offers the power to improve or destroy our environment, to kill or cure the human race. The difference between the two extremes may only be a matter of quantity. How much energy and radiation is released, over what period and in what way? These are important questions which this chapter is designed to help answer. You will study how the important aspects of radiation are quantified, measured and controlled.

LEARNING OBJECTIVES

After studying this chapter you should be able to:

1. define, know the units of, and use in calculations, the following terms: activity, exposure, absorbed dose, dose equivalent, exposure rate constant;
2. use the following terms correctly: committed dose equivalent, effective dose equivalent, collective dose equivalent, linear energy transfer, quality factor, relative biological effectiveness;
3. recall the background levels of radiation, and give examples of some common medical doses, such as chest X-ray;
4. discuss, in general terms, and by giving examples, the relationships between radiation levels and the incidence of damage or disease;
5. recall the general principles under which the use of radioisotopes is permitted, and give examples of maximum permitted dose levels.

11.1 BIOLOGICAL EFFECTS OF RADIATIONS

All types of radiation carry energy. Their biological effects are caused by the way that they deposit that energy in tissue. Radiations in the lower frequency regions of the electromagnetic spectrum dissipate their energy at the surface of the body: in strong sunlight we experience a heating and a reddening of the skin. The radiations we are concerned with in this theme can penetrate the body tissues to varying degrees and cause ionisation. The general way in which this leads to damage was described in Section 9.2 and a summary of the effects at different levels of biological organisation was given in Table 9.1. As a preliminary to quantifying these effects, let us consider what factors will affect the damage which these radiations cause.

Type of radiation

The interaction of α, β and γ with body tissue was described in Section 10.2 and schematically in Fig 10.3. The significant difference is in the extent of the ionisation produced, which affects the penetration into the tissue. The penetration of alpha particles into the skin is only about 50 μm whereas a high proportion of gamma radiation will pass completely through the whole body.

QUESTION

11.1 **(a)** Explain why sealed α sources, such as those used in physics laboratories, are unlikely to cause damage, while an open α source, such as a liquid or gas is potentially very hazardous.
(b) Why is the interaction of γ radiation with tissue very similar to that of X-rays?
(c) In what ways is the interaction of β radiation with tissue intermediate in its effect, between α and γ? Explain why this is so, with reference to Fig 10.3.

Activity of the source

As the activity is the measure of the number of disintegrations per second, this is clearly important in determining the effect of a particular source. Radiopharmaceuticals will always have a quoted or reference activity, as do laboratory sources. This represents the maximum activity; decay will result in a falling activity with time according to the equation

$$A = A_0 e^{-\lambda t}$$

It may be very difficult to determine the activity in a particular situation since all the relevant information may not be readily available. Take the example of radioactive fallout from the Chernobyl nuclear reactor, which reached Britain on May 2–3 1986. Estimates made of the activity of the radioactive contamination of Glasgow a few days after the Chernobyl accident, varied by a factor of four times, largely because of the assumptions made about rainfall (Fig 11.2).

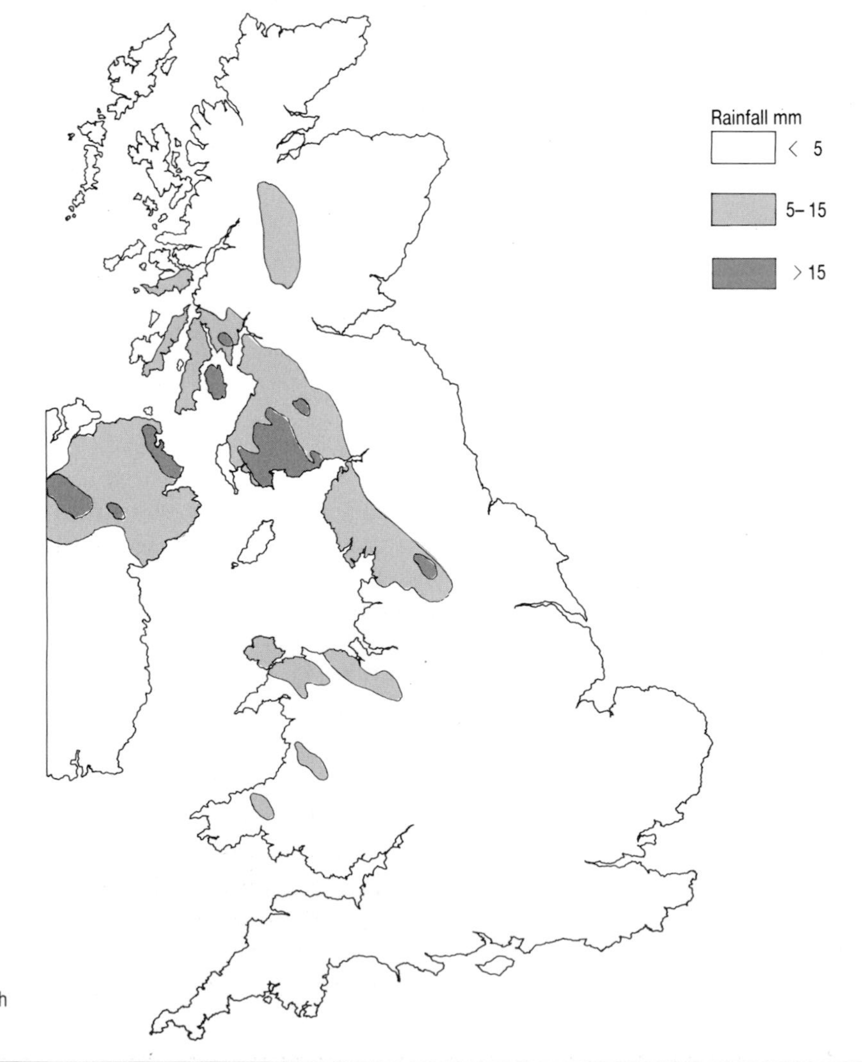

Fig 11.2 Rainfall in the UK during the passage of the Chernobyl plume 2–4 May 1986. This is not the total rainfall for that period, only that which intercepted the plume.

QUESTION

11.2 Briefly explain how the areas of Britain most polluted by the radioactive fallout from Chernobyl were those where it rained whilst the radioactive cloud was passing over. How would you expect this activity to cause exposure to humans? What would be the best way to measure it? How do you think a difference of a factor of four might arise in the estimates?

Time of exposure

For a given source of a particular activity, the exposure to radiation will be directly proportional to the period of time. With sources of short half-life, clearly the activity does not remain constant over long periods, and this will have to be taken into account. The biological effect of the exposure does depend on the period over which it is received. A number of small doses, or a continuous low level of irradiation is less damaging than the same dose given over a short period of time. This is because cells are able to recover from the damage caused when this is small. As a consequence of this, there are different dose limits for short and long periods of exposure.

Type of tissue irradiated

It was discovered as long ago as 1906 that cells differed considerably in their sensitivity to radiation. As a result of their studies, Bergonie and Tribodeau concluded that rapidly dividing or mitotically active cells are radiosensitive; mature, differentiated cells which are unlikely to undergo cell division are not. This means that those tissues which most rapidly replace themselves are the most susceptible to damage; the skin, bone marrow and reproductive organs. It also accounts for the ability of radiation to cure cancers as well as cause them. What distinguishes cancerous cells from normal ones is that they divide more rapidly. The least affected cells are those of the brain and bones.

11.2 DOSIMETRY

The quantification of radiation effects is complicated for three reasons:

1. The effects of the radiation's interactions with tissue is complex and not fully understood.
2. There are many relevant quantities which we need to measure.
3. There are at least two systems of units; in practical dosimetry the SI system of units has not yet completely replaced earlier systems.

In this section we will consider the quantities to be measured and the various units in common use. Some of the complexities and uncertainties will be considered in the next section.

Activity

The curious (!) conversion from curie to becquerel arises because the curie is defined as the activity of one gram of radium. What does this tell you about the activity of radium? How did the curie get its name?

This is simply a measure of the number of nuclei that disintegrate in one second. The unit the becquerel, Bq, is one disintegration per second. This is a very low activity, typical values of sources used in the teaching of physics are about one MBq, medical sources are often in the GBq (10^9 Bq) range. The old unit of activity is the curie, CVi which is equivalent to 3.70×10^{10} Bq.

Absorbed dose

The origin of radiation damage is in the energy given to the irradiated material, so the absorbed dose is simply the energy absorbed per unit mass:

$$D = \frac{E}{m}$$

The unit is therefore J kg^{-1} which is called the gray, Gy. The old unit the rad derived from the metric 'cgs' system; 1 rad = 0.01 gray.

Exposure

With X- and γ radiation it is found that the total ionisation produced is a good measure of the total energy absorbed. Exposure is defined as the total charge of one sign (+ or –) produced in a unit mass of air:

$$X = \frac{Q}{m}$$

The unit is therefore C kg^{-1}. The old 'cgs' unit of exposure is the roentgen, R; 1R = 0.01 C kg^{-1}.

Charge is usually easier to measure than energy, so this is a useful quantity. It applies, strictly speaking, only to X- and γ radiation in air. The relationship between exposure and absorbed dose in other circumstances can however be compared to this.

Relationship between exposure and absorbed dose

An exposure in air of 1 C kg^{-1} means the creation of $1/e$ electrons per kilogram of air, where e is the electronic charge. Each of these requires an energy, the ionisation energy, of 34 eV, so the total energy released is:

$$34e/e = 34 \text{ J kg}^{-1} = 34 \text{ Gy}$$

so for air:

$$D \text{ [Gy]} = 34 \times X \text{ [C kg}^{-1}]$$

and in general:

$$D \text{ [Gy]} = f \times X \text{ [C kg}^{-1}]$$

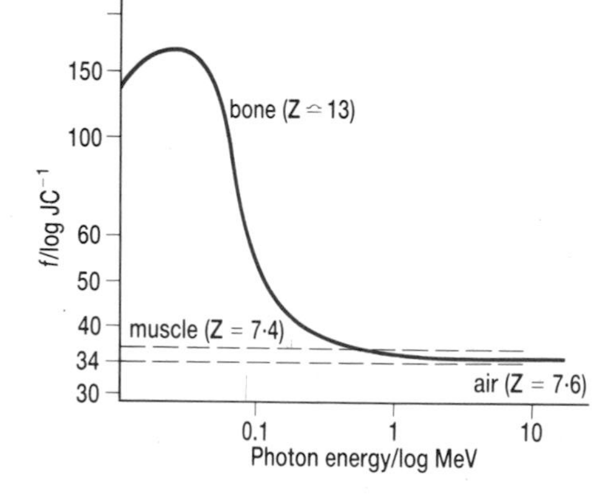

Fig 11.3 Relationship between exposure and absorbed dose for various materials.

where f is a conversion factor which depends on the absorbing material. The atomic number Z of soft tissue is similar to that of air (Z for air is 7.6, Z for muscle is 7.4), so $f \approx 34$ JC^{-1}. When the photon energy of the radiation is high, energy absorption is by Compton scatter which is independent of Z, so the relationship between exposure and absorbed dose is constant for all materials. Otherwise f varies with photon energy and the relationship can be computed with the use of a graph such as Fig 11.3.

Dose equivalent

Radiation damage depends on the distribution of the energy released, and this in turn depends on the type of radiation. Those with multiple charge and multiple atomic mass units, such as α, deposit their energy in a short path length. They are called high linear energy transfer radiations. This effect is expressed as a dimensionless constant the quality factor, Q, or relative biological effectiveness of the radiation. The values recommended by the International Commission on Radiological Protection, the ICRP, are given in Table 11.1. The dose equivalent of a particular radiation is defined as:

$$H = QD$$

The basic unit is J kg^{-1}, but it is called the sievert, Sv, to distinguish it from the absorbed dose. The old unit is the rem; 1 rem = 0.01 Sv. Fig 11.4 shows the relationship between these quantities, in a schematic form.

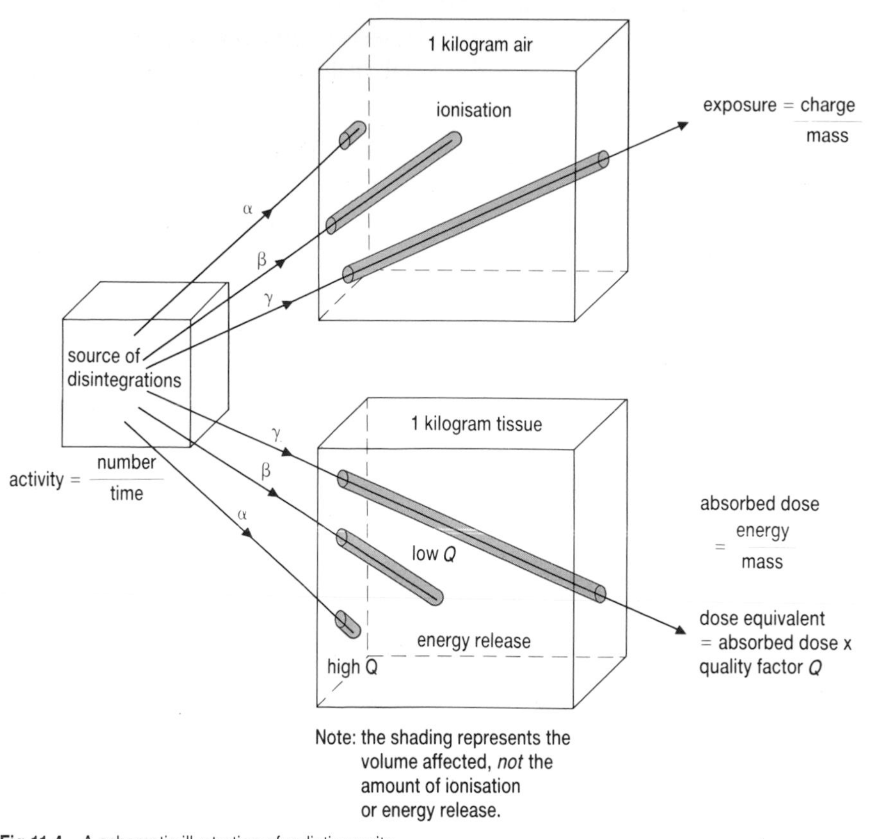

Fig 11.4 A schematic illustration of radiation units.

Table 11.1 Quality factors

Radiation	Q
X-rays, γ-rays, high energy β	1
'Thermal' neutrons	2.3
Other neutrons and singly charged particles	10
α particles and multiply charged particles	20

The linear energy transfer LET, of a charged particle is the average energy transferred to the absorbing medium, per unit distance travelled.

Some UK values

Annual doses from radiation of natural origin. Figures are averages for the whole population and are approximate (from NRPB 1987).

Effective dose equivalent	1870 mSv
Genetically significant dose	1000 mSv
Collective effective dose equivalent	103 kman Sv

A number of other measures are in common use. The **committed dose equivalent**, is the total dose equivalent over the period of time that a radionuclide is in the body. It will depend on the physical and the biological half-lives. The **effective dose equivalent** is a measure which takes account of the relative sensitivities of different parts of the body. Each is assigned a particular weighting so that a total can be made from whatever parts of the body are irradiated, which represents the total effect of the radiation, as if the whole body was irradiated uniformly. Of particular concern is that part of the radiation which may affect the reproductive organs, or gonads. This is called the **genetically significant dose**.

It is often necessary to calculate the total dose equivalent of a group of people, for example, those who live near a nuclear power station. The **collective dose equivalent** is simply the average dose equivalent, multiplied by the number of people exposed. It is usually expressed in man sievert, though the Friends of the Earth and others committed to anti-sexist language refer to the person sievert.

Exposure rate constant

The dose equivalent can be calculated directly from the activity of a source, by the use of this constant, *R* which represents the average energy release for the particular radiation of a particular nuclide. This is used for example in estimating the maximum dose during the use of sealed sources in physics laboratories. The constant is expressed in terms of the dose received, in mGy, at 1 metre over a period of 1 hour, from a source of 1 MBq activity. The constant therefore has a rather cumbersome unit, for example, γ radiation from Co-60:

$$R = 0.36 \text{ mGy m}^2 \text{ MBq}^{-1} \text{ h}^{-1}$$

Calculations

Here is a worked example using the rate constant.

A student is carrying out an experiment with a Co-60 source of labelled activity 5μCi (~0.2 MBq). She holds the source in tongs 25 cm long for the duration of the experiment which is 6 minutes. What is the maximum dose equivalent that she could receive?

$$D = \frac{R \times A \times t \text{ [hours]}}{(d^2 \text{ [metres]})^2}$$

$$D = 0.36 \times 0.2 \times 0.1 \times 1/(0.25)^2 \text{ mGy}$$
$$= 0.072/0.625 = 0.115 \text{ mGy}$$

This is the maximum dose because (i) there will be some shielding from the source container, (ii) only the nearest part of the body will have been exposed to this degree, (iii) the source will have decayed to below its maximum activity, $T_{1/2}$ for Co-60 = 5.3 years.

The dose equivalent is 0.115 mSv because the relative biological effectiveness of γ radiation is 1. We will see in the next section how this compares with a 'safe' dose.

QUESTIONS

11.3 (a) Suppose the student in the previous example held the source in her hand instead of tongs (this is against the regulations!), so that the distance from source to hand is reduced to 0.5 cm. What is her dose equivalent now, assuming the experiment still lasts 6 minutes?

(b) Suppose that a student was foolish enough to put such a source in his pocket for an hour; what dose would he receive if the distance between source and body is 1 cm?

11.4 How much energy is absorbed when a student of mass 60 kg receives an effective dose equivalent of 20 mSv, half from radiation of quality factor 1 and half from quality factor 2?

11.3 RADIATION LEVELS

Incidence of radiation

The responsibility for monitoring the extent of radiation exposure and setting recommended dose limits is with the ICRP, internationally, in conjunction with national organisations such as the National Radiological Protection Board (NRPB) in Britain. Their recommendations usually become the safety regulations, and these are discussed in the next sub-section. Both the monitoring and regulatory functions lead to changes in the estimated doses; in addition other organisations may contest the 'official' figures. Whenever you meet data on this subject, it is worth noting who published it, if it is their own work and when the work was carried out. This may help you to decide why data does not agree. Take as an example the pie charts in Fig 11.5. All were published by the NRPB and based on their monitoring over a period of about ten years. Before we try to account for these changes, let us review the sources of radiation.

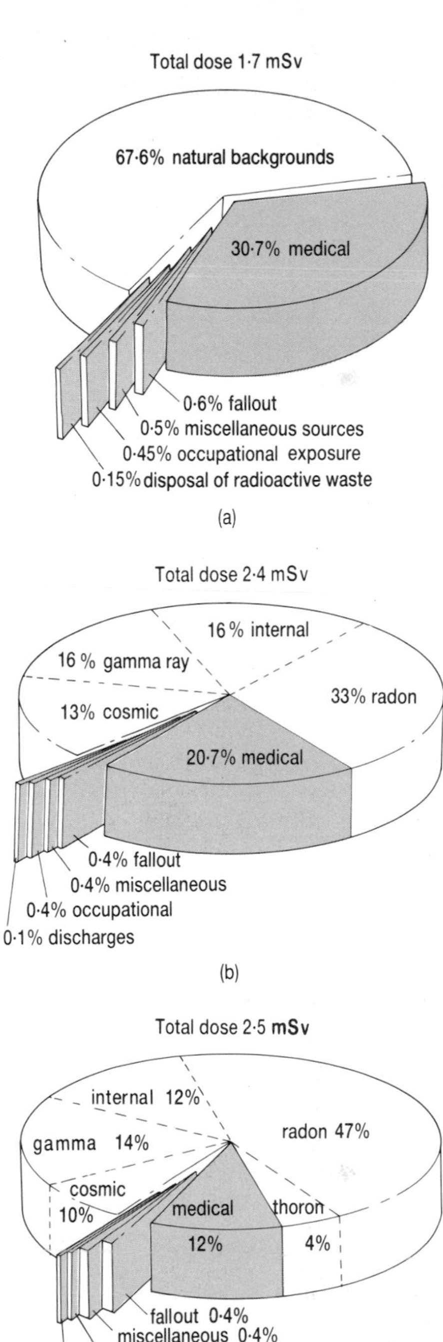

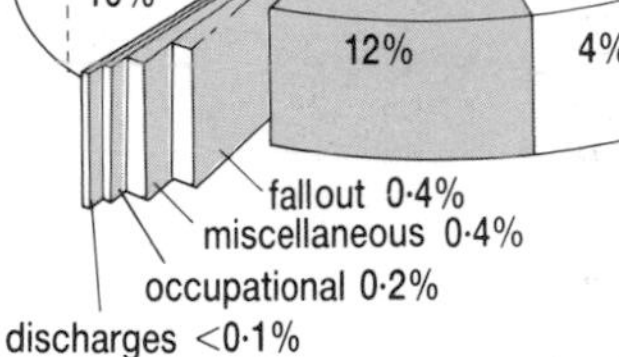

Fig 11.5 Annual UK population dose equivalent from all sources **(a)** NPRB 1977 **(b)** NRPB 1981 **(c)** NRPB 1988.

Natural background radiation

This originates from the action of stars which are vast nuclear reactors. Cosmic radiation is the direct result of this, mainly of course, from our own Sun. The Earth itself originates from this, so there are many naturally radioactive substances in the Earth, such as compounds of uranium, potassium and thorium. Some of these decay to form different radioisotopes which, in the case of radon and thoron pollute the atmosphere, particularly in buildings made from radioactive minerals. Concern about this has grown in recent years, and between 1977 and 1981 the NRPB included estimates for radon doses in its figures, for the first time and more accurate models are still being developed. Accurate estimates are difficult to make as the dose is very dependent on air flow, so monitoring is continuing. Potassium-40 is a common radioisotope in the body and the main source of the internal radiation.

Table 11.2 Annual dose of radiation of natural origin in μSv
Nature 16.3.1989

Source	UK (1988)	US (1987)
Radon and thoron from earth	1.30	2.00
Terrestrial gamma ray	0.35	0.28
Internal radiation	0.30	0.39
Cosmic radiation	0.25	0.27
Totals	2.20	2.94

The values quoted in Table 11.2 are averages for the country's population. The values are higher in regions of granite rocks (more uranium), in regions of higher altitude and latitude (less protection from cosmic radiation by the atmosphere), and in draught-proofed, centrally-heated homes. The variations in background radiation are, as you would expect, much greater worldwide. The maximum that is known for a large population is the 100 000 or so inhabitants in part of the Indian states of Kerala and Madras. As a result of the underlying rock they receive an average dose of 5–10 mSv per year; about four to eight times the world average dose. No significant increase in radiation-linked disease has been unequivocally associated with this population or any other, as a result of variations in background incidence (though it might be discovered yet!). Some worldwide figures are given in Table 11.3.

Table 11.3 Worldwide examples of annual doses from natural sources (various sources)

Region/City	Population	Dose/mSv
London, UK	11×10^6	1
New York, USA	17×10^6	1
Paris, France	10×10^6	1.2
Central France	7×10^6	~3
Parts of Cornwall, Yorkshire and Aberdeen, UK	4×10^5	~5
Kerala and Madras, India	1×10^5	5 to 10

Risks from radon

'Depending on the exact risk factors applied, radon exposure in the United Kingdom might be responsible for 6% or more of the annual incidence of lung cancer (41 000). Our advice in 1987 was that a lifetime exposure to radon at the 'action level' of 20 mSv yr^{-1} would give rise to about a 2% increase in risk of an individual dying of lung cancer. It is now clear that this figure has to be revised substantially to at least 5%, and that action to reduce the risks from radon are even more urgently required.'

R H Clarke & R Southwood of NRPB, Nature, 16.3 1989, p.197.

Artificial radiation

The largest source by far is medical uses as Table 11.4 shows. These values are averages for the whole population. In the case of medical uses this includes diagnostic X-rays, in which a large number of people will have

Table 11.4 Annual dose of radiation of artificial origin in μSv
Nature 16.3.1989

Source	UK (1988)	US (1987)
Medical uses	300	530
Fallout (including Chernobyl)	10	0.6
Miscellaneous sources	10	~100
Occupational exposures	5	9
Radioactive waste	1	0.5
Totals	**326**	**~640**

Table 11.5 Example of typical medical doses (NRPB)

Source	Dose/μSv
Chest X-ray	30
Pelvic X-ray	300
'Barium meal'	1000
Thyroid scan ^{99}Tc^{m}	480
Thyroid scan ^{131}I	16 400
Therapy ^{60}Co	40×10^6

small doses (0.1–2 mSv), and radiotherapy, in which a small number will have very large doses (30–70 Sv). In the 1980s the dose due to X-ray diagnosis has been reduced quite significantly by newer, more sensitive techniques and by its replacement with ultrasound examinations. Some specific medical doses are listed in Table 11.5. Other sources make a small contribution on average, but this may in reality be a large dose for a small number of people. For example the NRPB have estimated that the average additional dose from the Chernobyl accident is about 3 per cent of the total dose for the year from the date of the accident, whereas doses as high as 1.0 mSv per hour were recorded in some areas of Britain and Western Europe, a few days after the accident. Occupational doses are those received by people who work with radioisotopes, again a small proportion of the total population.

Miscellaneous sources include television sets, fire alarms which contain α sources and work like a cloud chamber, gas mantles containing thorium and liquid crystal watches which contain tritium. These latter have replaced luminous watches which had radium in the painted numbers. Those who painted the watches made a fine point on their brushes by licking. After some years it was realised that this caused high levels of radiation-induced diseases, from the ingestion of radium (Fig 11.6).

Fig 11.6 Workers at the Radium Luminous Materials Company – a view.

INVESTIGATION

Background radiation

Make measurements of the normal background radiation in your locality, using a Geiger–Muller tube and a scalar/counter. This will give you a result in counts per unit time. Now compare this with the radiation from other sources available. Possibilities are small quantities of uranium and thorium salts, a gas mantle (these are sold in camping shops), and about 500 grams of a potassium salt (this is about the quantity of potassium that you have in your body).

Try to obtain the best estimate of the background level of radiation for your area from reference materials available to you. You can then calibrate your first measurement so that you can obtain values of the other source doses in grays.

QUESTIONS

11.5 Answer the following questions with reference to the preceding section, or from any additional reading you may have done.

(a) What changes occurred in the period 1977 to 1988 which would account for the differences between Fig 11.5 a, b and c?

(b) Explain how the natural background radiation varies within a single country. Why would you expect the worldwide variation to be much greater?

(c) Find out why the miscellaneous sources mentioned are radioactive and list any others you know of.

11.6 **(a)** Work out the average additional dose of radiation received in the UK from Chernobyl in the year following the accident (data from Fig 11.5(c)).

(b) In the heavily polluted area mentioned, how many hours would it take to receive this dose, assuming no decay.

(c) The principal isotopes involved were iodine-131 and caesium-137. Find out their half-lives and adjust your answer if necessary.

(d) Is your own dose of radiation in the past year likely to be more or less than the average given? Think about the various sources and your own experiences, such as any medical X-rays, holidays in areas of high natural radioactivity, and so on. Can you make a quantitative estimate?

The Chernobyl accident provided a situation of activity deposition that was well characterised in time and in geographical distribution, and measurements along environmental pathways will refine our models. This accidental deposition reinforced the importance of some effects that we know about – such as the importance of wet deposition – and will cause us to consider the need to take account of specific situations that we had not considered previously in adequate detail – in particular, the behaviour of radionuclides in upland ecosystems. FA Fry from NRPB in the 1987 Mayneord Lecture

What level of radiation is safe?

The answer you gave to the last question was probably no, but if you could estimate the dose, how would you know if it was safe? This is not an easy question as you no doubt realise! The fundamental problem is that so many parts of the damage mechanisms described in Section 9.2 seem to depend on probabilities, not certainties. Like the process of radioactive decay, the results can only be described with any accuracy if there are a large number of instances. This is not yet the case for damage at high or moderate levels of radiation. It has been said that a benefit from the Chernobyl accident will be the large amount of new data.

The population of most of the northern hemisphere (about 3×10^9 people) had some additional exposure to radiation as a result. There was the opportunity to monitor this in some detail, but the effects will take many years to show themselves. A further complication in estimating the effects is that these do not cause disabilities or deaths which are uniquely identifiable with radiation exposure. Cancers for example are caused by smoking, dietary actions, atmospheric pollutants and other environmental causes, to which low-level radiation only adds a small proportion. The results of gathering a vast amount of data worldwide, and using the imperfect models of the causes of cancer were summarised in an authoritative survey by several eminent American radiation experts in the journal *Science* of 16.12.1988 (Table 11.6) Their general conclusions are shown in the box below.

The Global Impact of the Chernobyl Reactor Accident

Global impacts to health from the Chernobyl accident may be characterised as acute (non-stochastic) effects or as delayed, stochastic effects predicted on a probabilistic basis. No acute effects have occurred outside of the Soviet Union where 237 cases of acute radiation sickness, including 31 deaths, were reported.

Outside of the immediate Chernobyl region, the magnitude of radiation doses to individuals is quite small, leading to extremely low incremental probabilities of any person developing a fatal radiogenic cancer over a lifetime. Given present dose data, a doubling of leukaemia risk might be expected for the period 1988 to 1998 in the highly exposed Soviet populace in the 30-km zone around the reactor site. Some possibility also exists of a few added cases of severe mental retardation in recent progeny of this exposed group. No adverse genetic effects are expected to be observed in the entire group.

Probably no adverse health effects will be manifest by epidemiological analysis in the remainder of the Soviet population or the rest of the world. Projections of excess cancer risk for the Northern Hemisphere range from an incremental increase of 0% to 0.003%. An upper bound estimate would range from 0% to about 0.01%, still undetectable. Projections of other adverse health effects such as severe mental retardation or genetic disorders are so low as to be unobservable, compared to natural or spontaneous incidence.

The major global impacts from Chernobyl appear to be economic and social. On the basis of extrapolations from initial Soviet estimates, direct and indirect monetary costs may reach $15 billion, 90% of which would be in the Soviet Union. The social consequences are more difficult to quantify, but public concerns, whether justified or not, have increased, necessitating attention by medical, public health, and other authorities. Estimates of health and environmental effects, as well as economic and social impacts predicted in this article, are reasonable but early projections. Their evaluation and, where possible, validation in highly exposed populations and the environment will require study for some period of years.

Table 11.6 Projected health effects from Chernobyl
(various sources, quoted in *Science* 16.12.88)

Region	Population /millions	Collective lifetime dose /thousands of person-Gy	Fatal cancers (lifetime) natural or spontaneous /thousands	Fatal cancers (lifetime) radiation induced /thousands	Excess over natural or spontaneous /%
USSR	279	326	35 000	6.5	0.02
Europe (non-USSR)	490	580	88 000	10.4	0.01
Asia (non-USSR)	1900	27	342 000	0.5	0.0001
USA and Canada	250	1.2	48 000	0.02	0.00004
Northern Hemisphere	2900	930	513 000	17.4	0.003

Most of the existing data has come from the survivors of the Second World War bombing of Hiroshima and Nagasaki. A recent re-evaluation of this has led the NRPB to revise its advice on safe doses (see next section).

11.4 RADIATION PROTECTION

The system of protection, which is to a large extent internationally agreed involves a set of principles leading to regulations covering the use of radioactive substances in all circumstances.

Principles

The radiation protection agencies set maximum permitted dose levels (MPLs) for various situations. The principles on which these are based are:

- **There is no threshold, or safe dose** (a linear relationship is usually assumed for stochastic effects).
- **Optimisation** All additional doses should be as low as is reasonably achievable, (called ALARA).
- **Justification** Any use should show a positive net benefit.

Each of these is difficult to quantify and involves some value judgments. Thus it is to be expected that people will disagree about the extent to which radiations should be used in any particular circumstance. One of the more straightforward choices is in the use of radiotherapy in treating a fatal disease. Perhaps only one individual's risk and benefit has to be considered. The case reported in the margin indicates how agonisingly difficult even this can be.

A man in his late sixties was in terminal care from a tumour called pineoblastoma. His conventional radiotherapy treatment had blinded him; he couldn't move, he had lost weight. He was dying. Three weeks after a new treatment involving a targeted antibody carrying radioactive iodine he was able to walk out of hospital, free of pain but still blind. He survived two years with no problems. He relapsed and though he could have had another injection he was so depressed by his blindness that he refused further treatment.

The Guardian 20.4.1988.

One of the consequences of this approach is the setting of the *different* MPLs. For a radiation worker this is 20 mSv per year for the reproductive organs or the whole body, rising to 750 mSv per year for the limb extremities. For the lay person it is 1 mSv per year and for children under the age of 16, 0.5 mSv per year. As more data becomes available and as understanding of the damage mechanisms increases it is to be expected that MLP's will change.

QUESTION

11.7 (a) What are the risks and benefits of a dental X-ray procedure? How can the dose be made ALARA?

(b) What is the justification for setting a higher limit for radiation workers?

(c) Why are these workers permitted a higher dose to their hands and feet?

Regulations

Any medical use of radioactive substances or equipment producing radiations, must comply with the relevant regulations. These are usually set by national government departments on the advice of bodies such as ICRP and, in Britain, NRPB. Radionuclides are divided into classes by their toxicity (ability to harm). In addition to setting limits on the amounts, the regulations for each class deal with three aspects of safety: people (what are the necessary qualifications, duties and responsibilities), places (for which buildings, services and purposes), and procedures (how to do things safely).

Radiotoxicity The first distinction to be made is between sealed and open sources, as the hazard of contamination is absent from sealed sources. The sealed sources have the highest activity, cobalt-60 sources for teletherapy can be as high as 300 TBq (3×10^{14} Bq). The highest levels of open sources encountered in hospitals are iodine-123 and -131 for the therapy of thyroid cancers, having activities of about 1 GBq. Some hospitals require nuclides with activities of only tens of MBq. The nature of the radiopharmaceutical is important in determining the toxicity, as well as the activity, since this determines how it affects the body chemically.

Radiation more hazardous than previously thought

RADIATION is two or three times more dangerous than was previously thought, the National Radiological Protection Board said yesterday. The board, a Government watchdog, wants the limits for maximum legal radiation doses to be lowered accordingly.

The recommendations could affect about 2,000 people whose work takes them near the present dose limits. It will increase pressure on British Nuclear Fuels and the UK Atomic Energy Authority to lower radioactive discharges from the nuclear reprocessing plants at Sellafield and Dounreay.

Dr Roger Clarke, the director of the Protection Board, said a radiation worker exposed to the maximum legal dose of 50 millisieverts (mSv) a year had been thought to face a one in 2,000 risk of contracting fatal cancer each year. Now, the risk is put at one in 700.

"Continued exposure near the (current) dose limits represents a level of risk which verges on the unacceptable." Dr Clarke added.

Such exposure equates to roughly a one in 18 chance of contracting a fatal cancer during a 40-year working life.

Dr Clarke said people working near the dose limits included some maintenance workers at the Sellafield plant, Ministry of Defence staff working on nuclear weapons, industrial radiographers checking pipe welds and surgeons and theatre nurses exposed to radiation beams during operations. "They form about one per cent of those workers whose exposure to radiation is monitored – about 2,000 in the whole country," Dr Clarke added.

The NRPB suggests that employers should lower the maximum dose for workers to an average of 15 mSv a year while the Government's Health and Safety Commission considers reducing the legal limit.

The board's stance follows a decision by the International Commission on Radiological Protection to accept the findings of a definitive study on survivors of the Hiroshima and Nagasaki atom bombs. The indicated radiation was more likely to cause fatal cancers than thought previously. While the Commission decided a change in the dose limits was not necessary, the NRPB disagrees and is breaking the international consensus.

The board recommends that no member of the general public should be exposed to an artificial radiation dose of more than 0.5 mSv a year. The present legal limit is 1 mSv. Under the new risk estimates, this equates to a one in 33,000 chance of contracting a fatal cancer each year.

The Independent
19.11.1987

QUESTION

11.8 Read the passage in the box and answer these questions on it.

(a) Who are the people whose jobs put them at a risk of a one in 700 chance of contracting a fatal cancer each year?

(b) Do the reductions in the MPLs recommended in the article correspond with the opening statement?

(c) Try to find out the total risk in the general public of contracting a fatal cancer each year, hence work out what proportion of this would be added by a dose of 1 mSv.

People In general, regulations will say who is responsible for what, when radioisotopes are being used. In Britain, the Health and Safety at Work Act gives some responsibility to the individual worker as well as the employer. The regulations for work with each class of radioisotope identify a Radiation Protection Advisor (RPA), whose responsibility it is to draw up local rules. These are implemented at a given workplace, by the Radiation Protection Supervisor (RPS). Each person who initiates an activity involving the use of radiation, for example, the radiologist, is also required to take additional responsibility. So the chain of responsibility goes from the general to the particular.

Places Approval for the use of radioisotopes is given to particular places, subject to the appropriate people being available to carry out the required procedures with suitable facilities. For example, a hospital with a reasonably large diagnostic facility would have a specialist medical physicist as RPA with specialist radiographers as RPS. The required facilities might be those listed in Table 11.7. Areas containing any sources have to be labelled, and some areas have limited access to staff and patients. Monitoring of exposure levels in these sites, and of the personnel entering them is required (Fig 11.7).

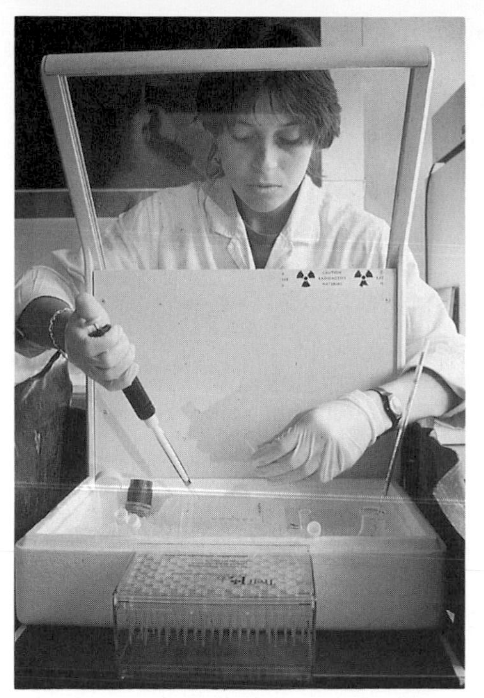

Fig 11.7 Safety procedures with radioactive tracers, here used for labelling in immunological research.

Procedures These are the detailed accounts of what is expected of those people with the various levels of responsibility described above. Those with the more widespread responsibilities will have the most detailed procedures to follow. The general purposes are to establish and maintain practices which keep radiation exposure as low as possible, and within any limits set. For the individual employee, this will mean following rules similar to those you have when using sources in physics. These may be summed up as:

> When using radioactive sources, increase your **distance** to a maximum, reduce your **time** to a minimum, use **shielding** where appropriate, and with open sources, avoid or contain any **contamination**.

A supervisor's procedures will include instructing and monitoring all staff, monitoring the performance of equipment, keeping records of staff, patients and equipment, and perhaps assessing new equipment and evaluating the changes needed in the safety procedures as a result of this.

Table 11.7 Facilities of a laboratory for handling Radionuclides

Specifically set aside for handling radionuclides.
Protective clothing used. Gowns or overalls, gloves and overshoes.
Negative pressure. Extraction fans to outside of building to prevent spread of contamination inside building.
Fume cupboards.
Floors, walls and benches should be smooth and without cracks. Must have non-absorbent and washable surfaces.
Benches strong enough to support lead shielding.
Wash basin with elbow and/or foot operated taps.
Sink with direct access to main drain for disposal of radioactive liquid waste.

SUMMARY ASSIGNMENTS

11.9 Find out the procedures governing your use of radiations in physics. Write a set of rules, or procedures for a medical student who is on a clinical attachment to the nuclear medicine department of a hospital. You should use the school/college regulations as a guide, as well as the information in Section 11.4.

11.10 **(a)** Define the unit of radiation exposure. Describe the mode of action of an ionization chamber and explain how it is used to measure exposure.

(b) Define the unit of absorbed dose and calculate the absorbed dose in air when the exposure is one unit.

(c) A neutron beam and a beam of gamma rays each produce the same absorbed dose in a certain body. State whether the dose equivalents are the same and give a reason for your answer.

The charge on an electron = -1.6×10^{-19} C
Energy required to produce one ion-pair = 34 eV

(JMB 1983)

Further reading

IAEA *Facts about low-level radiation*, International Atomic Energy Agency, 1986

A readable summary of the origins of radiation and the system of protection. Includes a summary of the different views of risks from low-level radiation.

NRPB, *Living with radiation*, National Radiological Protection Board (from HMSO), third edition 1986.

Authoritative, clearly presented guide to the origins, effects of, and system of protection from radiation. Particularly useful on units and quantities.

Harrison, R. Radiation protection, *Physics Education*, July 1989

Page, R.A, Environmental issues: what people think, *Chemistry in Britain*, p.559 1987

Environmental hazards – data and opinions, drawn from a wide variety of examples including Chernobyl and leukaemia clusters.

Chapter 12

DETECTORS AND IMAGERS

The measurement and imaging of radiations is required for a range of purposes and in a variety of situations, as the previous chapters have shown. This has led to the development of many devices and techniques. In this chapter we will study the principles of these and look at examples of the commonest types in use in hospitals. The photographic, ionising and exciting effects of radiation form the basis of all methods of detection and imaging. The chapter describes each of these in turn and this is followed by an account of some of the current developments in imaging techniques. Table 12.1 summarises the methods and devices.

Table 12.1 Methods and devices for detecting radiation

	Photographic films	Ionising gas chambers	Exciting solid state
measuring	film badge	free air thimble pocket Geiger	semiconductor thermoluminescent scintillation counters
imaging	radiograph	CT scanner	fluoroscopes rectilinear scanner gamma camera CT scanner

LEARNING OBJECTIVES

When you have studied this chapter you should be able to:

1. explain how the physical effects of photography, ionisation and excitation can be used to detect radiation;
2. describe common measuring and imaging devices which use these effects;
3. identify detectors which are particularly suitable for measuring and/or imaging each of the radiations X, α, β and γ for special applications;
4. give an outline of the use of computers in image processing, with particular reference to tomography and digital techniques.

12.1 PHOTOGRAPHIC METHODS

The radiograph

The remarkable properties of ordinary silver bromide photographic emulsion, make it the most common method of recording images from radiation. It consists of gelatin in which are suspended minute crystals of silver bromide, in X-ray use these have an average linear size of 1 mm. This is supported on a sheet of cellulose acetate, which gives it the name film. In use this is always surrounded by a fluorescent screen, the intensifying screen, which is described in Section 9.3. The X-rays, or other radiation, are absorbed by this screen, and visible and ultraviolet light is produced. This light is absorbed by the silver bromide; when a certain number of quanta have been absorbed, one AgBr molecule is ionised. The silver ion, Ag^+, forms a latent image centre which can then be developed into a black silver particle by chemical processing. The density of these black particles is dependent on the intensity of the incident quanta, which therefore produce an image in shades of grey. The processed film is called a *negative* because the darkest parts are those most exposed to the radiation. The processed film is usually viewed in this form, with the use of an illuminated screen (Fig 12.1). Radiographs are thus similar to slides rather than the prints which most amateur photographers view.

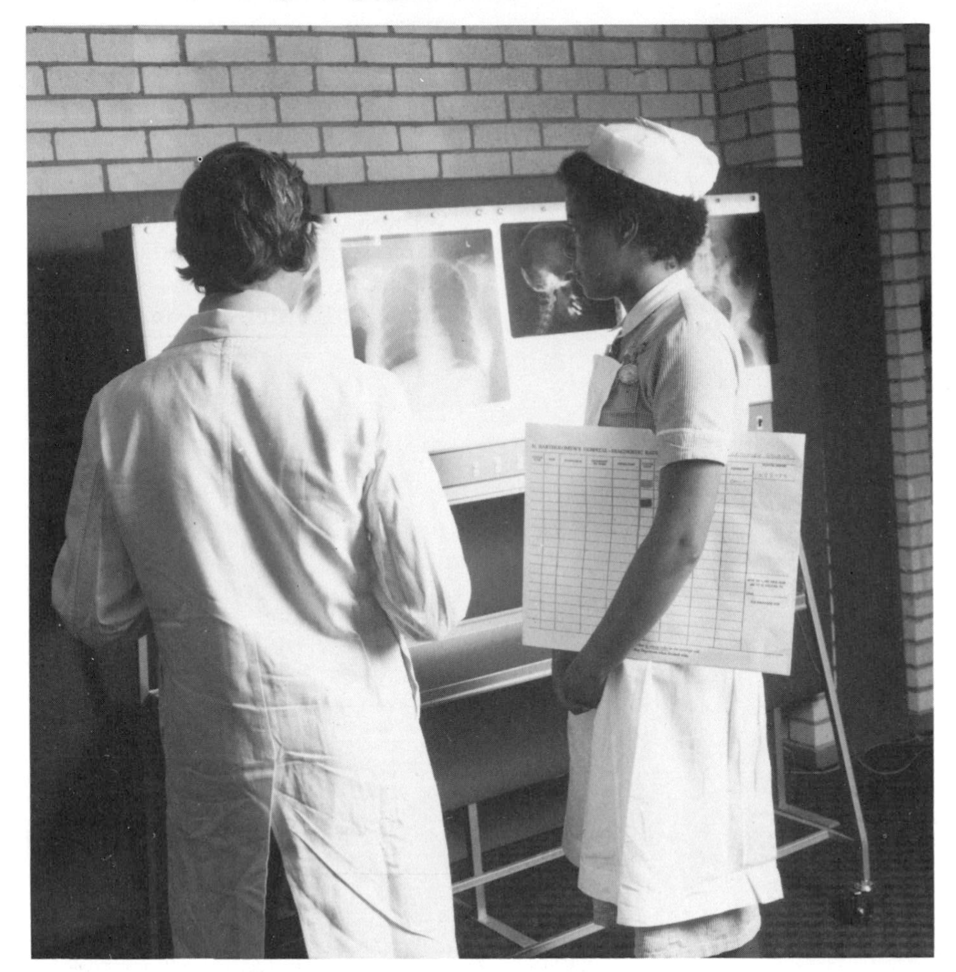

Fig 12.1 Radiologist and nurse examining radiographs.

INVESTIGATION

Film detection

Use dental X-ray film to detect the radiations from sources available to you in your laboratory.

Two types of film are generally available; one contains the developer and the fix for processing within the packet (like polaroid film). The other which is much less expensive, can easily be processed in the laboratory in dim light, as it incorporates a yellow filter. One side of the film is shielded with lead, so ensure that you use the other!

For sources try the sealed α, β, and γ; the most active of these will require an exposure time of about ten minutes. Other naturally occurring sources can be used, but the exposure times will need to be lengthened accordingly. Try taking a radiograph of a metal object, and try out the effect of filters of plastic and thin aluminium, such as are used in the film badge monitor, described in the next paragraph.

The film badge

This is the most common type of personal dosimeter. It consists simply of a piece of photographic film in a special holder. It is pinned to the clothing for a period of 1–4 weeks and then processed under standard conditions and the film blackening measured. The film is similar to a dental X-ray film but is a double emulsion type, having a sensitive (fast) emulsion on one side and a slow emulsion on the other. This enables a wide range of doses to be measured. The holder contains a number of filters, which enables different types of radiation to be identified (Fig 12.2).

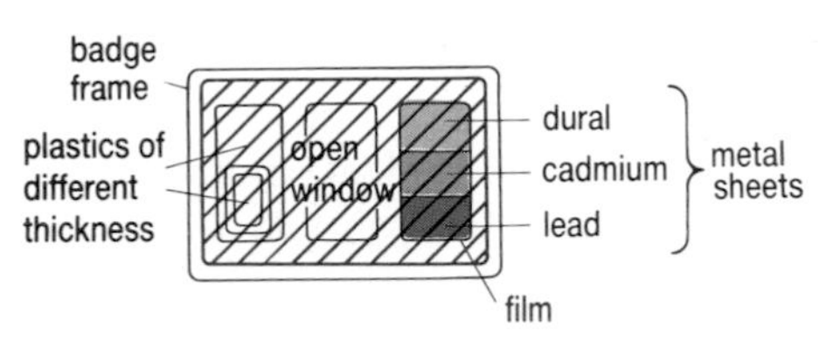

Fig 12.2 The film badge.

The beta dose is measured by comparing the blackening through the open window with that through the plastic filters. These enable the energies of beta to be estimated. The metal filters absorb beta and permit differentiation between photons of different energies. Dural, being an alloy mainly of aluminium, absorbs only low-voltage X-rays, whereas the lead alloys will attenuate all energies. Thermal neutrons (from a reactor), interact with the cadmium alloy, producing gamma radiation and additional exposure

The fast emulsion enables doses in the range 50 μSv to about 50 mSv to be measured; for higher doses this emulsion is stripped from the film and the slow emulsion then allows measurements of up to 10 Sv. The accuracy is only about 10 to 20 per cent, but this is sufficient for its normal role of low-level monitoring. The indium panel is there to detect higher than 10 mSv doses of thermal neutrons. These doses can then be measured more accurately with a Geiger–Muller tube.

Film badges are cheap and easy to use; they require no maintenance and can be sent away for processing. They provide a permanent record of exposure from a wide range of radiations. The limitations are that the film is sensitive to changes in ambient temperature and humidity, and the results are known only later, which may make it difficult to determine the origin of any unexpected exposure.

QUESTIONS

12.1 Explain briefly how the developed film will show patterns of blackening as a result of exposure to:
(a) low energy β radiation only,
(b) high energy X-radiation,
(c) direct contamination by a drop of liquid, emitting γ radiation.

12.2 **(a)** Why is it particularly important to measure the neutron dose accurately?
(b) Why will the film badge not measure α radiation, and why is this omission unimportant?

12.2 IONISATION METHODS – GAS DETECTORS

The result of radiation interacting with a gas is commonly the production of ions, positively and negatively charged particles. These can be collected at electrodes if a potential difference is set up across the ionised material. If the applied voltage is low, many of these ions will recombine. As it increases a point will be reached when all those produced will be collected. This is known as saturation, and the chamber is said to be operating in the 'ionisation region'. The electrical current resulting from this flow of ions is extremely low (about 10^{-12}A), but an electronic circuit known as a d.c. amplifier can be used to measure it. This current represents a mean value of the interaction of many charged particle or photon radiations. This type of device is called an **ionisation chamber**, and is shown schematically in Fig 12.3. Chambers are designed for a range of applications. Significant parameters are the gas with which it is filled, the material of the electrodes and the potential at which it is operated. Those used for health physics monitoring are usually air filled and made of low Z materials.

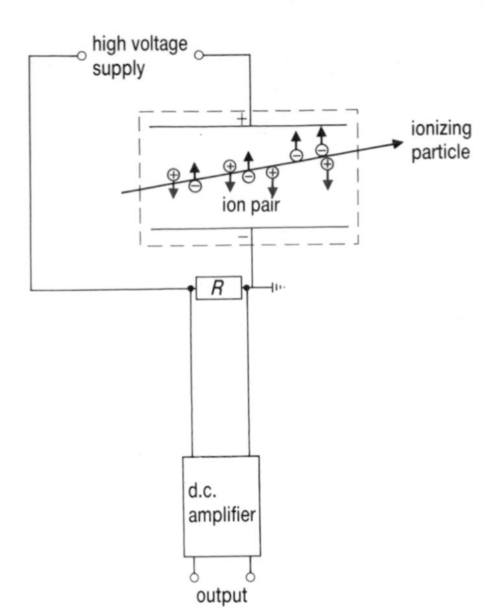

Fig 12.3 Operation of the ionisation chamber.

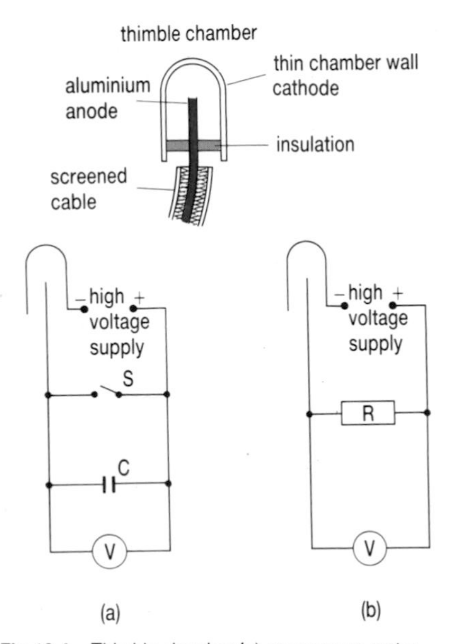

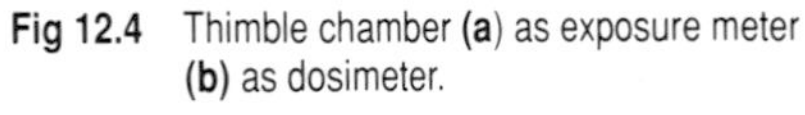

Fig 12.4 Thimble chamber **(a)** as exposure meter **(b)** as dosimeter.

Free-air chamber

The major advantage of the ionisation chamber over other detection methods is its accuracy (~0.1 per cent at best). The free-air chamber is designed to make use of this. It is a device for setting the primary standard in the direct measurement of exposure. It is too cumbersome for medical use but is used in, for example, the National Physical Laboratory (UK), to calibrate other instruments. The design of the device is similar to that shown in Fig 12.3, with careful attention being paid to the dimensions and accuracy of all parts of its operation. For example, the radiation is allowed in as a narrow, collimated beam, approximately parallel to the electrode plates; these plates are constructed to produce a uniform field; the charge is collected from the ionisation of a known volume of air; and it is measured with an electrometer. All these features improve the accuracy of the performance of this instrument.

Thimble chamber

This is a small version of the ionisation chamber, suitable for clinical use. The volume is usually less than 1 cm , and it accepts radiation from a wide range of directions. It is used to monitor exposure or exposure rate, at a point in or on a patient, or simulated patient (called a phantom). It is shown in Fig 12.4, as a small volume of air surrounded by a solid wall which is made of a material with the same mass absorption coefficient as air. This is known as an air-equivalent wall; the material is usually a graphite-plastic mixture, and it forms one electrode of the chamber. The other is a thin aluminium rod in the centre, insulated from the wall by a material such as amber.

The wall is made thick enough not to allow any secondary particles from outside to penetrate it. The ionisation measured is then mainly due to that occurring in the 'air' wall with a small contribution from the air inside. It is operated in the saturated condition, so that all the ionisation charge is collected. This may be measured as a voltage across a capacitor,

$$Q = CV \quad \text{or} \quad V = Q/C$$

and the result is a measure of the total exposure in C kg^{-1}. This instrument, an integrating dosimeter, or exposure meter, is shown in Fig 12.4(a). The alternative is to measure the voltage across a resistor R (typically $10^{12}\Omega$),

$$Q = It = tV/R$$

or,

$$V = QR/t$$

giving a reading of the exposure rate, C $kg^{-1}s^{-1}$. This instrument, a dosimeter, is shown in Fig 12.4(b).

The thickness of the wall is critical, for accurate readings, and dependent on the energy of the radiation to be measured. If the thimble is to be used to measure radiations of a higher energy than that for which it was designed, the wall can be thickened by the use of a 'build-up cap' such as a sheath of about 3 mm perspex. For lower energies a thinner walled chamber is required.

Pocket ionisation chamber

For monitoring personal doses it is inconvenient to have the cable connection to the electrometer. This device avoids the problem by using a **capacitor** (or condenser) **chamber**. In this the thimble is adapted to allow the chamber to be charged by contact with the central electrode. Attached to this is a quartz fibre which can move over a scale. In the fully charged condition it reads zero. When the chamber is exposed to radiation there is a gradual discharge as the ions give up their charge, and the fibre moves

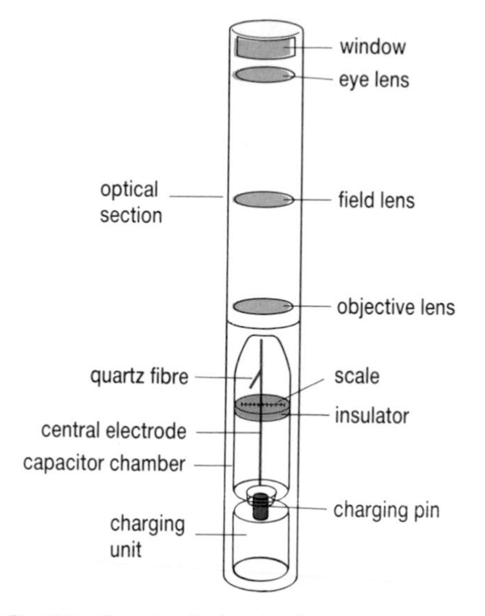

Fig 12.5 Pocket ionisation chamber.

across the scale. This is calibrated directly in C kg^{-1} or Gy, since these both are proportional to the loss of charge, and

$$Q = C\Delta V$$

where C is the capacitance of the chamber and ΔV is the voltage change causing the movement of the quartz fibre.

Because of the small size of the movement, the pocket chamber incorporates a built-in microscope (Fig 12.5). The instrument is a convenient method of obtaining a direct reading (usually the range is 0 to 2 μGy). It is accurate and independent of energy of the radiation over a wide range. Its disadvantages compared to the film badge, are that it is expensive, provides no permanent record and gives no information as to the type of radiation measured.

Counters

As the voltage applied across the ionisation chamber increases, the charge or current produced also increases as shown in Fig 12.6. This is because the original ions are given sufficient energy to cause secondary ionisation of the gas molecules by colliding with them on the way to the electrodes. This avalanche or 'gas amplification' effect can increase the current greatly and make it possible to detect individual ionising particles as pulses of current. Such an instrument is called a counter. At voltages just above saturation the pulses are proportional to the energy of the original ionising particle. Operating in this proportional region it is possible to identify the original nuclide that produced the ionisation, with the use of a **multichannel pulse analyser**, which can record both the number of pulses and the total charge of the pulse. Further increases in the applied voltage produce a greater avalanche, such that a constant pulse size is obtained, whatever the energy of the original radiation. This simplifies the electrical circuitry needed, but the type of radiation cannot now be identified. This is called the Geiger region and the instrument that operates in this way is the **Geiger (or Geiger–Muller) counter**.

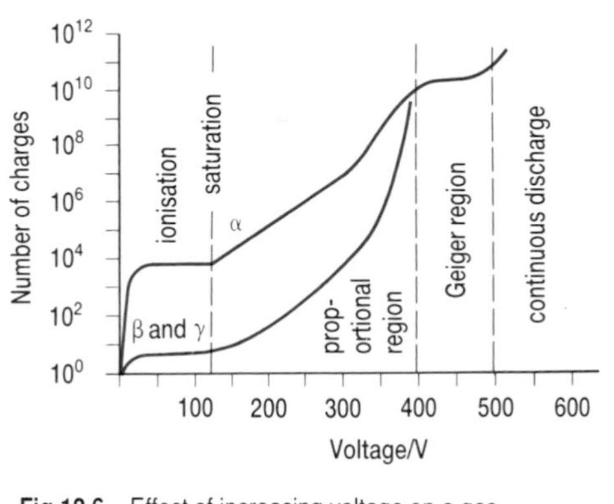

Fig 12.6 Effect of increasing voltage on a gas ionisation chamber.

The Geiger counter

You may be familiar with the structure and operation of the Geiger counter from your investigations of the properties of ionising radiations. You may have heard that these are expensive because they are almost exclusively used nowadays in physics education! The counter does have some applications in medicine because of its ability to measure very low activities. It also has the advantage of producing a stable, reliable output with no need for amplification, so the device can be small and portable. Its major disadvantage is the 'dead' time during which it cannot register a pulse since it is recovering from the effects of the previous one. This is about 1 ms, and so puts a severe limit on the maximum activity that can be accurately measured.

It is used in medicine to measure β radiation, including phosphorus-32 inside the body, by means of a special miniature counter, about 20 mm by

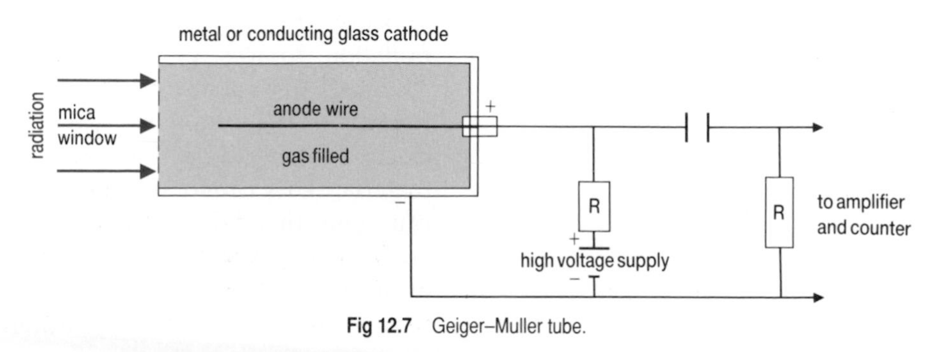

Fig 12.7 Geiger–Muller tube.

2 mm in diameter. Because it is rather insensitive to γ radiation it can be used to count charged particles in the presence of γ. It is also used in the monitoring of equipment to detect contamination.

The structure of the counter is in principle like other ionisation chambers. The gas is a mixture of argon and a 'quenching agent', of a halogen gas to reduce the dead time. The central anode is tungsten, and the body is metal or glass with an inner conductive coating, to make the cathode. The operating voltage is usually between 500 and 1000 V (Fig 12.7).

QUESTION

12.3 With reference to the preceding text, or a core physics textbook, make notes on the following aspects of the Geiger counter.

(a) Explain how it produces a pulse of current from an avalanche of charge, with the aid of a labelled diagram.

(b) Explain the cause of the 'dead time'.

(c) Estimate the maximum activity which can be measured, from the value of the dead time quoted above.

(d) Why is background radiation a particular source of error in Geiger counts, and how may this be avoided?

12.3 EXCITATION METHODS – SOLID STATE DETECTORS

The effect of radiation on many crystalline substances is to raise the energy of electrons in it. The highest level in which the electrons normally exist is the valence band, above which there is a forbidden band. The transfer of energy may raise the electron through this to the conduction band or the exciton band, or the presence of impurities may give rise to the electron being trapped in the forbidden band (Fig 12.8). In the conduction band electrons are free to migrate, when a potential difference is applied. In the exciton and forbidden bands they are not, but the trapped energy can be released to provide a measurable signal. In the exciton band this occurs spontaneously; in the forbidden band it must be stimulated. Each of these effects give rise to detectors, which will now be described.

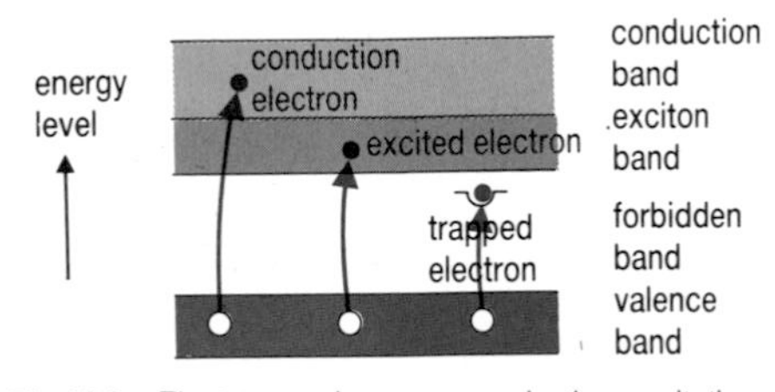

Fig 12.8 Electrons and energy : conduction, excitation and trapping.

Semiconductor detectors

In substances, such as cadmium sulphide, silicon and germanium, the energy of the radiation excites an electron from the filled valence levels to the conduction levels. This provides a pair comprising a conduction electron, and an electron vacancy which is called a positive hole. The energy to produce this is about 1eV, which is low, so the potential sensitivity of the detector is high. If a potential difference is maintained across the semiconductor the charge carriers will move and a current pulse will be observed. The devices have been used in place of Geiger counters, though they are expensive and rather insensitive to γ radiation.

Thermoluminescent dosimeters (TLDs)

These devices are replacing film badges for monitoring personnel, and are also used for monitoring patients' doses during therapy and calibrating radiation sources in hospitals. The principle of the TLD is electron trapping. Certain materials such as lithium fluoride (LiF), and calcium fluoride (CaF_2), can, when exposed to radiation store part of the energy in electrons trapped in the isolated level resulting from impurities in the material. They cannot migrate and conduct, but if the crystal is heated they can gain thermal energy which then enables them to lose all their additional energy as light. The amount of light emitted is a measure of the number of trapped electrons, and hence the original exposure.

Substances with the ability to store and subsequently release light energy

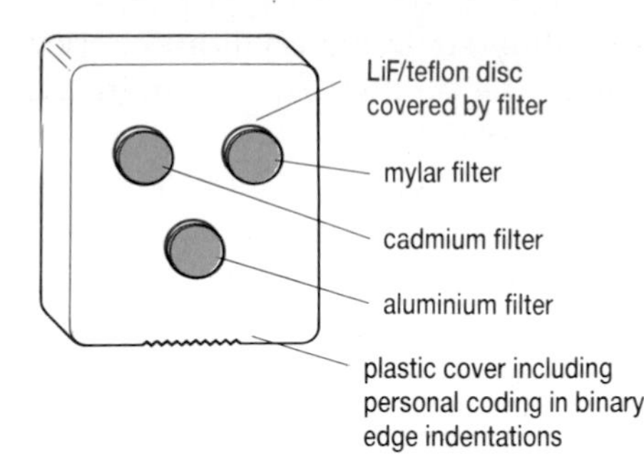

Fig 12.9 Thermoluminescent dosimeter

are called **phosphors.** LiF is the most commonly used in TLDs as it has the following features:

- it can detect β, γ, X- radiations and thermal neutrons,
- it responds to a very wide range of total energy and rate of energy deposition,
- it has a similar mass absorption coefficient to tissue and air, so that dose estimates can readily be made,
- it is sensitive and gives reliable results,
- the light is released on heating to a suitable temperature (~200 °C).

The phosphor is usually incorporated into the plastic Teflon, and produced as a disc. This can be attached, in a light-proof plastic container, to the body or clothing. Fingers and hands may be measured directly in this way, for example by taping a plastic pouch to the appropriate part – that which has been contaminated. Alternatively the discs are incorporated into a badge, worn over a period of time like a film badge (Fig 12.9). Different filters in front of the thermoluminescent discs make it possible to measure the effective dose to different parts of the body, for example, mylar for skin and aluminium for the gonad dose.

TLDs are inexpensive, requiring only small quantities of LiF, which can be re-used, and they can be made very small to measure exposure within the body at sites such as tumours. The necessary heater and light emission analysis unit is expensive however; to be economic their use needs to be widespread.

Fluoroscopes

A fluorescent substance is one in which the energy of radiation moves electrons into the exciton band. They then immediately re-emit this absorbed energy of radiation, as light. Dynamic imaging of the body with X-rays was carried out for many years by use of a glass screen coated with zinc-cadmium sulphide, the fluoroscope. Direct viewing is no longer practised, because of the high exposure to X-rays required. Modern fluoroscopy uses an image intensifier tube as described in Chapter 9. The fluorescent screen is backed with a photocathode which changes the light into electrons. These are then accelerated, and focused onto another fluorescent screen, to produce a gain in brightness of about 5000. The output can then be viewed directly, still- or cine- photographed, or televised, using a special camera tube. Direct viewing is rarely used; the use of television processing offers computer image-enhancement possibilities as well as recording facilities.

Scintillation counters

These operate on the same principle as fluoroscopy; the emission of light as an electron returns to the valence level. In this case the purpose is to count the emissions or scintillations, and therefore measure the exposure.

There are a number of **scintillator** materials in use including those in solution, and large organic molecular crystals, which are used for detecting β particles. Those most widely used are simple ionic crystals such as zinc sulphide (to detect α), lithium iodide (for thermal neutrons) and sodium iodide (for γ).

The effect of the light emission can be magnified with a **photo-multiplier tube**, which is a development of the image intensifier discussed earlier. Light from the scintillator ejects electrons from a photocathode. This is a thin layer of semiconductor material coating the inner surface of the tube. These photoelectrons are then accelerated and multiplied by a series of electrodes called dynodes, held at increasingly positive potentials along the tube. The dynodes are made from an alloy such as beryllium-copper, which causes several secondary electrons to be emitted by the energetic incident electron. This process is repeated several times, causing multiplication each time. The electrons are collected by the anode and output through a further amplifier. The complete device is called a scintillation counter (Fig 12.10), and it is able to amplify the original light output by a factor of between 10^5 and 10^8, but strictly in proportion to the original energy. By using a multichannel pulse analyser it is possible to obtain the energy spectrum of the incident radiation.

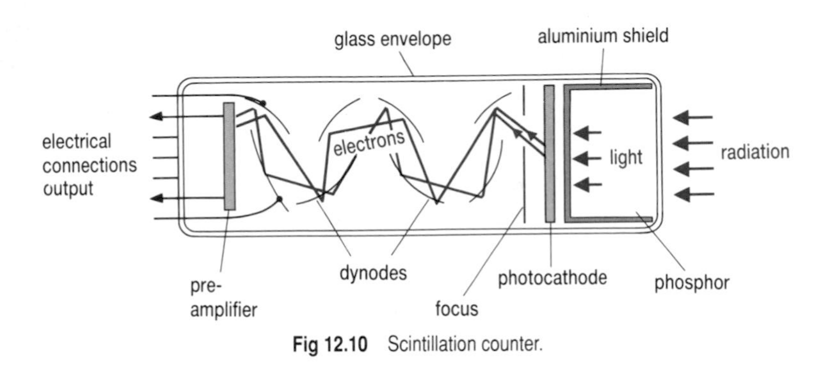

Fig 12.10 Scintillation counter.

The most common scintillator is sodium iodide, activated with about 0.5 per cent thallium iodide, so it is usually abbreviated as NaI(Tl). The substance is hygroscopic, so it has to be kept in sealed containers. These are usually made of aluminium, with a glass or quartz window which is coupled to the photomultiplier tube with silicone oil or grease. It has the advantages of

- high efficiency in converting about 10 per cent of the incident energy into light,
- rapid scintillation,
- excellent γ detection ability, due to its high density, and relatively high proton number.

The scintillation counter has widespread use in hospitals, as it can measure all types of radiation commonly encountered, with the selection of a suitable scintillator. It is particularly favoured for the detection of X- and gamma radiations, and can be used to analyse the energy spectrum of γ, and thus identify the source nuclide. It is also used to monitor low intensity tracer levels in investigations using the gamma camera (see next section). The main restrictions on its use are its relative expense and its size.

QUESTIONS

12.4 **(a)** Describe the physical principles underlying the action of a scintillation detector and photomultiplier, i.e. a scintillation counter.

(b) Why is a thallium-doped sodium iodide crystal [NaI(Tl)] often used as the scintillator material for the detection of γ rays?

(ULSEB 1988, part)

12.5 Table 12.2 summarises the most commonly used detectors for various purposes. List the factors which are important in choosing a detector, and account for the particular choices listed.

Table 12.2 Most common uses of detectors

Type/situation	Film badge	TLD	Thimble chamber	PIC	Geiger counter	Scintillation counter
low energy X-rays			✔			
high energy X-rays						✔
α particles					✔	
low energy β					✔	✔
high energy β					✔	
γ rays						✔
personnel (general)	✔	✔				
personnel (protected areas)				✔		
low activity					✔	
local and internal body sites		✔				

12.4 RADIOISOTOPE IMAGING – SCANNERS AND CAMERAS

The process of creating a two dimensional image from the emissions of radionuclides within the body, is quite different from that for X-rays. Earlier techniques included the laborious moving of a Geiger counter backwards and forwards across the body. This process of scanning can be automated and the scintillation counter is used as this is a more suitable detector for the gamma radiation.

The rectilinear scanner

This imaging device was developed in 1950, using these principles, as shown in Fig 12.11(a). The detector is collimated so that it receives radiation from a particular place, when in a particular position. Interference from emissions from different depths can be reduced by a special 'septa', a lead collimator which is 'focused' on a fixed plane (Fig 12.11(b)). The movement of the detector is replicated by a light source which produces an exposure on film to match the intensity of the radiation. Such scanners have been widely used in hospitals, but for most applications they are now superseded.

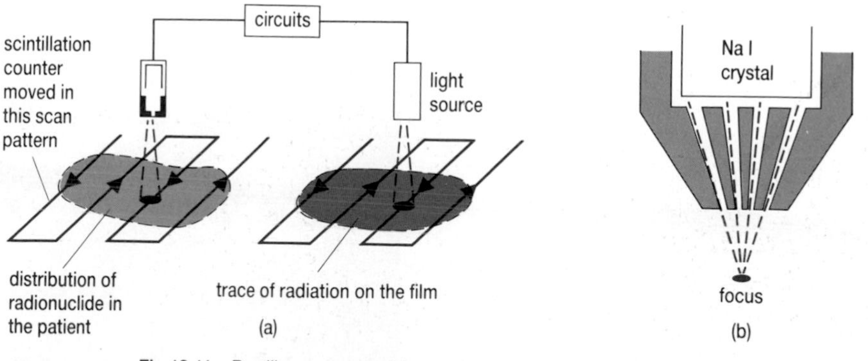

Fig 12.11 Rectilinear scanner **(a)** scanning method **(b)** focusing method.

The gamma camera

This was invented in 1957 by Anger, which gives it its alternative name. The purpose is the same as the rectilinear scanner, but the detector is designed so that no movement is needed. Essentially it consists of a large disc of sodium iodide, typically 40 cm in diameter and 1 cm thick (Fig 12.12(a)).

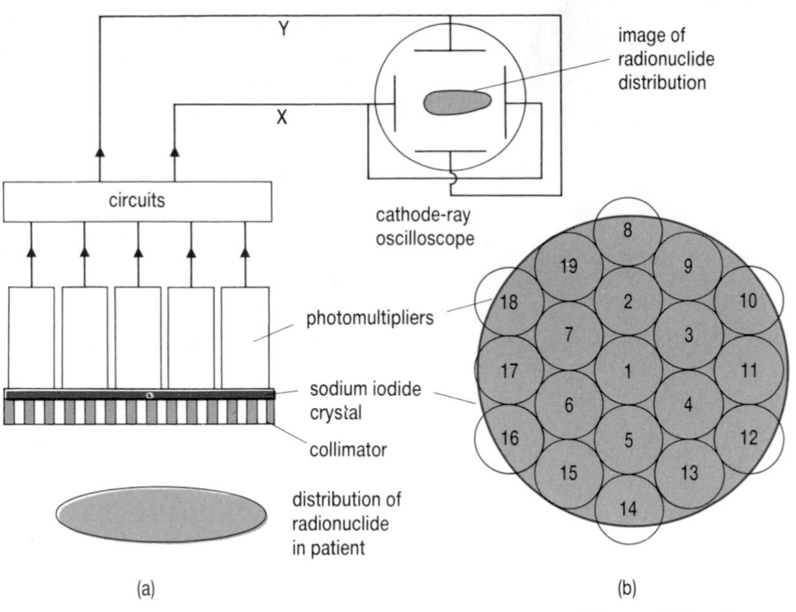

Fig 12.12 Gamma camera **(a)** scheme of operation **(b)** arrangement of photomultiplier tubes.

To this are attached an array of photomultiplier tubes, just like those of the scintillation counter. These are arranged in a geometric pattern. Fig 12.12(b) shows a common arrangement of 19; more advanced machines use up to 75. These tubes will receive signals from each flash of light in the crystal. Because they are different distances from it, the intensity will differ. The result is nineteen output pulses of varying size, from which, with the help of some clever electronics, the position of the light flash can be deduced. This is then converted to signals representing its x and y coordinates so that it may be displayed on a cathode ray screen, and a Z coordinate for its brightness.

A lead collimator is used as in the rectilinear scanner, to restrict the directions from which the gamma radiation is received. Here it is a circular slab of lead pierced with many holes perpendicular to the face. This ensures that only those gamma rays are received which are emitted at right angles to the crystal and the image is not degraded by those received from other directions. Sometimes collimators are used with non-parallel holes, for example to reduce the field of view, or with a single 'pinhole' for small organs such as the eye or thyroid.

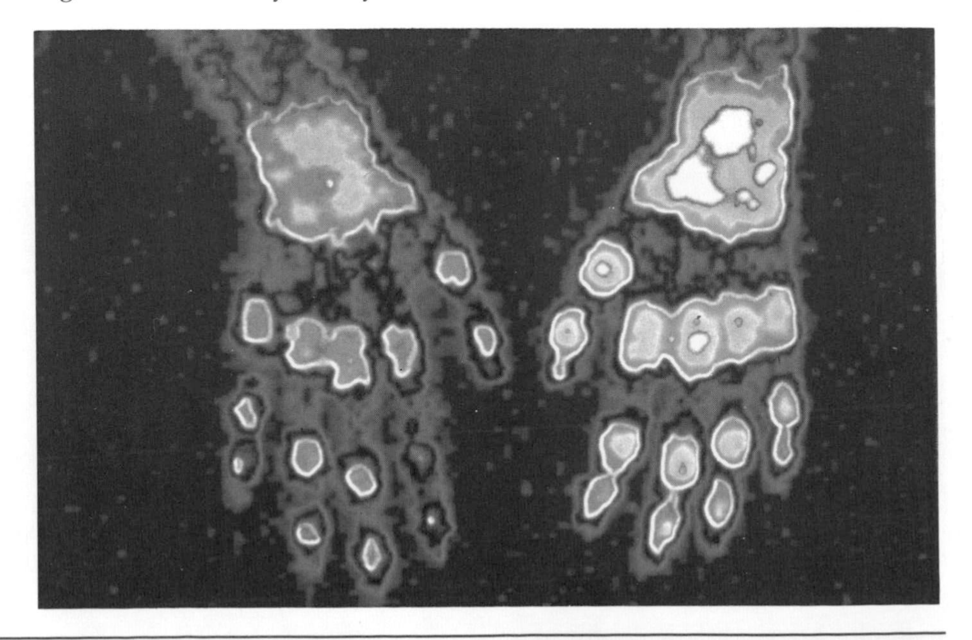

Fig 12.13 Gamma camera image of arthritic hands. The radionuclide concentrates in the inflamed joints, showing that the hand on the right is severely affected.

The isotope most commonly used is $^{99}Tc^{m}$ with a γ photon energy of 140 keV, an alternative is ^{123}I with a photon energy of 160 keV. A change in the collimator thickness allows for the energy difference between the two gamma emissions.

The gamma camera image is composed of many small dots (up to one million), the frequency of which is proportional to the region with the greatest concentration of the radionuclide (Fig 12.13). Examples of their wide applications in diagnostic imaging were given in Section 10.5. One of the most significant features of the image is the way that it is computer processed, which means that it is in a digital form. This permits a range of additional processing to be carried out, examples of which are given later.

Variations on the standard camera include a larger crystal for a larger field of view (Fig 12.14), a mechanical scanning method to cover the whole body, and a rotating detector to produce tomographic images.

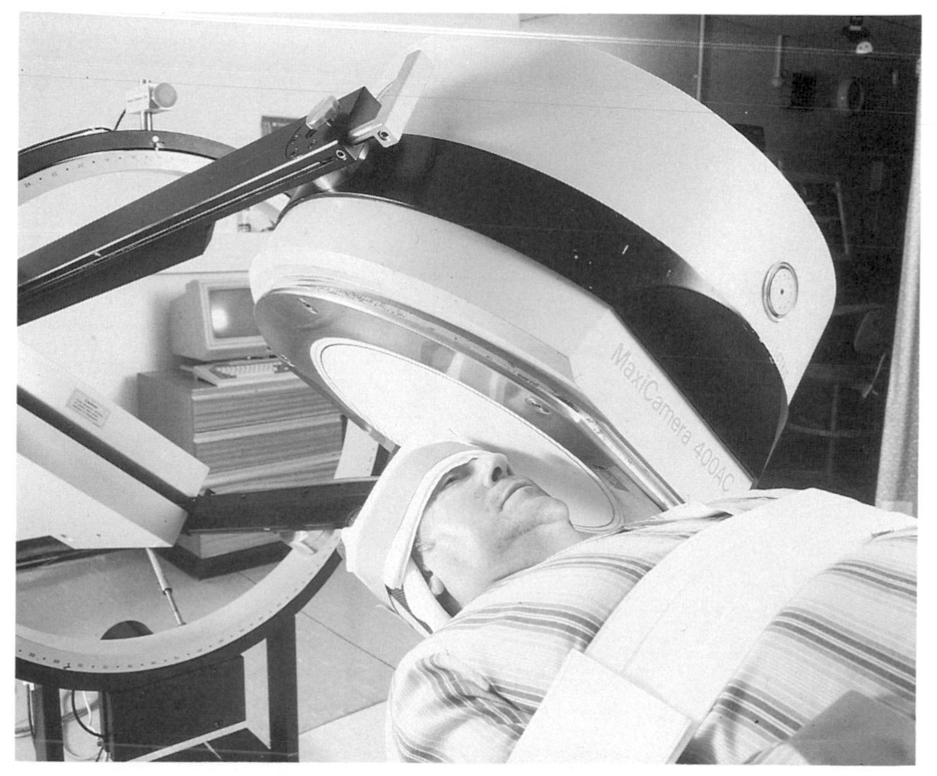

Fig 12.14 A large field of view gamma camera.

QUESTIONS

12.6 Give reasons why the gamma camera has superseded the rectilinear scanner.

12.7 **(a)** State **three** factors which are important in choosing a suitable radionuclide for nuclear imaging of particular parts of the body. Name **one** radionuclide in common use for this purpose.
(b) Draw a labelled sketch of the main component parts of a gamma camera. Explain how it can be used with a radionuclide to obtain the desired image of a particular part of the body.
(c) Explain how information obtained from such an image differs from that obtained from a diagnostic X-ray image.

(JMB 1987)

12.5 TOMOGRAPHY – CT SCANNING

This method of producing images of a slice of the body was introduced in Chapter 9. It enables detailed pictures to be produced of particular planes of the body. Those with conventional radiography are only longitudinal, computer techniques introduced in the last 20 years have made it possible to scan axially, or at any cross section required. This technique is called computed (axial) tomography, or **CT scanning**. Similar methods can be applied to radionuclide emissions where the technique is called emission computed tomography or **ECT**. The method which uses the rotating gamma camera to detect photons in the way described above, is called single photon ECT, or SPECT. A second method is more complex. It uses two rotating detectors to detect the annihilation radiation emitted from the patient who has been injected with a positron emitting radionuclide. This is known, not surprisingly, as positron emission tomography, or **PET**. The latest method to be introduced into hospitals uses entirely different physical phenomena, that of magnetic resonance, or **MR**. It is therefore not strictly part of a theme on ionising radiations. It will be considered as an alternative to the other tomographic imaging methods in the next chapter.

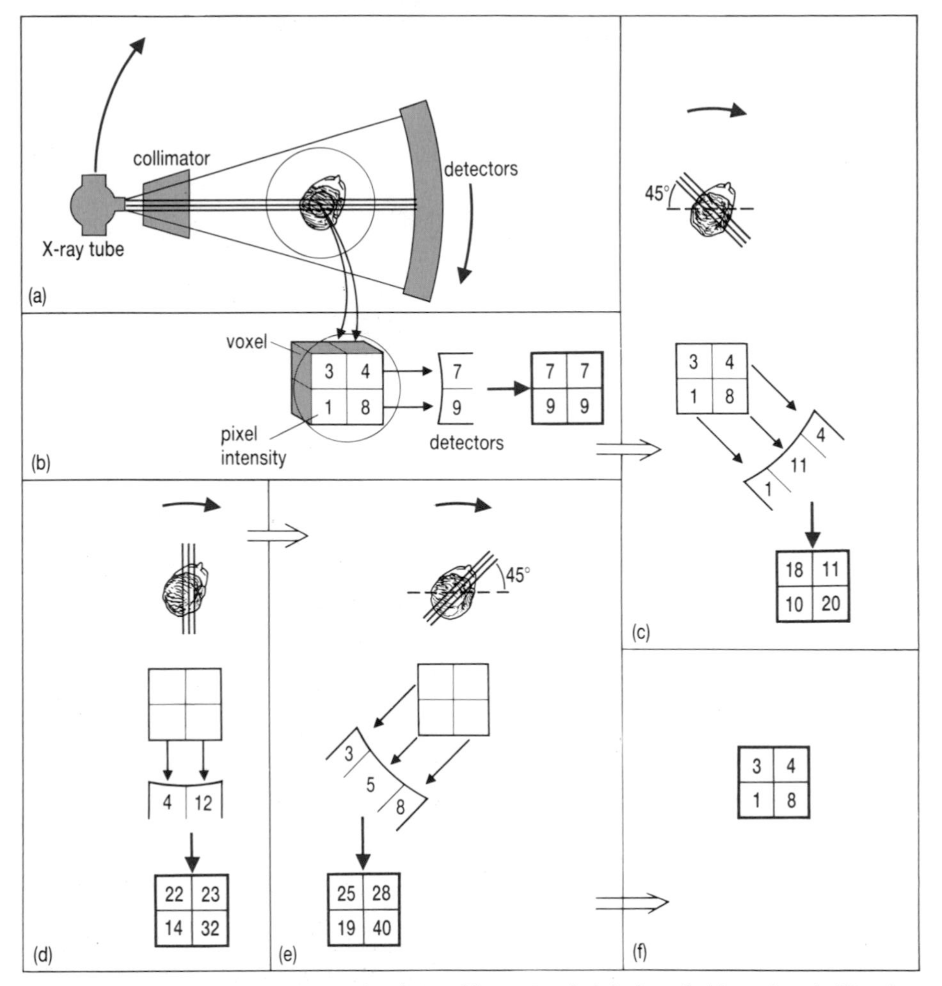

Fig 12.15 Tomography **(a)** arrangement of equipment **(b)** summing pixels horizontally **(c)** turn through 45° and add new intensities **(d)** repeat **(e)** repeat **(f)** subtraction of background and reconstruction.

Principles of computed tomography

The task is to make an image of a section of the body, from measurements made around its axis (Fig 12.15(a)). The plane is divided into small volume units called voxels (it is assumed that the slice has a finite thickness). The intensity from a given voxel, called a pixel, is reconstructed from the measurements made along a series of strips across the whole plane. The principle may be illustrated by back projection which is illustrated in Fig 12.15(b), though more sophisticated mathematical methods are used in practice. The total intensity measured along any strip is simply allocated to each pixel in that row. For example, viewing the sample array of four in (b) from the right, gives totals of 7 and 9. A second view at 45° to this adds 1, 11 and 4. These are added in (c) to give new totals. The procedure is repeated in (d) and (e), to give totals of 25, 28, 19 and 40. The final step is to remove a background of 16 from each and divide by 3 because of the duplication of view. The original values of 3, 4, 1 and 8 reappear, as if by magic! As we said this is an illustration of the principle, in practice the computer takes many more steps and uses rather more complex formulas.

CT equipment

The **X-ray source** must be highly collimated to produce the precise paths required, typically each strip of image is only a few millimetres wide. This means most of the radiation energy is wasted. The tube voltages and exposures are greater than for conventional radiography, typical values being 130 kV and 300 mA s^{-1}. Therefore a great deal of heat has to be dissipated from the anode. Because of the tube movement during scanning, they tend to fail after a few months in service and replacements are a large expense in the running of a CT department.

Filtration is very important, to ensure the beam is homogeneous and that the patient skin dose is reduced. In the earliest models the patient was surrounded by a water bag to absorb stray unattenuated beams. Present models can avoid this problem. An array of several hundred scintillation counters act as the **detector**, mounted so as to receive the beam as the tube rotates (Fig 12.15) or able to match the movement in earlier models. The traditional scintillator for X-rays is NaI(Tl). This has some limitations in the scanner and is being replaced by caesium iodide thallium-doped crystals coupled to silicon semiconductor photodiodes. Other models use xenon gas ionisation chambers.

The X-ray tube and detectors are mounted within the scanning **gantry**. This is a frame with an aperture in which the plane of the body to be scanned can be positioned (Fig 12.16). It contains the driving mechanisms and the sequencing sensors. The frame can be set at different angles for alternative sections. The **couch** for the patient is motorised to allow accurate positioning of the body in the scan plane. A single scan takes between 2 and 8 seconds.

The collection, reconstruction and display of the scan data requires a powerful **computer** and sophisticated peripherals and **software**. The image reconstruction of each pixel intensity value is the result of about one million computations. The image is displayed on a TV screen to produce a picture of the slice scanned with variable contrast and brightness. It can be stored

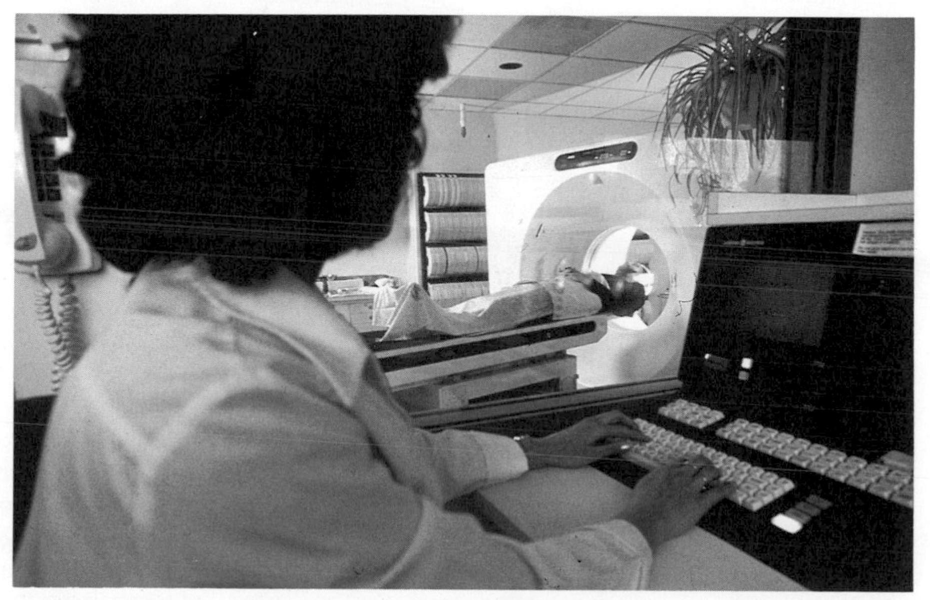

Fig 12.16 CT scanner in action with the patient in the gantry aperture and the radiographer at the control consul.

in the computer memory and used to construct a slice in a different plane in combination with all the other pixel data. You can learn more of the growing use of CT scanning in the next chapter.

12.6 DIGITAL SUBTRACTION

Subtraction principles

The technique of 'subtracting the background' is often used to increase the contrast of an image, so that more detail can be detected. For example blood vessels, being soft tissue, are often obscured in radiographs by the greater contrast of bone. A contrast medium can be injected into the bloodstream so that it changes the intensity of the image. If two pictures are taken before and after the injection, only the blood vessel intensity will vary. The contrast of one picture is now reversed, by making a negative, i.e. dark for light. When the two pictures are combined together the common parts such as the bone will be reduced in intensity (one dark and one light) whereas the blood will be increased in intensity (both light, or both dark). This has been a well established 'analogue' technique in X-radiography. The advent of computing power has enabled this to be developed. In this respect it is similar to computed tomography.

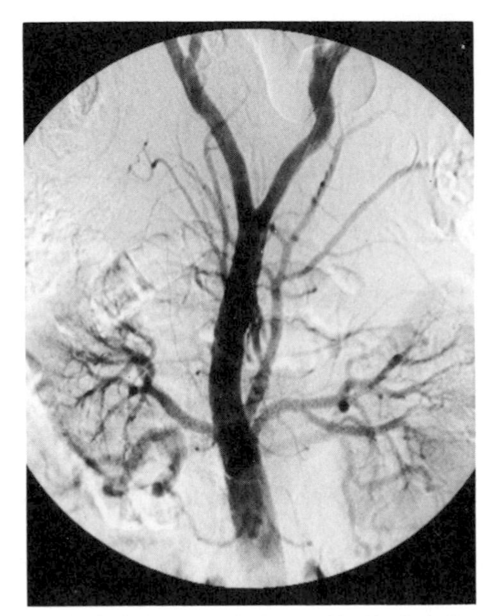

Fig 12.17 Digital subtraction image of the blood supply to the kidney.

Digital subtraction angiography (DSA)

The original analogue image is fed through a TV camera to an analogue to digital converter which turns the shades of grey into digital information. This can then be processed by the computer, compared with other contrast pictures and stored. Angiography, the displaying of the blood vessels is at present the most important application of digital subtraction (Fig 12.17). However, the increased availability and lower cost of computing power, is certain to make digitising of radiological images more widespread in the near future. One method recently announced uses a fine laser beam to 'read' the specially processed film, so that pulses were produced for digitising.

QUESTIONS

12.8 How could a laser reader be used to transmit the radiograph by telephone?

12.9 Read the article on DSA in the box, and answer these questions on it.

(a) Why is it particularly useful to image the blood vessels?

(b) What is a catheter?

(c) Why does the fluorescent dye show up under X-rays and how is this use different from the contrast effects from a barium meal?

(d) What is misleading about the fourth paragraph?

New X-ray equipment for safer treatment

People who once would have bled to death can be cured by doctors manipulating a tube inside their arteries. The doctors are guided by watching an X-ray image replayed on a television screen.

Although the technique to show doctors what is happening in blood vessels has been available for several years, new X-ray equipment is making the process safer. It is also extending the range of complaints that can be treated.

With the technique, called digital subtraction angiography, doctors can pinpoint the position of tumours and blocked or damaged blood vessels.

To see the blood vessels, the doctor first takes an ordinary X-ray, which does not show soft tissue. The image is converted to digital information. The doctor then inserts a tube, or catheter, into one of the patient's veins or arteries, helped by the X-ray image on the TV and a knowledge of anatomy.

When the catheter is in position, a fluorescent dye which shows up under X-rays is pumped into the patient, and more X-rays are taken and stored as digital information recorded by the conventional X-ray. From the image recorded with fluorescent dye in the blood vessels, the doctor obtains a picture of the blood vessels. And from the picture a diagnosis is possible.

The usefulness of the X-ray equipment and digital subtraction angiography does not stop with diagnosis. For example, orthopaedic surgeons carried out a biopsy on the fibula of a patient and damaged an artery. Guided by the X-ray image, doctors were able to insert steel coils in the artery to block it off and stop her bleeding to death.

New Scientist 31.1.1985

SUMMARY ASSIGNMENTS

12.10 **(a)** Explain what is meant by *dose equivalent* and why it is important in radiation dosimetry. Calculate the energy delivered to a person of mass 70 kg by a dose equivalent to the whole body of 30 mSv (3 rem), half being acquired from radiation of quality factor one, the remainder from radiation of quality factor three.

(b) A film badge used for personal radiation monitoring contains various filters through which radiation must pass before reaching the film. Explain how this helps in making an estimate of the dose equivalent received by the wearer of the badge.

(c) Describe the principle of operation of the thermoluminescent dosimeter.

(JMB 1986)

12.11 **(a)** Mention three devices used in estimating radiation doses received by personnel using radioactive material or other sources of radiation.

(b) Describe one such device. Indicate how, using the chosen device and other simple apparatus, the type as well as the quantity of radiation may be estimated.

(COSSEC AS part, specimen)

Further reading

You will find further examples of developments in Medical Imaging in the next chapter.

Theme **4**

THE MEDICAL IMAGING DEPARTMENT

Those hospital signs which still say 'X-ray department' are becoming increasingly out of date, as alternative techniques replace or complement the radiological methods which have dominated the internal examination of patients for so long.

The changes are of three broad types; there is the improvement of existing techniques, such as better resolution of ultrasound images; the transformation of a technique by the use of computer aided digital processing, for example tomography; and the invention of a completely new method or modality, of which magnetic resonance imaging is the most important at present. Most of these changes are based on the physics you have been studying, but the technical details can be very complicated.

What this theme aims to do is to concentrate on the *effects* of these changes on the practice of this area of medicine, rather than on the technical details of the changes themselves. The purpose is not to try to teach you the skills of the hospital physicist or medical practitioner – as a short study within a physics course, that would be rather ambitious! It will, however, bring medical considerations to the fore.

Prerequisites

Before starting this theme you should have some familiarity with chapters 8-12 of this book.

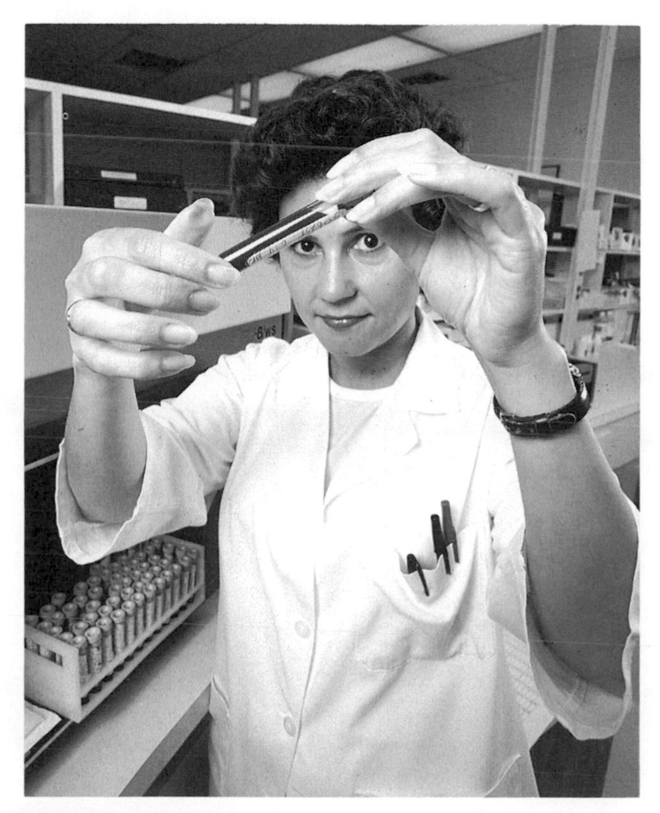

If you visit a medical imaging department you are unlikely to see a physicist at work – except by appointment. All the sophisticated equipment is operated routinely by medical staff – radiographers, radiologists and nurses – for whom the physics is very much in the background.

Chapter 13

A SIMULATED CASE STUDY

'Shall I refuse my dinner because I do not fully understand the process of digestion'

O. Heaviside

This chapter aims to put you in the position of those who work with medical imaging. You, like they, will have to evaluate situations on the basis of incomplete knowledge and understanding, and make recommendations for future practice. We hope that you will work with your fellow students who will have different roles and views, to reach a common conclusion, in the same way that those working in a hospital would need to discuss and decide on policies for the future. This activity is called a simulated case study, because you are simulating (acting out) the role of people in a situation, given relevant information taken from real life cases. The aim is rather different from the preceding themes. In previous chapters you have been presented with ideas and information on a range of physics topics, within the realm of medicine in the background. In this chapter the emphasis is reversed.

LEARNING OBJECTIVES

In this chapter you will have the opportunity to:

1. review your knowledge and understanding of the existing diagnostic imaging techniques of ultrasound, X-radiography and radioisotopes;
2. familiarise yourself with new developments in diagnostic imaging, for example computer tomography (CT), and magnetic resonance imaging (MRI);
3. compare the advantages and disadvantages of different techniques and modalities, from given information briefs and additional personal research;
4. identify the different roles of those involved in providing medical imaging services, and interpret a brief to present the views of one of these, on a physics related issue;
5. contribute to a group decision-making process.

13.1 BACKGROUND TO THE CASE STUDY

We first set the scene with a summary of relevant information on the hospital department to be simulated: the equipment, staff and its standing in the locality. The particular activity of this chapter is set in the context of a ten year strategic plan for the district in which the hospital is situated; an outline of this concludes the introduction. You will be concerned with the development of diagnostic imaging services in a district general hospital (DGH).

For the purposes of this simulation, much of the detailed description has been abbreviated and many of the circumstances simplified. But care has been taken to retain the essential features of the real case on which this example was based.

St. Katherine's Hospital

St Katherine's is the major hospital in the Exhampton district health authority (DHA). The DHA is the unit of administration of health services; most planning is done at this level. Exhampton is a metropolitan district

Organisational notes

To achieve these objectives, it will be necessary to work on this case study in a rather different way from your usual studies. The two most important factors are allowing enough time, and taking this work seriously. The fictional nature of the situation does not reduce its significance; it may help to think of the simulation as a model of reality, rather like a mathematical equation models the relationship between physical variables. The allocation of the amount and distribution of time will need to be discussed with your teacher/tutor.

The following is recommended:

Preparation: read the whole of this chapter quickly and decide which of the roles you wish to adopt (about 1 hour).

This should occur at least a few days before the next stage. During this time it is particularly helpful to visit a medical imaging department, or have a presentation from a specialist. Literature is readily available from equipment manufacturers and professional associations, which can be very useful at this stage.

Planning: re-read the relevant sections of this chapter, make your decisions and write a draft report. Discuss this with the others in the group and incorporate this in your final report (about 3–4 hours).

Presentation: present your plans to the group (10 minutes), arrive at a group consensus plan, and decide how to present this to the whole class (about 1.5 hours). Present as a group (15 minutes), answer questions and receive an 'expert' evaluation of your group's proposals (about 1.5 hours).

This is summarised in Table 13.1.

Table 13.1 Organisation of timetable

		Timing/Hours	
	Stage	**Classwork**	**Homework**
Week 1	*Preparation*		
	read chapter 13		1
	visit or visitor	~3	
Week 2	*Planning*		
	decide roles	$\frac{1}{2}$	
	draft plan		~1
	discuss plans	1	
	finalise individual plan		~1
Week 3	*Presentation*		
	prepare group plan	1	1
	present group plan	$1\frac{1}{2}$	
	evaluation and decision		
	Total	**4 + 3**	**4**

with a population of approximately 250 000, part of the South Westshire regional health authority (RHA). The RHA is a larger administrative unit of the health service; some major decisions are made at this level. There are six other hospitals in the Exhampton district, all small, specialised and without any diagnostic imaging facility. The revenue, or running costs of the district is currently £50M per year, of which about 75 per cent is spent on staff and 25 per cent on other costs. There is also an allocation for minor capital expenditure of £250 000, for minor developments and the purchase of expensive equipment, and the region holds an allocation for major capital developments, which is usually spent several years in advance! The DHA employs approximately 2500 staff and is generally reluctant to increase the numbers.

St Katherine's is a moderate sized hospital of some 500 beds, serving the mixed population of part of the inner and outer suburbs of a major city. Its orthopaedic department has a national reputation, and students from the local medical school are placed in various departments during their clinical training.

The medical imaging department at St Katherine's has a total annual budget of £900 000. It has a staff which includes the full-time equivalent of 15 radiographers (specially qualified technicians) and 3 radiologists (specialist doctors). The department comprises six rooms for general X-ray investigation, which also include special equipment for imaging teeth, breasts, skulls and the blood system. There are two ultrasound machines and a relatively new gamma camera in its own suite of rooms. The department is fully stretched by its present workload. It does not have equipment for computed tomography or magnetic resonance imaging, and its ultrasound and radionuclide imaging facilities have scope for extension(Fig 13.1).

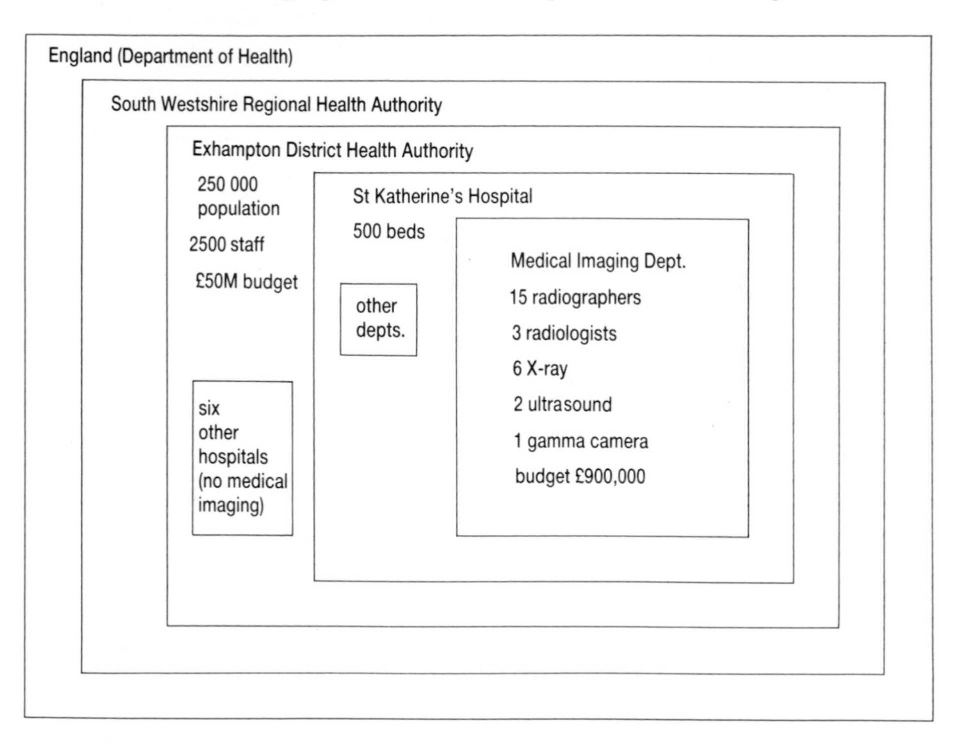

Fig 13.1 The health service structure.

A ten-year plan for the district

The district health authority is currently engaged in compiling a ten-year development plan. As part of this the medical imaging department is required to meet national policy and show an increase in 'cost effectiveness', which can mean either a lower cost for individual treatments, or a greater volume of patient throughput, for the same overheads. In addition the department has the opportunity to review the possible development of its service in the district in the next ten years. The following are options which need to be explored:

1. The upgrading and extension of the ultrasound facilities, combined with a rationalisation of the X-ray provision, to provide a more effective and less hazardous service with the most up-to-date equipment.
2. The addition of a tomographic facility to the gamma camera, or its replacement with a PET system, to maximise the capability of this modern computer-aided system.
3. Supplementing the X-radiology with a CT facility, to improve the scope of the investigations that can be carried out.
4. The construction of a new MRI facility to enhance the reputation of the hospital as a centre of medical research.
5. The computerisation of patient and departmental records.

ACTIVITY

You are to take on the role of **one** of the key people with a contribution to make to the ten-year plan. With four or five of your fellow students, you should allocate the following tasks, and then meet together to devise a combined response which can be presented to your teacher, the other groups in your class, and possibly a wider audience, for critical evaluation.

1. You are the senior radiologist in the department, with a major interest in the efficient running of the department and the welfare of your staff and patients; you wish to propose option 1. You should read Section 13.2, review Chapter 8, and present your arguments.
2. You are the radiographer who recently established the present radionuclide imaging facility and you are committed to option 2. You should read Section 13.3, review Sections 10.4, 10.5, 12.4, 12.5 and 12.6, and make your case.
3. You are the technical sales representative of Seiverts, the leading manufacturer of computed tomography equipment. Section 12.4 gives you a briefing on your equipment, use this, Section 12.5 and any other information – and your sales skills, to persuade the district to purchase your CT scanner.
4. You are a radiologist with a strong interest in research and in teaching medical students. You are convinced that the hospital should be at the forefront of medical practice. You have had some experience of MRI at the research stage and consider that this department should adopt option 4. Read Section 13.5 and use this and any other evidence you can find to persuade the RHA to make its major capital equipment funds available.
5. You are the hospital physicist. Your responsibilities include radiological safety in the department and the installation of new instrumentation including computer facilities. You should read Sections 13.2–13.6, review Chapter 11, decide which developments you support, and prepare a case for the DHA.
6. You are the hospital's development officer. Your responsibility is to ensure the best possible use of the hospital's very limited resources for development, consistent with national and regional health policies. You should read Sections 13.2–13.7, decide which developments you support, and play a leading role in formulating the department's ten-year plan.

13.2 ULTRASOUND

A widespread change in the routine examination of patients has come with the development of ultrasound techniques, in particular the use of **'real-time'** and **Doppler** investigations, which had become standard by the mid 1980s. The underlying physics and examples of applications, were given in Chapter 8 of this book. Recently available options and additions to equipment include **digital computing** of the Doppler image and **duplex scanning**. This latter is the presentation of the Doppler waveform of the blood flow alongside a two-dimensional image of the blood vessel under scrutiny. Doppler is an inexpensive, though time-consuming screening technique for the ever-growing need to investigate the circulation system, with a view to surgery.

The present provision of ultrasound within the hospital has evolved in an unplanned way. Its frequency of use has grown greatly in the past ten years, and is expected to continue to grow. The ten-year plan offers the opportunity to review these developments, to rationalise the provision within the medical imaging (MI) department, with a consequent increase in efficiency, and the provision of additional equipment and staffing, to permit an extension of ultrasound into new areas of treatment. The following papers contain the information on which the changes should be planned.

Obstetric ultrasound

Memo from head of obstetrics dept. to head of MI dept.

The continued provision of ultrasound examination by this department is problematic. The current use is approx. 200 k units and expected to rise rapidly if better provision can be arranged. The following deficiencies have been identified:

a) Our two machines are unsatisfactory, one is unreliable and the other has a low resolution only. Both require considerable medical time to produce reasonable scans.

b) We have no satisfactory record system for the volume of information that ultrasound has produced. It seems more sensible to incorporate this in your system which I know you are planning to revise with computer processing.

c) Our radiographer staffing is inadequate and inefficient, depending as it does on regular borrowing of one of the two staff from MI, and the constant need to refer to MI for supervision and specialist advice.

(Just between ourselves, this ultrasound business is getting too specialised for we obstetricians, and we'd be pleased if you'd take the whole thing over!)

Interventional radiology

A review by the consultant radiologist

This is perhaps the most exciting and certainly the most rapidly growing branch of radiology. The range of minor surgical techniques within the competence of the radiologist is now extensive, a few examples only are given here. They are all, of course, more time consuming than mere examination, some will require additional equipment.

The general procedure is the carrying out of a physical intervention, usually with a catheter (tube) inserted through the skin (percutaneously), whilst viewing the operation with real-time ultrasound or with reference to radiographs. It can be illustrated with the now well established procedure of percutaneous transluminal angioplasty(PTA). This term is simply medical jargon for 'replacing part of an artery through the skin, whilst looking through at it' (Fig 13.2). A part of the artery which is damaged is located by the imaging method, into it is introduced a so-called balloon catheter, via a minor incision under local anaesthetic. The balloon is then inflated at the correct site, until the artery is restored to its original shape.

The long-term efficacy of angioplasty compared with conventional surgery is at present under scrutiny, but early results have shown comparable success rates. The potential savings are large:

PTA, including all materials and overnight stay	£300
Arterial bypass surgery	£3000–£4000
Amputation	~£30 000

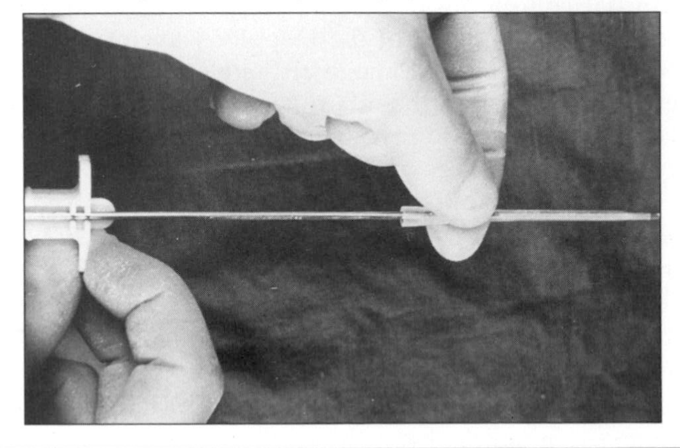

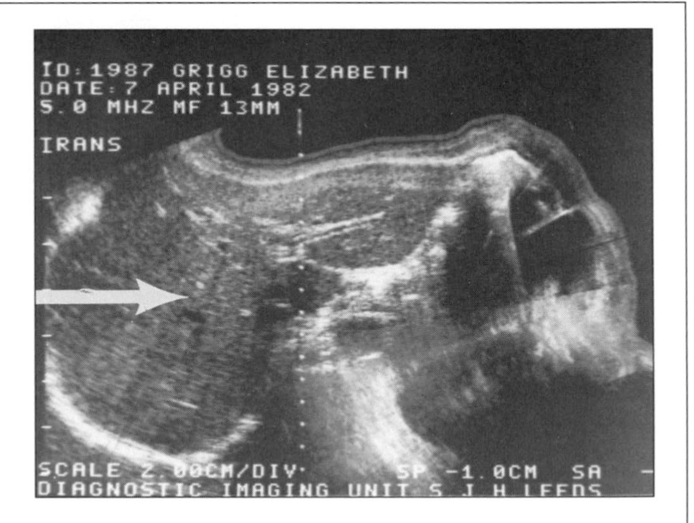

Fig 13.2 Ultrasound intervention technique **(a)** scan showing intended pathway for needle **(b)** assembly of needle within sheath.

Other examples include the removal of calculi, such as kidney stones, and of small foreign bodies, the draining of cysts and abscesses and the taking of biopsy specimens from internal organs.

The MI department could carry out some 3000 of these interventions with the addition of 2 ultrasound machines, the replacement of one fluoroscopic X-ray machine, and 2 additional radiographers and 1 radiologist.

Departmental data

Current level of ultrasound use is 200 k units, using 2 machines and 2 full-time-equivalent radiographers. Duplex scanning and digital processing are not available.

New equipment and staff: capital and annual costs:

- Ultrasound scanner, including Doppler, high resolution images, integrated thermo-printer for hard copy and full range of transducers — £50k

  add on duplex scan with Doppler — £15k

  computed sonography system, with Doppler — £80k
- Fluoroscopic, general purpose X-ray machine — £200k–£300k
- Qualified radiographer salary — £15k
- Consultant radiologist salary — £30k
- Support staff, clerical, porters etc. — £10k
- Accommodation at present full; additional rooms cost: serviced — £20k

  unserviced — £10k

Note that a sum of £30k may be available for computerising the departments records system.

DECISIONS ABOUT ULTRASOUND

1. As senior radiologist, what are the advantages of taking over obstetric ultrasound? What costs need to be incurred to ensure effective provision? Does this result in savings elsewhere?
2. What use will be made of the improved Doppler and intervention techniques? How will you ensure that the planners see that this is cost-effective?
3. Write a report of no more than 600 words (2 sides A4) which is the senior radiologist's contribution to the ten-year plan. You may wish to discuss with your colleagues, their proposals, and their reactions to yours, before you complete this.

13.3 NUCLEAR MEDICINE

The use of radionuclide imaging has become increasingly common with the development of the **gamma camera** which had become a standard piece of equipment by the 1980s. Its mode of operation was described in Chapter 12 of this book, along with the two currently available emission computed tomography (**ECT**) methods – single photon ECT (**SPECT**) and positron emission tomography (**PET**). There are many reasons for the use of nuclear medicine. In some cases it is the primary diagnostic tool, in others it is an adjunct or follow-up investigation, to X-rays. The advantages of radionuclide imaging are its sensitivity, compared to X-radiography, which reduces the dose, and its ability to make quantitative measurements and record the dynamic functioning of the part of the body under investigation.

The facility at St Katherine's was established three years ago by the radiographer currently in charge. It consists of a gamma camera under fully automated computer control and a range of collimators/filters, enabling a variety of radionuclides to be used, though the vast majority of investigations use technetium-99^{m}. The ten-year plan can be expected to include an increased use of the existing facility, requiring additional staffing and running costs, and the provision of a tomographic facility. The following papers contain the information on which the proposals should be based.

Survey of the current status of SPECT

Extract of an article in 'RAD Magazine', April 1988, by Andrew Todd-Popropek

The first tomographic systems were introduced into nuclear medicine some years before the first (X-ray) CT systems were used clinically, pioneered in the States in 1964. From these small beginnings, it is now estimated that the majority of all systems purchased for use in nuclear medicine have the *capability* of performing SPECT, even if not always used. It is interesting to look at how the current generation of SPECT systems perform and to try to predict in what ways SPECT might evolve. This article is concerned with the questions of 'instrumentation', the design and performance of systems, rather than of their clinical application.

Rotating gamma cameras and dedicated slice systems

The great majority of all SPECT systems in current use comprise a conventional gamma camera, mounted on a gantry, rotating about a patient, as illustrated in Fig. 13.3(a). The camera moves through a small angle, a conventional planar image is acquired and the process repeated until the camera has completely rotated through 360°. Many variations exist, but the basic principles are the same. The set of planar images at various angles about the patient are acquired and are then reconstructed using algorithms similar to those of (X-ray) CT. However, in SPECT there are additional problems, notably those of attenuation and scatter correction. Early systems were also plagued by artefacts, but these have now been largely eliminated.

In competition, a small number of 'dedicated slice' systems also exist. Fig. 13.3(b) shows a diagram of such a system. These are systems where a series of detectors are placed in some geometry around the patient and data collected from one or several slices, usually by means of moving the detectors by translation and rotation. Such devices can normally *only* perform SPECT and can generate only a limited number of slices. They have a major advantage in that they are much more sensitive than rotating gamma cameras and images can be acquired in a much shorter time. Although not many such systems exist, they have passionate advocates.

The rotating system produces a complete set of 3D data, whereas the slice system does not. Resolution of the image is dependent on how near the patient is to the detector. The slice system is superior to the rotating camera, with resolutions of less than 10 mm; it also has a greater sensitivity.

Summary and conclusion

From the point of view of instrumentation, SPECT systems have considerable potential advantages over conventional planar systems. They can produce images with improved contrast to noise ratio and facilitate reasonable estimates of *relative* concentration of activity. Although some improvement is still possible given advances in collimator design, the lack of sensitivity and limited resolution of rotating gamma camera systems means that dedicated slice systems and other systems with 'unusual geometry' are potentially important. For these developments, additional investment is needed by manufacturers. This will happen only as nuclear medicine is demonstrated to be clinically viable in competition with other techniques, something that seems to be occurring given the recent advances in radiopharmaceuticals. Fortunately, there seems recently to have been a substantial increase in the numbers of clinically valuable SPECT studies, and such developments may actually happen.

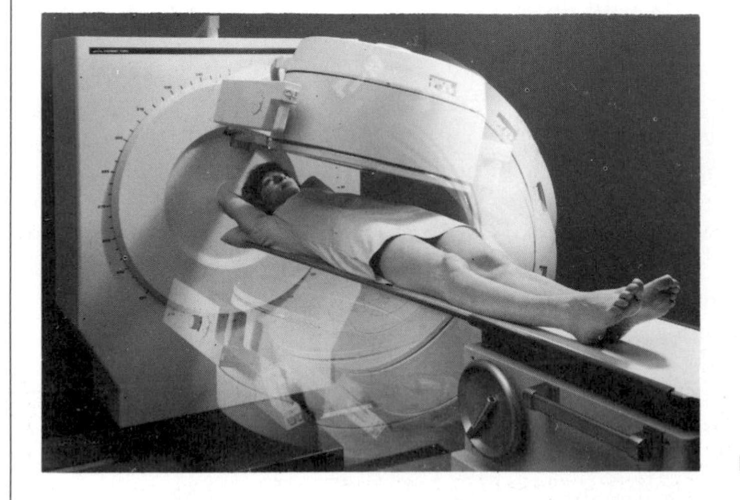

crystals and PMs

rotation

front

side

Fig 13.3 SPECT **(a)** rotating gamma camera **(b)** dedicated slice system.

MUPPET - A New Low-cost PET
News report in 'RAD Magazine', May 1987

Like gamma camera imaging now in widespread use as a diagnostic tool, Positron Emission Tomography (PET) is based on the use of short life radioisotopes.

But it is up to 50 times more sensitive and able to detect lesions as small as 5 mm, deep in the body, compared with 20 mm using a conventional gamma camera.

PET imaging can also measure physiological function quantitatively, making possible the study of such things as cerebral and coronary blood flow.

Despite these obvious advantages, two major obstacles have prevented its introduction to clinical use – a price tag of up to £2 million and the need for an expensive cyclotron to generate essential radioisotopes.

A solution?

Now scientists in the Department of Physics at the Royal Marsden Hospital and the Institute of Cancer Research at Sutton, Surrey, in collaboration with the Rutherford-Appleton Laboratory of the Science Research Council, believe they have solved both problems.

They have designed and built a computer-based PET clinical prototype (Mark II) incorporating a radically different gamma ray detection system that dramatically cuts cost – bringing the estimated market price in line with conventional gamma camera computer systems at about £100,000–£200,000.

The clinical prototype is up to 10 times more sensitive than gamma cameras in use today.

Alternatives

Work carried out at the hospital on an earlier small prototype (Mark I) has shown that cyclotron-generated isotopes can be replaced for many nuclear medicine applications by in-house radiopharmaceuticals, such as Gallium 68.

The new PET imaging system, which incorporates a computer-controlled rotating gantry with couch for positioning patient and camera, has been designated MUPPET, for MUltiwire Pro-portional chamber Positron Emission Tomography.

Breakthrough

The MUPPET system has achieved its cost breakthrough by developing a new method of gamma ray detection on multiwire proportional chambers.

A proportional gas ionisation counter has been modified by the addition of multiple lead photon converters. When rays strike the lead they produce electrons which are then detected by the gas proportional counter.

By using fine anode wires and lead cathode strips inside each detector, it is possible to obtain values of the x, y and z positions of the conversion points of both gamma rays.

Encouraging

To date, doctors at the Royal Marsden have scanned over 100 patients with the Mark I detector with most encouraging results.

Their work has shown that it is possible to carry out highly sensitive PET imaging at a hospital site where there is no radioisotope-producing cyclotron. The Mark II detector has a larger area of view and increased sensitivity.

Fig 13.4 MUPPET installed at the Royal Marsden Hospital, Sutton.

Departmental data

	running cost	capital cost
• Gamma camera (existing type)	£40k	£100k
• SPECT Rotating upgrade to existing gamma camera (new mount and computer hardware and software)	minor	£40k
• SPECT Dedicated slice system	minor	£150k
• MUPPET (complete system and radioisotopes)	£80k	£200k
Additional building and services		£50k
• Current staffing is 3 radiographers		£45k
Each new machine would require 1 extra radiographer + 1 technician		£25k

DECISIONS ABOUT NUCLEAR MEDICINE

1. As the radiographer in charge of nuclear medicine will you recommend an upgrade to the existing gamma camera, the dedicated SPECT system or the new low-cost PET system? Consider the relative advantages and costs of each.
2. Justify the extra expenditure in the ten-year plan – how will nuclear medicine be more cost effective?
3. Write a report of no more than 600 words (2 sides A4) which is the nuclear radiographer's contribution to the plan. You will probably need to discuss this with those working on the CT scanner and MRI system (it is unlikely that the DHA could support all of these), before finalising this.

13.4 X-RAY TOMOGRAPHY (CT SCANNING)

The major recent development in X-radiography is **computed tomography**, the principles of which were described in Section 12.5. The method is particularly good in examining soft tissue, as it can detect small differences in attenuation of X-rays. The tomographic technique removes the interference from other structures which are out of the chosen plane. As it is measuring the amount of X-ray absorption quantitatively, this information can be stored digitally in computer memory, recalled and processed to give a selection of reconstructed images. This imaging modality is becoming common in large hospitals, as its cost is falling to within the range of a high quality fluoroscopic X-ray room. It has undoubted potential within the MI dept., though the radiation dose is quite high and it may in time be supplanted by the radiation-free MRI modality.

At the present time the purchase of CT represents a major investment for the Exhampton district, but there is already an increasing demand for CT scans using facilities outside of the district. This growing cost is likely to make the purchase of either CT or MRI cost effective within the next decade. Use the information in the following extracts to put the case for purchase of a CT scanner.

The Seivert XJS CT scanner

Extract from a sales brochure

As a pioneer of medical imaging equipment (founded in 1910) and with an abundance of knowledge and experience, Seivert's advanced technological capability has brought forth its finest fruit in the XJS CT scanner. The revolutionary design of this whole-body, large throughput machine takes CT scanning beyond the fourth generation, with unique features found in no other comparable system.

***Patient throughput** – *up to 25 per day*
The high speed processing unit, high quality software and large capacity X-ray tube have reduced typical examination time by 50%. Slices can be scanned, reconstructed, displayed and filmed in as little as 15 seconds.
(This doubles the number of patients that can be seen in a day.)

***Scan times** – *down to 0.5 seconds*
High X-ray pulse rate offers range of scan times as low as half a second upto 5 seconds for more detailed images. This minimises artifacts due to patient movement, and reduces exposure without loss of valuable information.
(This is about one quarter of the scan time of a typical CT scanner.)

***Resolution** – *down to 0.4 mm*
Over 2000 highly sensitive solid state detectors produce unparalleled image sharpness and sensitivity.

***Slice thickness** – *down to 1 mm*
This is essential for the accurate imaging required of

eye, ear, nose, spinal cord and for neurosurgery. Slices of 10, 5 and 2 mm are offered as well as 1 mm. (This matches the best of our competitors.)

***Large viewing field and automated gantry operation**
These features allow the fullest range of examinations, from the head of a child to the abdomen or lung of an adult, without loss of resolution and with maximum convenience to the patient.

***User friendly console**
This includes all controls ergonomically designed and the facility for an additional, remote console for separate image analysis and processing.

***Software** – *to provide an extremely wide range of image processing*
For example: 3D image construction, simultaneous display of several projections or sections, variable contrast enhancement, overwriting of numerical intensity values or information labels etc. Optical disc archiving for rapid storage and retrieval. Compatible with other models (e.g. for radiotherapy planning).

***High reliability**
New design of X-ray tube with dramatically increased life, guaranteed for 40 000 slices.

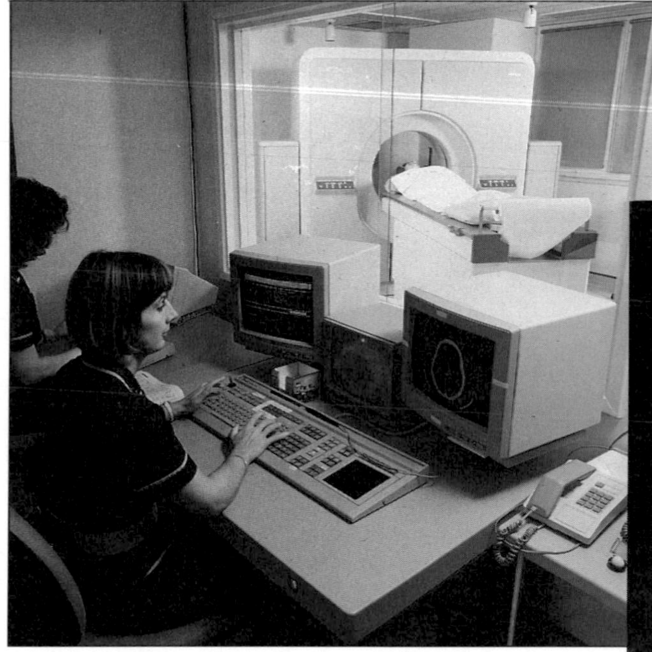

Fig 13.5 Dual console, triple screen console for CT scanner.

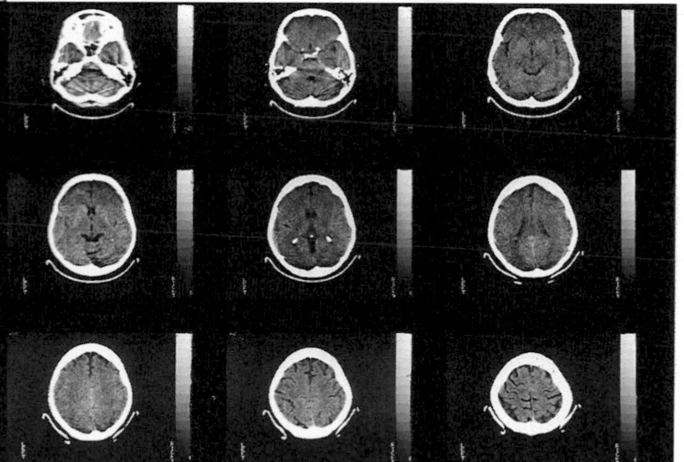

Fig 13.6 A series of axial sections through the brain by CT scanner.

Income and expenditure for CT scanner

Extract of district working party report

Present situation

Scans have to be purchased outside the district at £100 each patient. The present level of patients is 500 per year and rising rapidly.

Estimates of use

Following the purchase of a scanner, demand is expected to rise as follows (either from the district's own patients or from referrals from other districts).

Year	*number of patients*	*income/nominal cost*
1	1000	£100k
2	2000	£200k
3	3000	£300k

Remaining at 3000 thereafter.

Costs

Item	*Capital cost*	*running cost per year*	
Basic equipment	£260k	upgrades	£5k
Computer hardware & software	20k	updates	5k
Site preparation, services	20k	services	10k
Insurance, depreciation			10k
Interest on loan of capital			25k
Staffing			45k
Supplies			50/patient
Totals	*£300k*	*£100k + £50/patient*	

Raising the money

Article from the 'Radlet Weekly News'

Community appeal funds CT scanner

Just one year after the appeal was launched, the St Mary's hospital CT scanner appeal has reached its target of £250000. The money raising campaign has inspired local people to jump out of planes, play marathon tennis, darts and music sessions, and hold numerous jumble and bring and buy sales. Local businesses and industry undertook, through the chamber of commerce, to double whatever was raised by donations from the community. The appeal was launched after a request for the equipment, by St Mary's chief radiologist, Dr Waltham-Thompson, was turned down by the Dulminster health authority. Dr Waltham-Thompson's own department played a leading role in the fund raising, with projects such as the sale of ultrasound photographs to expectant mothers and of personal radiation dosimeters following the accident at the Newfield nuclear power station.

Presenting the cheque at the celebration ceremony, the patron of the appeal, Lady MacBath paid tribute to all those who had helped. "It was a wonderful community effort", she said, "which shows a spirit of enterprise and commitment by a core of dedicated people, supported by all sections of society. Not only have your efforts saved the Health Service the cost of buying the scanner, but it will provide a future saving in its operation. This is because it will replace other, more expensive types of examination, requiring longer stays in hospital. Most importantly," she added,"it will provide a safer, more effective and less traumatic service to those in need, of which we can all be proud".

DECISIONS FOR THE CT SCANNER SALES REPRESENTATIVE

1. As sales representative you will have to persuade both the radiologist, and the planner to support your case. Use the technical information and the financial information appropriately for each. For example, you could work out from the data given, how many years it will take for the scanner to pay for itself.
2. What financial support can you or your company offer to the district or region, to ease the burden of the capital outlay or interest repayments? You could quote the example reproduced here, or offer an extended repayment scheme, a lower interest loan arrangement, or reductions for multiple purchase within the region.
3. Anticipate the objections and alternatives raised by the other participants, and write a brief report of no more than 2 sides of A4, presenting 'the facts' and outlining your strategy as sales representative.

Be creative, your personal commission if you clinch the sale is 5 per cent of the capital cost!

13.5 MAGNETIC RESONANCE IMAGING

Magnetic resonance imaging (MRI), is a completely new method of imaging, or modality, which is finding its way into hospitals in the early 1990s. It is based on the phenomenon of nuclear magnetic resonance (NMR), a process in which protons interact with a strong magnetic field and radio waves to generate electrical pulses that can be processed in a similar way to the computed tomography for X-rays. The medical application of nuclear magnetic resonance (NMR), began in the 1950s, but the first images of live patients were not produced until the late 1970s.

During the 1980s the technique slowly emerged from research laboratories, to become the most promising imaging and biochemical analysis tool of the 1990s. The images produced are in many ways similar to those of the CT scanner, but without the radiation hazard. Although the equipment presently available is considerably more expensive to install and to run, it must be considered as an alternative to CT scanning over the period of the ten-year plan. The case to be made may require this longer time scale to be successful. It will also be helpful to provide those involved in the decision making, both an introduction to the physics of the technique, and an account of its value as a research tool. Use the following papers to make this case.

The physics of MRI

This phenomenon is rather complex, involving a number of steps. You will probably need to read this more than once to take it in!

- Atoms which have an unequal number of neutrons and protons in their nucleus, spin.
- Because the nuclei are charged, and moving charges produce a magnetic field, they act like tiny magnets.
- In a magnetic field they will align themselves, just like iron filings do on a larger scale.
- This alignment is not exact however, and the nuclei rotate around the direction of the field, while spinning. This motion is called **precessing**, and it is exactly the same as the motion of a top spinning in gravity.
- The frequency of the precession depends on the magnetic field and on the nature of the nucleus. It is called the **Larmor frequency** and is found to lie in the radiofrequency (RF) part of the electromagnetic spectrum.
- So, by applying a pulse of radiation of this frequency, it is possible to disturb the precession, causing **resonance**.
- When the pulse ends the nuclei return to their equilibrium state with the emission of RF radiation. This occurs over a short period of time, called the **relaxation time**.
- There are in fact two different relaxation processes, the times of which can be measured. It is these which form the basis of MR image formation.

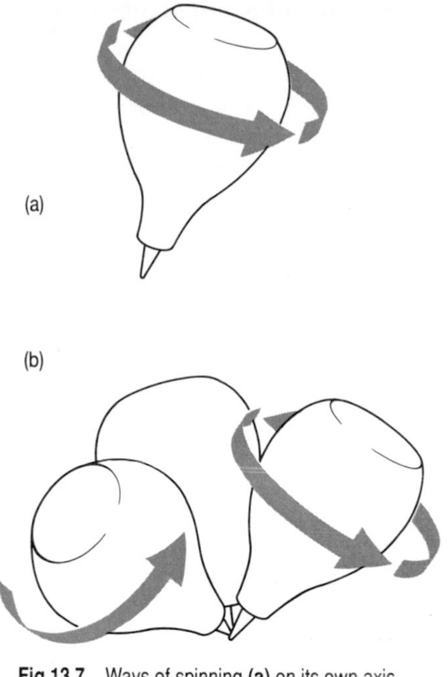

Fig 13.7 Ways of spinning **(a)** on its own axis **(b)** spinning and precessing around the vertical axis.

Examples of nuclei able to resonate in this way are hydrogen, phosphorus and carbon. Because of its abundance, in body fluids in particular, hydrogen is the nucleus used in imaging.

An MR scanner requires the following (Fig 13.8):

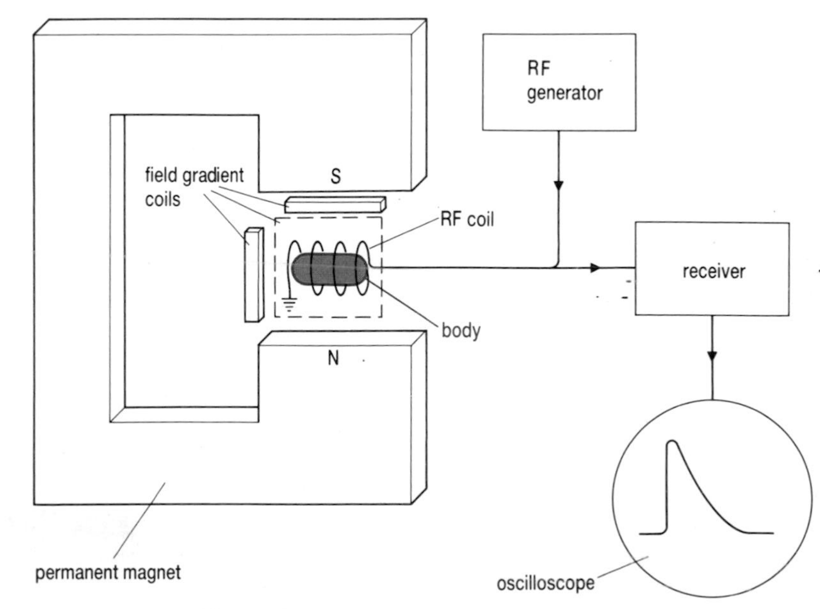

Fig 13.8 Schematic diagram of an MR system.

Fig 13.9 MR scanner magnet – with legs!

- A field strength of up to 2 tesla which is uniform and constant, over a volume large enough to accommodate a patient (Fig 13.9). Very large and powerful electromagnets are used, made of superconducting niobium–titanium alloy. It is only superconducting at a temperature of –269 °C, so needs to be immersed in liquid helium. The replenishment of this is the major cause of the high running cost of this equipment. If conventional copper coils are used, a heavy current is required to generate the high field. This produces a great heating effect, which is inconvenient and results in power costs similar to the superconducting type,

- A method of creating a gradient in the static field, so that the signals detected from the nuclei can be located in position. This is usually done by three separate coils under computer control, so that the image can be constructed,
- Radiofrequency pulses which are supplied by yet more, (RF) coils. The pulses are transmitted to the patient in special sequences, which produce the required contrast from the tissue,
- Radiofrequency receivers, linked to a CRT for display as an image or trace,
- Rooms of suitable size and construction, because of the weight of the magnet, to house the scanner and computer equipment. Screening from the magnetic field is required; this is usually by enclosing the machine in a metal cage.

There have been no reports of any adverse effects from the magnetic or RF fields, though care must be taken if patients have metal implants.

The potential of MR imaging and spectroscopy

Extract from the appeal literature of another hospital

Magnetic Resonance Imaging

The study of the behaviour of body molecules in a powerful magnetic field, combined with scanning technology has given another revolutionary form of body scanning – Magnetic Resonance Imaging (MRI). For certain parts of the body, including brain, spinal cord and heart, and in certain diseases, MRI is superior to CT scanning. It does not utilise X-rays, has no dangers or complications and is non-invasive. It offers an early and easy diagnosis of many conditions, including cancer, and helps to plan the most effective treatments.

At St. George's we already have a research team with pioneering experience of MR technology and we are hoping as a result of the Appeal to be able to buy, install and run an MR scanner.

It is not just in imaging, however, that MR is creating a revolution.

Magnetic Resonance Spectroscopy (MRS)

Modern MRI instruments of the highest power have an extra capability. Using the technique of Magnetic Resonance Spectroscopy (MRS), they can be used to detect and measure chemicals within the body without having to take any samples – not even blood samples.

We have never before been able to monitor biochemical changes at the site of the disease. The possibilities for solving practical medical problems, such as the following, are exciting:

Has the heart muscle been irreversibly damaged after a heart attack?
How is a transplanted heart or kidney being damaged by the process of rejection?
Is a cancer responding to radiotherapy or chemotherapy?
Are anticancer drugs given to a patient actually being taken up by the cancer?

Many of these questions cannot be answered at present, but our research teams will be able to tackle them directly using the new instrument. The results will be applied to clinical practice within the new St. George's.

MRS is also a pure research tool of immense promise. Because of its ability to measure the chemical changes in normal volunteers or patients we will be able to study the effects of many diseases or physiological processes.

St. George's is already in the forefront of research into the medical applications of this new method, particularly in the field of cancer research. In 1981, Dr. John Griffiths was the first to apply MRS to the study of cancer, and he and his research team have remained world leaders. Indeed, the Cancer Research Campaign have pledged £450,000 to support this research for the next few years, but we have to rely on this Appeal to provide the instrument itself.

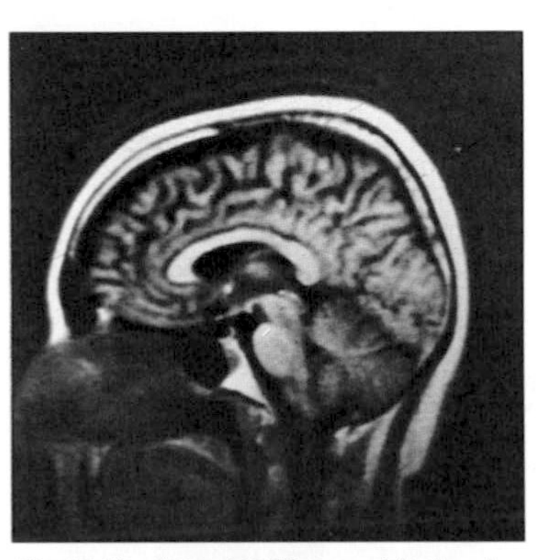

Fig 13.10 Lateral NMR scan of the head.

Fig 13.11 Research workers at the computer terminal of a magnetic resonance spectrometer used in determining chemical composition.

Costs

	running cost	capital cost
MRI complete unit, all staff and services	£400–£800k	£600–£1000k
MRS research unit, additional staff and services	~£100k	~£100

DECISIONS FOR MRI

1. As consultant radiologist, explain to your non-physicist colleagues how NMR produces images of body tissue.
2. Give a brief report on the value of the MR technique in medical diagnosis and biochemical investigations. Mention especially the advantages it has over CT scanning.
3. Make a case for the hospital to be a regional centre for MRI and MRS based on the present expertise of the staff (especially yourself!), and the teaching functions of the hospital.
4. Include these three aspects in your contribution as consultant radiologist to the ten-year plan, as a report of about 600 words (2 sides A4). You will need to try to gain the support of the senior radiologist and anticipate the opposing arguments of other contributors to the plan, before completing this.

13.6 THE ROLE OF THE HOSPITAL PHYSICIST

The hospital physicist works in the department of medical physics, which is staffed with a small number of graduate physicists and engineers, and a larger number of technicians (Fig 13.12). The department has four roles, and the extent to which a particular physicist is responsible for these will depend on the size of the department and the experience of the person.

Fig 13.12 Hospital physicist at work.

There are legal requirements for **health and safety** duties to be carried out, as described in Section 11.4. These will be predominantly concerned with the use of ionising radiations in a medical imaging department. There are in addition, hazards due to the use of high voltages and currents, in the equipment used; other possibilities include non-ionising radiations, such as laser light and the strong magnetic fields used in NMR.

The physicist will have a key **scientific service** function in relation to the medical investigations being carried out, and the equipment which is used. The particular duties may range from routine advice to radiologists and radiographers on the radiological properties of a pharmaceutical, to the monitoring of detection and measuring equipment.

Research and development are an important aspect of work, which may include new or improved procedures, equipment, and instrumentation such as computer use, or aspects of medical research projects.

In the areas of **management and teaching**, the physicist will provide:

(a) advice, training and education for staff in the physical, health and safety aspects of radiation and instruments for measuring and imaging,

(b) direction of technical and administrative staff carrying out the above duties,

(c) evaluation and selection of new equipment, and collaboration in the manage-ment of its installation.

The department at St Katherine's hospital is small, with only one physicist and five technicians. The physicist is mainly concerned with radiological safety, and the technicians with aspects of physical measurement, including electronic and mechanical devices, major equipment and 'anything attached to the patient'. The job is therefore rather routine, with some collaboration with physicists in other hospitals on development projects. Any ten-year plan must have the safety implications examined by the physicist, details of which are contained in the following extract. It also offers the opportunity to increase the developmental capacity of the department (which would mean increasing its size), so that it can respond more effectively to the technical demands of this rapidly changing field.

Duties of the physicist in radiological protection

Extract from policy statement of the Regional Health Authority

The physicists' role is to apply scientific expertise to the assessment and development of imaging equipment, the monitoring and control of radiation doses to patients, staff and the general public and the determination of the overall efficacy of imaging techniques. This involves liaison and co-operation with members of the other relevant professions involved, i.e. radiologists, radiographers, equipment manufacturers, service engineers and Health Authority engineers and administrators.

In compliance with the Ionising Radiations Regulations 1985, (IRR), every Health Authority which makes use of ionising radiation should have appointed an appropriate Radiation Protection Adviser (RPA). The RPA may be a named individual or a corporate body such as a medical physics department with individual physicists appointed to give advice in specified fields.

Some specific duties of physicists in support of the advice of the RPA are indicated below.

DIAGNOSTIC RADIOLOGY

From the point of view of ensuring the safe use of ionising radiations in this field, the physicists would normally be responsible for:

a) proposals for protection measures to be incorporated in X-ray rooms and subsequently surveys of rooms and equipment, and confirmation that the measures have been executed effectively (IRR Regs 6.2 and 32),

b) performance tests at installation, routinely and when faults are suspected, to ensure that any equipment used in an X-ray examination operates so as to restrict the radiation dose to the extent consistent with the clinical objective (IRR Reg 33.1) thus optimising the benefit of the examination,

c) measurement and calculation of doses to patients – including those following irradiation of an undisclosed pregnancy; advice to the radiologist on the magnitude and distribution of doses received,

d) calibration of monitoring and test equipment used in (a) and (b) above.

NUCLEAR MEDICINE

In nuclear medicine it would be expected that the physicist would be responsible for:

a) proposals for the design, construction, or adaptation of premises and equipment for work with unsealed radionuclides (IRR Reg 6.2),

b) proposals for designation of areas requiring special supervision or control and review of the areas designated in response to changes in the work (IRR Reg 8),

c) tests of the operation of safety features and warning devices against external radiation and against spread of radioactive contamination (IRR Reg 6.2),

d) drafting a plan for disposal of radioactive wastes for submission to the Department of the Environment,

e) ensuring that unsealed radioactive sources are prepared in accordance with good radiation protection practice (IRR Ref 6); protecting the public from irradiation by radioactivity administered to patients both before and after those patients leave hospital (IRR Reg 6.1); ensuring that equipment used, the physical aspects of techniques employed and the products administered in diagnostic or therapeutic radioisotope procedures restrict exposure of the patient to the level necessary to achieve the desired clinical result (IRR Reg 33.1); ensuring that the radiation exposure and radioactive contamination of staff, the public and the environment does not exceed the limits laid down (IRR Reg 7),

f) verification of records relating to work with unsealed radionuclides and accounting for all radioactivity including physical confirmation of stock inventories (IRR Reg 19).

In addition the physicist has responsibility for the overall supervision of safety measures which arise under the Health and Safety at Work Act etc. including measures related to the potential biological hazards of ultrasound, and nuclear magnetic resonance.

DECISIONS FOR THE PHYSICIST

1. What are the radiological protection and instrumentation implications of the changes outlined in Sections 13.2–13.5? Will these result in a safer environment for patients and/or staff? Will there be any financial savings or additional staffing or resource requirements?
2. Which of the changes proposed do you support as giving the hospital physicist a more satisfying role? Are there others which you feel should be included, such as the increased use of computers in processing images and records?
3. Write a report of not more than 600 words (2 sides A4), explaining your views on the ten-year plan proposals. You will need to obtain some information on these from the other participants, or make your own evaluation of Sections 13.2–13.5.

13.7 PLANNING CHANGE

The preparation of the ten-year plan which is the focus of this case study, is a national requirement, to be carried out at the regional and district levels. The purposes nationally, as expressed by the government Department of Health (DH) are:

- To identify inefficiencies in the service, remove them and produce an overall improvement in the cost-effectiveness of the health service.
- To target the scarce additional financial resources in key areas of need in health care, and on major proven technical developments which may be expected to produce savings in the future.

There are clearly implications for medical imaging in each of these.

More specifically, guidance from the DH has indicated that:

- savings are expected from the rationalisation of MI departments,
- money should be made available at district level for the increased use of computing in the processing, storing and accessing of medical information of all kinds, and
- regions should establish a centre where the complete range of currently available imaging modalities may be used, and that substantial additional finance may be provided for this.

The following papers interpret these guidelines for the Exhampton district, and should form the basis of the overall plan by the development officer.

Strategic planning: implications for the Exhampton district

Part of a memo from the regional planning officer

In compiling your plan for the district, please give urgent consideration to the following priorities:

General

- In pursuit of greater efficiency, ensure that there is no duplication of effort, for example the provision of the same or similar services in more than one department, as this has often been found to result in waste. Rationalisation of such provision is expected to produce savings.
- To improve the cost effectiveness of the service it is imperative to consider alternative ways of carrying out routine investigations and minor interventions. Considerable savings can follow from an increase in outpatient treatment and the consequent reduction of patient overnight stays.
- The district should aim to be more self-sufficient in its health care, and thus spend less on purchasing services from other districts and regions. this may require the approval of additional capital expenditure.
- Finance for building works is severely limited, any expansion of a particular department or service should be by internal reallocation of space.

Specific

- Capital grants are available for the purchase of a limited amount of major equipment, such as new imaging modalities. It has to be noted that in the South Westshire region, the Exhampton district is not at present the obvious choice for a major investment of this kind. St Katherine's, the district general hospital, is rather small and is close to to a major teaching hospital in the centre of the city. On the credit side, the present department has successfully established its nuclear medicine facility recently, and has a staff that is relatively well qualified and highly regarded by others in the region. The case for any major extension of MI provision will have to be made very convincingly, however, and may need the support of all the specialists in the region.
- A review of all existing major capital equipment is required in the ten-year plan. The opportunity should be taken to plan the replacement of outdated and inefficient equipment, within the general financial constraints.
- Finance is available specifically for the computerisation of patient records. In view of the nature of the records in the MI dept., and the current level of computer expertise, it is expected that this department would take a lead. (I attach details of the experience of another hospital.)

Computerisation of an X-ray department

extract from an internal report of the Highbury General Hospital

The advantages we have found in the computerisation of our X-ray department may be summarised as follows:

Patient services

1. Speedier, tidier and clearer booking-in procedures, arrangement of appointments and communications with patients.
2. Increased reliability of records and improved availability of recall from files.
3. Improved communication with other departments in the hospital, and other sites. (This department gave such a successful service in this area that we have advised others in the implementation of compatible systems.)

Records

1. The opportunity was taken to revise the system so that it takes account of recent changes in practice in the department, and is flexible enough to cope with future changes, such as correlations between new imaging methods.
2. Information on patient throughput, use of equipment and consumable supplies, staff workloads, etc. can be computer logged for processing to provide the statistics increasingly demanded.

3. Financial spreadsheet programs can calculate the components of cost of various examinations, both fixed and variable.
4. Stock control of consumables and regular reordering can be computer processed (this is some way off for this department at present).

Management

1. The workload of this department has increased by 100% over the last ten years with the only increase in clerical staffing over that period being directly involved in work with the new CT scanner. Without the introduction of extensive computer processing, this would be impossible.
2. The detailed statistics which are routinely produced are invaluable for forward planning. These can include cost projections (though these will undoubtedly involve personal assumptions of the planners, not necessarily shared by all!).

Designing the system was easier than anticipated, as we were very fortunate in having the advice of an experienced software house. We would recommend any department considering such computerisation to review their requirements with software designers at a very early stage. This will probably result in them modifying existing programs to meet precise needs, at the lowest cost.

DECISIONS FOR THE PLANNER

1. As development officer for the hospital, you need to review all the possible plans, by reading Sections 13.2–13.7. Check each possible proposal against the criteria in the regional memo, and list your own priorities for approval, with reasons, in a report of no more than 600 words (2 sides A4). Pay particular attention to the computerisation plan, which you are expected to implement.
2. Convene the group and decide which plans are to go forward to the region for approval. Arrange who is to present what at this meeting.
3. Present your group's plans, calling on members of the group to present their 'expert' contributions, as appropriate. The presentation should not exceed 15 minutes, and there will be an opportunity for questions from the other groups. When all groups have presented their plans and been questioned on them, the regional planning officer (your teacher or an outside 'expert'), will give a public assessment of each, and decide which, (if any!) are approved.

ANSWERS

Chapter 1

1.1 See Fig 1.18 (a) and (b)

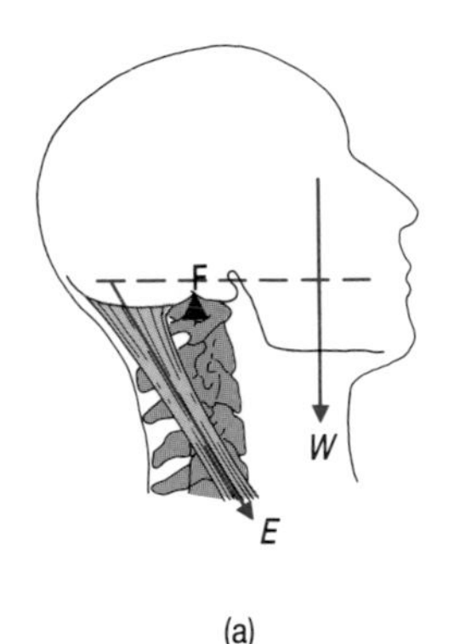

(a)

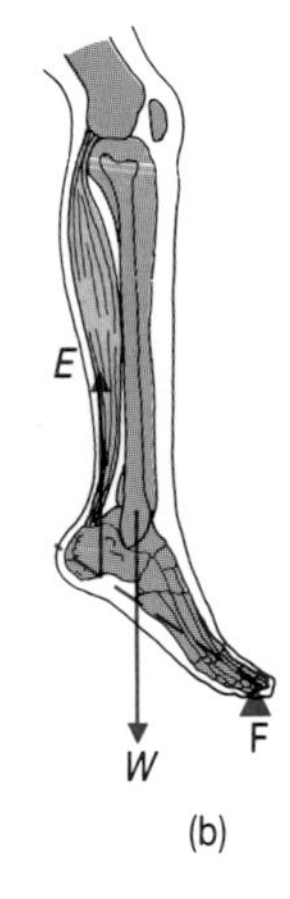

(b)

Fig 1.18 **(a)** Looking up, **(b)** Standing on tiptoe.

1.2 Biceps force = 9036 N Elbow reaction = 7756 N
(assuming each arm supports half the weight at the palm of the hand)

1.3 Compressive stress $= 0.7/2\ W \times \cos 50° = 0.35 \times 700\ \text{N} \times 0.6428 = 157.5\ \text{N}$

Shear stress $= 0.7/2\ W \times \sin 50° = 187.7\text{N}$

1.4 From the figures given in the examples in the text and answer to question 1.3:
157.5 : 2385 : 4886 or 1 to 15 to 31

1.5 **(a)** (i) $F = \mu V$ where μ for shoes is 0.7 and V = body weight W.
i.e., $F = 0.7W$
For walking horizontal force is about $0.2W$, easily opposed by friction.

(ii) For a given body weight and ground surface if θ is smaller F is smaller.

(b) We suggest you do a practical investigation on this one!

1.6 **(a)** see text

(b) high initial centre of gravity

(c) increasing forward momentum

(d) keeping centre of gravity low in relation to body position. It is possible for it to stay lower than the bar, yet for the body to clear the bar.

1.7 **(a)** About 1 kg per hour

(b) (i) A large area of the skin is exposed, other methods of loss are not sufficient to produce the rate of cooling required.

(ii) The body is almost completely covered, any sweat produced will not evaporate.

(c) The amount of energy loss is still low but the other mechanisms are hardly operating.

1.8 Read off from the graph, the value of BMR in kJ m^{-2} h^{-1} and convert this to watts using values of surface area, from Table 1.2.

1.9 Example of a day's energy expenditure:
Average value from the range given in Table 1.3 and a moderately active day: 8 h sleeping + 6 h sitting and standing + 9 h moving and $\frac{1}{2}$h climbing and carrying + 1 h walking and housework + $\frac{1}{2}$h moderate sports = 14.3 MJ

1.10 **(a)** 363.06 kJ h^{-1}
(b) 100.85 W

1.11 Panting increases the rate of supply of oxygen required for the higher rate of respiration. Massaging brings oxygenated blood to oxidise the lactic acid in the fatigued muscles.

1.14 **(a)** The kinetic energy of the run up can be used in the vault with the help of the pole.
(b) Glass fibre poles came into use in the 1960's, which were much more effective in absorbing and releasing the runner's energy – usually! (see photo, in the theme introduction).
(c) 5.6 m

Chapter 2

2.2 There is a greater density of rods.

2.3 About one minute, depending on the brightness.

2.4 It stimulates the red and green receptors equally;

red + green = yellow.

2.5 Study Fig 2.5(b).

2.7 Cyan is the complementary colour to red. The white light from the wall will stimulate all the receptors. If the red cones have been depleted of their photosensitive agents, in the region of the ball's image, then only blue and green will operate. One test would be to try different colours of wall.

2.8 The refractive index of water is about 1.33, close to that of the cornea, reducing the refractive power from ~ 46 D to ~ 6 D.

2.10 10 m, which is probably rather further than you can see. Can you explain the difference? Remember this is only a distinction between two points, to recognise an object, the brain requires more information.

2.11 **(a)** Diverging, –0.5 m
(b) (i) –5 D (ii) 0.11 m (iii) 0.25 m (normal vision).

2.13 **(a)** Longsight, use lenses of power +3 dioptres
(b) Presbyopia, use bifocals, with the lower half of power +2 D and upper half –0.2 D.
(c) Astigmatism, use a lens with a horizontally oriented cylindrical correction.
(d) Diplopia, possible correction by exercising the rectus muscles to strengthen them.

2.14 **(a)** Chromatic aberration has caused the eye to be unable to superimpose the two images.

(b) The image of the right hand object has fallen on the blind spot of the right eye.

(c) Reducing the size of the viewing aperture has increased the depth of field. Reducing it to 1 mm has caused significant diffraction effects.

2.15 **(a)** Refer to Section 2.2 and Fig 2.5
(b) $0.5 \times 0.3 \times 10^{-3}$ m = 0.15 mm

2.16 **(c)** A, myopia, for a far point of infinity a lens of –2 dioptres.
B, hypermetropia, for a near point of 0.25 m a lens of 2 dioptres.

2.17 **(b)** (i) person (ii) needs a converging lens with positive power.
(iii) find from the graph the person's near point (n), and substitute this in the formula

$$P = \frac{1}{n} + \frac{1}{0.02}$$

to calculate the power of the eye. Subtract this from the power required for a near point at 0.25 m which is 54 dioptres.

Chapter 3

3.1 Fluid fills the Eustachian tube so that the pressure across the eardrum cannot be equalised.

3.3 $I_x = I_0 e^{-kx}$

3.4 **(a)** 90 dB
(b) 10^5

3.5 **(a)** 10^{-12} Wm^{-2} from 1 kHz to 3 kHz
(b) 120 dB, 140 dB, no
(c) ~50Hz to 14 kHz

3.7 **(a)** 10^{16}
(b) 10^4 Wm^{-2}
(c) from 10^{-10} to 100 W m^{-2}

3.8 **(a)** For diffraction the aperture must be of the same size as the wavelength of the sound. It causes a change in direction of the sound;
(b) Speech frequencies are low;
(c) Consider the time period of a sound wave (=λf) compared to the time between separate sounds;
(d) Consider the path lengths for sound from each speaker to each ear.

3.12 **(b)** 5×10^{-3} W m^{-2}, 10^{-2} W m^{-2}, increase 3 dB

3.13 **(b)** 78.4 dB

Chapter 4

4.1 **(a)** You may have found that the range is too large and consequently the sensitivity is too low for this instrument, which is usually used to measure greater temperature differences.
(b) This will depend on how you did the experiment. Possible sources of error include the contact between the thermocouple and the body; the voltage measuring instrument; the reading of the meter scale; the process of calibration.

(c) A constant reference temperature must be maintained; good thermal contact with the body must be maintained.

(d) If you read off the temperature from tables, these will have been obtained by calibration against a reference thermometer of some kind. If you used a commercial thermometer to calibrate your thermocouple, this will have also been calibrated against a reference thermometer.

4.3 **(a)** Taking drinks of water whilst running.

(d) 'He faces months of intensive rehabilitation'. His physical condition is still very weak and damaged (he will have to learn to walk with one leg). Also he has suffered brain damage, from which he can be expected to recover quite shortly.

4.4 There are illnesses which do not cause a rise in body temperature. Children, who are most at risk from rapid temperature rise, are not good at keeping the thermometer in contact with their bodies. The person's normal body temperature may differ from the accepted norm of 37 °C, so that changes may be hidden.

Chapter 5

5.1 See Fig 5.15 below

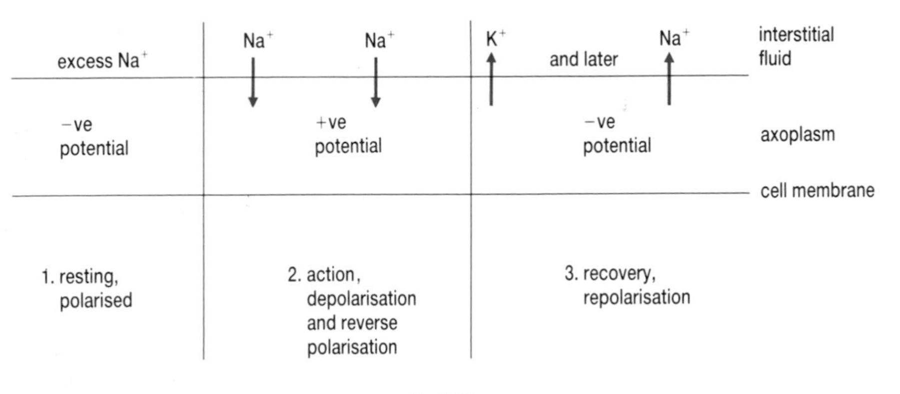

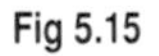
Fig 5.15

5.2 **(b)** This is the pump for the systemic circulation, which includes organs and fine capillaries all over the body, so a strong contraction is required to provide the pressure.

5.3 **(a)** It occurs at the same time as the much stronger ventricular depolarisation which hides it.

(b) About 1 per cent

5.5 **(a)** Immediate availability of information over an extended period, suitable for intensive care.

5.6 **(b)** 55, 120, 57–86, 67

5.7 **(a)** Heavy plastic handles on the paddles, good insulation for the wiring and power unit, possibly insulating gloves and mat for the operator to stand on.

(b) $20 \times 3000 \times 5 \times 10^{-3} = 300$ J

(c) (i) 150 Ω (ii) At the surface of the skin

(d) see Fig 5.16 on next page

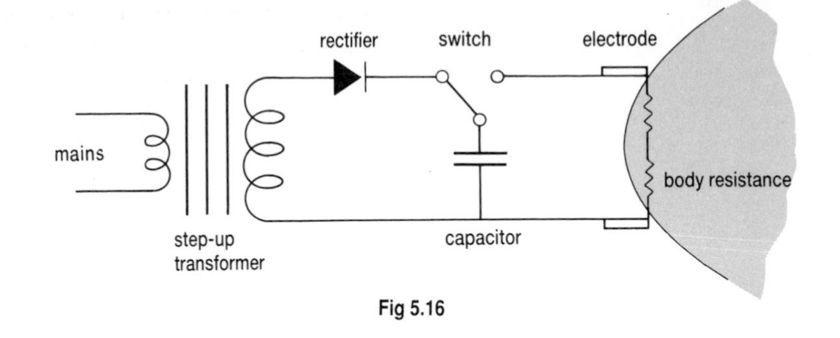

Fig 5.16

5.9 (a) Fig 5.3 shows the action potential for a nerve. The maximum potential is –90 mV, so over a distance of 10 nm, this gives a field strength of $90 \times 10^{-3}/10 \times 10^{-9} = 9 \times 10^{6}$ Vm^{-1}, so the membrane is a better insulator.

(b) number of sodium ions containing

one positive charge	$= 25.8 \times 10^{15}$
total charge	$= 41.28 \times 10^{-4}$ C
current density	$= 4.128$ A m^{-2}
current	$= 2.064 \times 10^{-11}$ A

Chapter 6

6.1 There is a pressure head to the saline solution to stop the blood leaking out. Heparin is added to the solution to prevent clotting. There are risks of infection both to and from the patient.

6.2 Use the data and conversion factors in Table 6.1, e.g. arterial blood

$$\text{pressure, maximum value} = 140 \times \frac{1.01 \times 10^5}{760} \text{ Pa} = 18.6 \text{ kPa}$$

6.3 (a) $F = pA, \dfrac{l}{t} = \dfrac{v}{A}$

(c) 1.38 W

6.4 (a) Operator practice will help to avoid some of the false results, such as those from the subject moving. Inaccuracies can originate from the fitting of the cuff; in the manual version, the reading of the manometer; in the electronic version, possibly the amplification of the signal.

(b) Pressure in the vascular system varies with height. A drop of 10 cm below the position of the heart will increase the pressure by approx. 10 cm H_2O or 7.5 mm Hg.

6.5 (a) 10^{-2} $\mu V\ V^{-1}$ mm Hg^{-1}

(b) 0.60 mm.

6.6 Peizoelectric, capacitance and resistance transducers can all measure displacement directly (explain how in your answer), inductance transducers measure velocity because the effect depends on movement, not simply position.

6.9 (b) (i) Refer to question 6.3 and Fig 6.3

(ii) Substitute for $\overline{v}_A^{\,2}, \overline{v}_P^{\,2}, \overline{p}_P$ from the information given in the paragraph between the two equations.

(c) (ii) $\dfrac{16 + 10.2}{2}$

(iii) compare the time interval between opening and closing, with the whole time period.

(d) $\frac{1}{2}Pv^2V = \dfrac{M}{L^3} \times \dfrac{L^2}{T^2} \times L^3 = \dfrac{ML^2}{T^2}$ = Force × distance

$$pV = \frac{ML}{T^2L^2} \times L^3$$

(e) 13.72 + 9.1 = 22.82W

Chapter 7

7.1 **(a)** 0.616 **(b)** 38°

7.2 **(b)** 48 per cent

7.5 **(b)** The parallel collimated beam can be conveniently used in endoscopy. The monochromatic light is used for particular tissue interactions. Coherence of the phase of photons is also mentioned, this is important in holographic imaging, where interference patterns are produced but not in the treatments discussed here.

(e) Argon and helium-neon, the energy is supplied electrically. A high potential difference across the gas filled tube causes the formation of ions by discharge.

7.7 **(a)** 10^5 W mm^{-2}

(b) (i) 1.27×10^{-3}

(ii) 1.27 (over an area of 10^{-3} mm)

(d) (i) Turn your head or blink your eyes immediately.

(ii) Very little. The area of retina damaged would be too small to notice, and there would be no pain sensation.

7.8 **(b)** (i) See Fig 7.14

$\sin i_c = 1/1.5 = R/(R + 0.1)$ so $R = 0.2$ mm

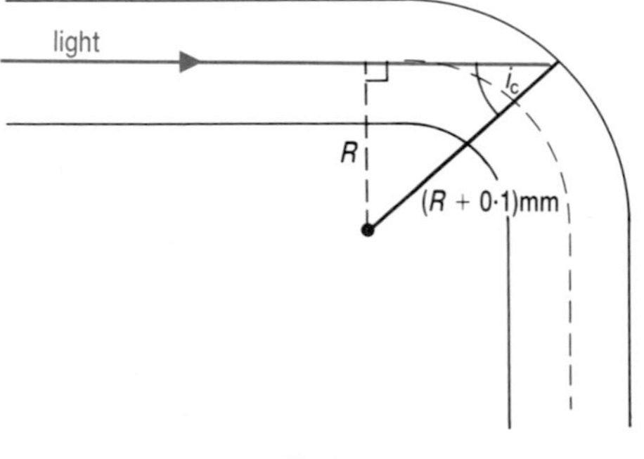

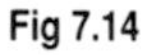
Fig 7.14

Chapter 8

8.1 **(a)** (i) 2.72 mm, 0.97 mm, 2.53 mm, (ii) 1.26 mm (half a wavelength)

(b) 10 cm (the pulse travels both ways)

8.2 **(a)** 0.26 – 0.44 or about 35 per cent.

(b) Almost 100 per cent reflection so there is no possibility of seeing tissue beyond.

8.3 Between 6 and 20 MHz to examine the eye at a depth of 1.5 to 5 cm.

8.5 See Fig 8.13.

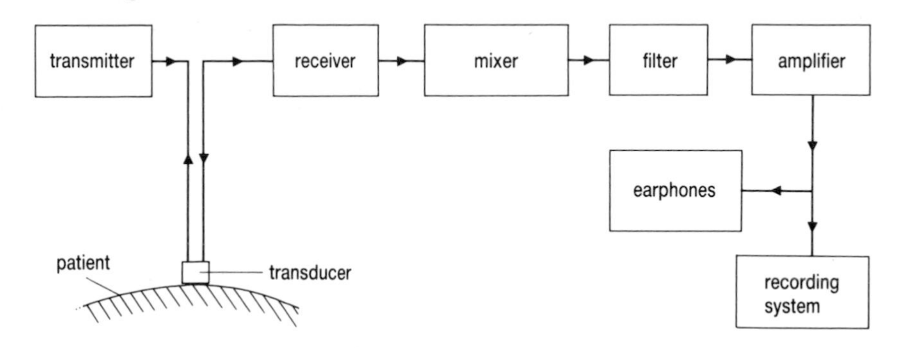

8.6 **(a)** The Doppler shift equations only show the difference, not which is larger.
(b) 4.5 mm s^{-1}

8.7 **(a)** 0.75 mm (half a wavelength).
(b) and **(c)** refer to Section 8.5.

8.9 **(a)** Substituting the values of Z in the equation gives $R = 0.999$ for the air-tissue interface, and $R = 1.7 \times 10^{-3}$ for the jelly-tissue interface.

8.10 **(b)** Substituting in the formula given, $v = 0.17$ m s^{-1}

Chapter 9

9.1 Refer to Fig 9.7

9.2 Complete Table 9.4

Table 9.4 Answer to question 9.2

change	E_{max}	λ_{min}	shape of spectrum	characteristic lines	total intensity
increase V_0	increases				
increase I_0		unchanged			
increase Z				shifted to higher proton energy	

9.3 Use the conversion given to transform the values of photon energies given in Fig 9.5 (a) and 9.6, as is shown in Fig 9.4.

9.4 **(a)** 4kW
(b) 2.5×10^{17}
(c) 1.6×10^{-14} J
(d) 0.0124 nm

9.5 See Section 9.2

9.6 See the preceding sub-section

9.7 **(a)** 21 MW m^{-2}
(b) 0.21 MW m^{-2}
(c) 0.026 MW m^{-2} (One eighth).

9.9 Electrons have a charge so can be focused electrically or magnetically, X-rays are uncharged.

9.12 Refer to Section 9.2
There is less simple scatter and photoelectric absorption of energy, with the higher energy photons, so there is greater penetration, greater deposition of energy at depth and smaller loss at or near the surface.
At MeV energy levels Compton effect predominates. This is independent of Z, so bone with high proton number, absorbs less of the energy than the mechanisms that predominate at lower energies.

9.16 These are two examples of how advances in a particular field are made as a result of the technology being used from a completely different area, in this case nuclear weapons development and fundamental particle research.

9.18 **(a)** see Fig 9.7
(b) the maximum tube voltage is unchanged
(d) 1/8 is three half-thicknesses so HVT = 1.0 mm.

Chapter 10

10.1 By taking potassium iodide tablets the thyroid of an irradiated person in which the radioactive iodine-131 collects can be 'flushed out' so that it contains only stable I-127.

10.2 Activity from 96.8 per cent U-238 = 11.96 MBq kg^{-1}
Activity from 3.2 per cent U-235 = 2.53 MBq kg^{-1}
Total specific activity of enriched uranium = 14.49 MBq kg^{-1}

10.3 ${}^{220}_{86}Rn \rightarrow {}^{216}_{84}Po + {}^{4}_{2}He$
The polonium also decays by α emission

10.4 **(a)** ${}^{14}_{8}O \rightarrow {}^{14}_{7}N + \beta^{+} + \nu$
(b) ${}^{19}_{8}O \rightarrow {}^{19}_{9}Fl + \beta^{-} + \nu$

10.5 **(a)** Magnetic and electric deflections: β great (high charge to mass ratio), α little (low charge to mass ratio), γ none (no charge).
(b) γ is em radiation $\therefore v = c$. Lower energy γ has a lower frequency ν and longer wavelength λ. $E = h\nu$, $c = \nu\lambda$.

10.6 **(a)** 1.8 m
(b) 21 mm of lead (four half-thicknesses).

10.8 ${}^{A}_{Z}X + {}^{1}_{0}n \rightarrow {}^{A+1}_{Z}X + \gamma$

10.9 ${}^{14}_{7}N + {}^{1}_{0}n \rightarrow {}^{14}_{6}C + {}^{1}_{1}p$, ${}^{6}_{3}Li + {}^{1}_{0}n \rightarrow {}^{3}_{1}H + \alpha\ ({}^{4}_{2}He)$

10.10 Completions for table 10.3 neutron, α, neutron excess, neutron deficiency, β^{-}, β^{+}, no, yes.

10.12 Refer to Table 10.6 and Section 10.4

10.16 **(a)** refer to Fig. 10.6.
(b) ${}^{99}_{42}Mo \rightarrow {}^{99}_{43}Tc^{m} + \beta^{-}$ (half-life 67 h)
(c) ${}^{99}_{43}Tc^{m} \rightarrow {}^{99}_{43}Tc + \gamma$ (half-life 6 h)
(d) see Table 10.6

Chapter 11

11.1 **(a)** Alpha's range in air is only a few centimetres, so it is unlikely to irradiate the body, as an external, sealed source. As a liquid or gas it can contaminate the body's surface, or be ingested and give a high local irradiation.

(b) The properties of X- and γ radiations are the same.

(c) The mass and charge of β have values between those of α and γ, which results in it having an ionising ability approximately midway between .

11.2 This is a question for discussion, to see if you can identify all the possible factors.

11.3 **(a)** 0.288 mSv (maximum dose permitted on any one occasion is 0.05 mSv)

(b) 0.719 mSv (maximum dose permitted in any one year is 0.5 mSv)

11.4 900 mJ

11.5 **(a)** The major changes are (i) The inclusion of radon after 1977 which has increased the proportion of natural radioactivity. (ii) Its contribution is larger again after 1981, perhaps as a result of the monitoring programme. (iii) Restrictions on the use of X-rays has reduced the medical proportion. There are probably other changes you could discover.

11.6 **(a)** Approximately 64.5 μSv

(b) 64.5 hours

(c) about 3 days

11.7 **(a)** The risk of radiation induced disease against the avoidance of toothache and other complications. Use the lowest possible doses, and have X-ray examinations only when really necessary.

(b) Their risk is set against the benefit of the wider population. In addition their actual exposure can be carefully monitored and controlled.

(c) These contain bone, muscles, etc., which are less radiosensitive.

11.10 **(a)** Refer to Section 12.2 for details of the ionisation chamber.

(b) 34 Gy

(c) Refer to Table 11.1

Chapter 12

12.1 **(a)** Dark under the open window, light in all other parts

(b) Moderately dark over most of the film

(c) A dark patch, corresponding to the shape of the drop; the intensity will depend on which filter it landed

12.2 **(a)** The RBE of neutrons is 20, so the dose equivalent is very high

(b) α cannot penetrate the wrapping of the film, which keeps out the light. α emitters are not used in medical diagnosis or therapy

12.3 **(b)** Ionisation has taken place throughout the volume of the tube, so a finite time is required for the ions to travel to the electrons, before this pulse can be distinguished from that from the next burst of ionisation

(c) 1 kBq.

(d) Because the counter is often taking measurements of the same order of magnitude as background radiation. The count may be corrected either by taking the background count in the absence of the source, and then subtracting it from the total count, or operating the counter within a shielded area

12.4 **(b)** i) low energies, and small differences in energies can be detected
ii) high count rates are possible (up to 10^4 per second)
iii) high density means a higher probability of the γ photon interacting with an atom in the scintillator

12.7 **(a)** See Section 10.4 and Table 10.6; $^{99}Tc^m$

12.8 A question with a number of possible answers!
Laser signals and optical fibres are now used for telecommunication; 'fax' machines send pictures by telephone; computers can download programs to draw pictures. British Telecom have developed a system called Imtran, specificially to transmit radiological images. It consists of portable transceivers which accept images directly from monochrome video sources, and transmit them over ordinary telephone lines. Pictures are displayed on monitors in 32 or 64 seconds time, depending on resolution and can be stored on tape. The running cost is limited to the cost of the telephone call.

12.9 **(a)** Images can show thickening of the blood vessels, clots in them and other damage can lead to fatal conditions
(b) A small tube which can be inserted into the body, through a small incision
(d) The first sentence says that the initial picture does not show blood vessels; in the last the doctor is using this image to help her find the veins and arteries!

12.10 **(a)** Refer to Section 11.2 for dose equivalent, 0.14 J
(b) and **(c)** refer to Sections 12.1 and 12.3

Appendix 

TECHNICAL INFORMATION ON THE INVESTIGATIONS

Chapter 1

The mechanics of jumping

This will require the participation of the P.E. or athletics department, if it is to be carried out with any degree of accuracy, though some rough estimates could be made with the use of laboratory measuring instruments, outdoors.

Chapter 2

Monocular and binocular vision

This is a series of simple investigations, requiring no equipment, which can be carried out at home.

Chapter 3

Sound and hearing

No specific investigations have been described in the text, but there are several possibilities, depending on the equipment available and the students' previous experience. Sound level meters are sold by most of the major school laboratory suppliers. For example the CS15C Sound Level Indicator from C.E.Offord (Microscopes), with a range of 36 dBA to 110 dBA, for which an inexpensive booklet of suggested projects is available. An alternative way of measuring is to use a microphone with a CRO. Possible investigations include a person's range of hearing under different conditions (as suggested in question 3.6), noise levels in various situations and the absorption and reflection of sound by different materials.

Chapter 4

1. Measuring body temperature with a thermocouple

It is intended that the apparatus required for this should be as simple as possible, in contrast with part (b) later in the chapter. A thermocouple can be made in the laboratory from 26 or 28 SWG copper and constantan wires. This will give a reasonable reading on a Scalamp galvanometer. An alternative is to use a thermistor in a bridge circuit with a high resistance (inexpensive bead thermistors are available from RS Components).

2. Temperature monitoring using a microcomputer

This is intended to be as 'high-tech' as available equipment allows. For example the Philip Harris temperature sensors can be used with their EMU data logger, or with the VELA.

Chapter 5

Measurement of ECG

Philip Harris offer an ECG interface for use with a BBC B or Master computer. Unilab have a Biological amplifier which can monitor a range of electrical signals from the body. These can be output to a CRO, chart recorder, or to a suitable computer through Unilab's Microcomputer Interface.

Chapter 6

Measurement of blood pressure

This can be carried out with the traditional manual sphygmomanometer, or by one of the large range of electronic equivalents now marketed by laboratory suppliers. These are easier to use but generally give less accurate results (which is why the traditional method is still used in medicine). If you are considering purchasing equipment for this, possibilities include Pulse Monitors from Educational Electronics, Griffin and Philip Harris.

Chapter 7

Optical fibres

No specific suggestions for investigations have been included in the text. There are a number of suppliers of non-coherent optical fibre, transmission kits and accessories, for example RS Components, Philip Harris and Irwin. These are useful to demonstrate transmission of signals. It is more difficult to find a reasonably priced source of coherent fibres; a hospital may be able to loan a fibre endoscope.

Chapter 8

The properties of ultrasound

A visit to a hospital will probably be neccessary to see ultrasonic imaging and Doppler measurements. Kits are available to investigate the nature of ultrasound (e.g. Unilab's Ultrasonics Kit), and to use it in measurement (e.g. Educational Electronics' Motion sensor which is a transmitter/receiver with software for the BBC microcomputer).

Chapter 9

1. Half value thickness

This investigation requires only a range of thicknesses of wood and lead and the standard cobalt-60 source in its storage box.

2. Shadows

This is a simple excercise which can be done at home with household items.

Chapter 10

1. Half life of radioisotopes

2. Range of radiations in air

These experiments could usefully be done at this stage, if students have not already performed them in their physics course. Details are not given in the chapter but may be found in standard physics texts.

Chapter 11

Background radiation

A range of everyday materials and substances can be measured using a standard laboratory G-M tube and a scalar or ratemeter. Suggestions include uranium, thorium and potassium salts, a gas mantle (these are coated with thorium oxide), granite, and brazil nuts!

Chapter 12

Film detection

This investigation makes use of the sealed laboratory sources, or uranium salts. It requires dental X-ray films which you may be able to beg from local dentists. The films and processing chemicals can be purchased from photographic suppliers; the processing can easily be carried out in a dimly lit laboratory.

Other possible practical work will depend on the availability of different detectors. It may be possible to obtain film badges and TLDs from a local hospital or laboratory for examination, though processing will not be easy to arrange. The performance of the Geiger counter at different voltages can be investigated, to determine the optimum conditions; details can be found in standard physics textbooks.

Chapter 13

Medical Imaging

This is a non-practical simulation excercise, but a prerequisite is a visit to a hospital department to see imaging equipment in action. Technical notes on the simulation are given in Appendix 2.

Appendix 

TECHNICAL INFORMATION ON THE CASE STUDY

Aims

The general purpose, and the specific objectives of this exercise, are stated at the beginning of the chapter in the usual way. You will realise that the presentation, content and activity generated by the chapter are unusual. They all owe more to the spirit of the syllabus option than to the letter. That is to involve students in the implications of physics in a field of intrinsic interest and topical importance, in which many students may have career ambitions. Thus the content is up to date, the presentation is realistic and the activity simulates that which students may be involved in, or affected by, in their future careers.

Planning

The activity should take about four hours class time, four hours homework and additional time for a relevant visit. The ideal programme for students is an introductory reading of the chapter and a hospital visit one week, with the planning and presentation work in the following two weeks, as shown in Fig 13.1. Arrangements can be varied considerably to suit circumstances, but experience has shown that three factors are particularly important in ensuring success:

1. The commitment to take the activity seriously; the different skills students will be using in this simulation are (at least!) as valuable in the practice of physics as the ability to recall and use formulae.
2. The allocation of sufficient time.
3. The provision of supporting material in non-book form. The best is a visit to a hospital imaging department for all students, following preparation with a member of staff. Some departments will not allow students under the age of 18 into protected areas, for safety reasons. This means thay may be unable to view 'live' examinations. There are still many advantages in a visit, and most departments are willing to help. Magazines and sales brochures can be obtained freely from manufacturers and other organisations; these add a verisimilitude to the simulation.

Tutor's Role

The advance planning is most easily carried out by the tutor, who will probably need to help the students with their own planning. Once the roles have been understood and allocated, the tutor need have only a supporting role, for example, providing additional information and encouraging creative responses, as required. The tutor has been given a role in the final presentation. This involves organising the procedure, chairing the discussion, evaluating all the contributions and deciding on 'the best plan'. This role could, with advantage, be shared with a member of staff of the school, college or local health service, who was unfamiliar to the students.

Notes on reports for presentation

It is clearly not appropriate to give answers to the problems set for each role. It would also be counter productive to make specific suggestions of proposals available to all students, since there is an element of competition in the excercise. The following are given to tutors as possible guidelines. For the success of the excercise these should be referred to by students, sparingly and only when in difficulties!

Role 1, Ultrasound

Planning guidelines and papers for the radiologist indicate a strong case for expansion of the ultrasound facility, and a new X-ray fluoroscope for the intervention radiology. The proposal might be:

Item	Running cost /k£ yr^{-1}	Capital cost Fixed /k£
4 ultrasound machines (2 each for obstetrics and intervention)		240
1 X-ray fluoroscope (for interventions)		250
3 radiographers (2 for intervention, 1 for obs.)		45
1 radiologist	30	
2 serviced rooms (1 for intervention, 1 for obs.)		40
1 unserviced room for records		10
2 support staff (porter and clerical)	20	
Variable		
Patients for intervention, 3000 @ £300	900	
Saving		
Patients not requiring major surgery, 3000 @, say, £5000	–15 000!	

So the intervention facility has a great potential for saving/making money, it remains to be argued where the patients would come from.

The establishing of obstetric ultrasound in the MI department can be argued in terms of general efficiency, effective and rapid handling of patients, and possible growth of throughput at no extra cost, all factors which would result in greater cost effectiveness in accordance with planning guidelines.

Role 2, Nuclear medicine

The radiologist's report could well consist of a comparison of the three tomographic options, with reference to performance, ease of change-over from existing procedures, staffing implications, running and capital costs. It would be an advantage to include a comparison of cost-effectiveness, if factors such as the speed of operation and the potential to sell the service to other districts is included. This will probably show that PET is 'better', but more expensive. Tactics in the presentation will depend on whether this or a less radical alternative is chosen. Whichever it is should be contrasted (favourably) with CT and MRI.

Role 3, CT scanning

The sales representative should be encouraged to use some enterprise and flair; the items for this contribution are designed to assist in this. For the radiologist, the advantages of CT over gamma cameras and MRI, and the advantage of Seivert's scanner over the competitors, need to be presented. The planner can be provided with an income and expenditure account over ten years using the data from the extract. Without any additional financial help, the scanner can show a small 'profit' over that period, (assuming costs given and ignoring inflation, which can be assumed to affect both the income and expenditure side of the balance sheet). It is then up to the salesperson to improve this to 'an offer they cannot refuse'.

Role 4, Magnetic resonance

The argument here needs to have a different flavour – that of the health of medical science being dependent on research and development. This part of the proposal calls for some imagination, whereas the rest needs a reasonable grasp of physics – and some chemical and biological knowledge could well be put to use in discussing the research applications of MR spectroscopy. The brief contains a short account of the complex processes involved, so the person in this role needs to be able to understand and represent this. The able student could supplement the case with some background reading, for example the review article by Christine Sutton 'A magnetic window into bodily functions', *New Scientist* 11.9.1986. The presentation of the case would benefit from reference to the local situation including factors such as: convenient location near city centre, with good services and pleasant surroundings; cooperative projects with St George's (the hospital with the reseach appeal extract); research experience of present St Katherine's staff, etc.

Role 5, Hospital physicist

The simplest way to approach this role might be to draw up a comparison of the four major equipment proposals, with respect to safety. This would not rule out any particular development, since suitable safety procedures could be implemented. These would cost, in staff time (make assumptions on the basis of the work to be done and the salaries quoted in the other sections), and installation requirements (see section 13.4). The potential for an increase in job satisfaction might lead the physicist to cooperate with the planner over installing the computers suggested in that brief. This could lead to an electronics development project, such as designing suitable analogue to digital converters to interface measuring equipment with the new computing capacity. Finding savings in the increased efficiency from computer systems, would enable such projects to be funded. The emphasis of this report might be on safety, but any opportunity to widen the brief, by applying some physics, by the student, should be encouraged.

Role 6, Planner

The person in this role has to evaluate all the other proposals against the planning priorities and guidelines given. Some of these are specific like the computerisation, for which a proposal should be made by the planner, perhaps with the help of the physicist. Others are more general and call for interpretation by the planner. Cost effectiveness is likely to feature highly in the decision making, but is not defined anywhere. It would help if the planner had a definition which she or he was able to use. The planner's initial proposals are to be modified by the group discussion, in the light of the case made by each 'expert'.

Group role, the ten-year plan

The planner convenes the meeting where all present their proposals and decisions have to be made. This will demand compromise and collaboration between individuals so that the efforts of all are directed to arguing, say, the advantages of the one tomography unit chosen (it is clear that a bid for more than one of SPECT,PET,CT and MRI will fail). A balance might be struck between the saving of money by the more efficient use of resources (eg. computerisation, obstetric ultrasound), and a high-profile, aggressive marketing of an attractive but expensive new facility (e.g. CT or MRI). Many combinations are possible in the presentation of the final plan to the whole class. The criteria on which it should be judged in the final evaluation are clarity, consistency, and the ability to meet the priorities and guidelines of the health authority. It is preferable that the regional planner, the teacher or outsider in role, gives a brief verbal report on each group's efforts, and is able to choose one plan to approve. This provides a suitable conclusion to the enterprise.

Appendix 

FURTHER RESOURCES

ORGANISATIONS

Institute of Physical Sciences in Medicine
Hospital Physicists' Association
2 Low Ousegate, York UO1 1QU Tel. 0904 610821

These two linked organisations publish general careers information and a range of titles of professional interest on the applications of physics and engineering to medicine.

College of Radiographers
Wimpole Street, London W1. Tel. 01 935 5726/8

Professional body which publishes *Radiography - a career.*

Royal College of Radiology
38 Portland Place, London W1. Tel. 01 6364432/3

Professional body which publishes *Diagnostic radiology as a career.*

National Radiological Protection Board
Chilton, Didcot, Oxon OX11 0RQ Tel. 0235 831600

A list of the board's publications and a range of free information sheets are available from the Board's information office. Most of the other publications are available from HMSO. The booklet *Living with Radiation* is particularly recommended.

MEDICAL EQUIPMENT

This is normally too expensive for schools to purchase. Some small or obsolete items may be donated, but generally the equipment is best seen in action in the hospital. An alternative is to arrange for a radiographer or hospital physicist to visit the school or college with items, or slides of equipment. This can be readily supplemented by suppliers, catalogues which are available on request and full of illustrations of the latest improvements. These are particularly recommended for Chapter 13.

Suppliers
This is a selection of some of the larger firms, many more are listed in *Radmagazine.*

Amersham International plc, Lincoln Place, Green End, Aylesbury, Bucks HP20 2TP
radioisotopes, radiotherapy equipment.

Brtitish Telecom, Mercury House, Bond Street, Bristol BS1 3TD
telephone transfer of medical images (ring freefone Imtran).

Diasonics Sonotron Ltd., 2 Napier Road, Bedford MK41 0JW
MRI and other imaging equipment.

DuPont (UK), Wedgwood Way, Stevenage, Herts SG1 4QN
radiological screens and films.

Elscint (GB) Ltd., Tower Road, Berinsfield, Oxon OX9 8LW
diagnostic imaging including ultrasound, X-ray tomography and gamma cameras.

GEC Medical Equipment Ltd., PO Box 2, East Lane, Wembley, Middlesex HA9 7P
gamma cameras and ultrasound.

IGE Medical Systems Ltd., Coolidge House, 260 Bath Road, Slough, Berks SL1 4ER
medical imaging systems.

Philips Medical Systems Ltd., Kelvin House, 63–75 Glenthorne Road, London W6 6LJ
ultrasound, linac, radiotherapy.

Scientific Medical Systems, Blackhorse Road, Letchworth, Herts SG6 1HL
medical imaging equipment.

Scintronix, 1 Drummond Square, Brucefield Industrial Estate, Livingston EH54 9DH
medical imaging equipment.

Shimadzu, Kingsbury Industial Estate, Church Lane, Kingsbury, London NW9 8AU
X-ray and other medical imaging systems.

Toshiba Medical Systems, Manor Court, Manor Royal, Crawley, W Sussex RH10 2PY
ultrasound.

JOURNALS AND AUDIOVISUAL RESOURCES

Journals

Professional medical and physics journals are often too technical or too detailed for student use. Some articles for tutors have been listed below. The most useful sources of topical articles for students are *New Scientist*, *Physics Education* (especially July 1989 issue) and *Radmagazine,* a free paper supplied to hospitals by Kingsmoor Publications Ltd., PO Box 3, Harlow, Essex CM19 4RF.

Audiovisual aids

The availabilty of suitable slides, films and videotapes is very limited. The best source is probably current TV broadcasts such as *Horizon, Equinox* and *QED,* for example 'Keyhole Surgery' broadcast in April 1989 in QED, which gave an excellent introduction to endoscopy, with support from X-radiography and ultrasound, and 'Invasion of the Body Scanners', from *Equinox* in October 1989 which concentrated on MRI.

BOOKS, LEAFLETS AND ARTICLES

Because this is a longer list than the preceding it is organised by theme, with an initial list of items of general use in medical physics. Those marked (*) are listed at the end of the appropriate chapter.

General

Aird, Edwin G.A. *An introduction to medical physics,* Wm Heinemann Medical Books Ltd. 1975.

> A useful supplementary text for tutors and students, though somewhat dated.

Cameron, J.R. and Skofronik, J.G. *Medical physics,* John Wiley 1978.

> An attractively written book which provides an excellent source of background and supporting information. Too detailed for general use and becoming dated.

Hallett, F.R. Speight, P.A. and Stinson, R.H. *Introductory biophysics*, Chapman and Hall 1978.

Provides a more technical treatment of many syllabus topics, for example sound and hearing, optics and vision, radiation, nerve potentials. Useful for tutors as the target reader is an undergraduate with a reasonable mathematical competance.

Institute of Physical Sciences in Medicine *Physical Sciences in Medicine* (*intro.)

(see also the 'Organisations' listing).

Jennett, Bryan *High technology medicine – benefits and burdens*, Oxford paperback 1986. (*intro).

Kane, J.W. and Sternheim, M.M. *Physics* , John Wiley, third edition 1988.

This textbook for American college students of the life sciences has a readable and well-illustrated style and an unusually large range of problems of different kinds. Throughout the book physics topics are related to human needs, often in medical applications. Particular reference is made to Chapter 22 (*3), Chapter 18 (*5), and Chapter 29 (*13), but it is a useful source throughout the course.

Physics Education, *Medical physics*, Institute of Physics vol 24, no 4. July 1989.

A special issue on medical physics, consisting of articles commissioned to meet the needs of students at this level, from practitioners in UK hospitals. General articles include:
Chadwick, R. *Medical Physical: an introduction.*
Chadwick, R. *Medical physics undergraduate degree courses at university.*
Clayton,L.B. *School - hospital links.*
Other articles are referred to in Chapters 4, 7, 9, 10, and 11.

Prentice, Andrew, Human energy on tap, *New Scientist*, p.40, 20.11.87.

A full description of the use of radioisotope labelling to measure human energy conversion. Hydrogen-2 and oxygen-18 isotopes label the water drunk by the subject. Collections are made of urine and saliva for about two weeks and measurements are made of the passage of H-2 and O-18 in water and carbon dioxide, through the body. This enables the breakdown of foods and the release of energy to be calculated.

Pope, Jean *Medical physics*, Heinemann Educational, 1984.

A concise account of the topics in A-level physics syllabuses. A useful source of additional questions.

Ronen, M. and Ganiel, U., Visiting a hospital *Physics Education*, vol. 24, p.18, 1989.

Ronen, M. and Ganiel, U., Physics in medical diagnosis – an optional unit for a high school *Physics Education* vol. 19, p.288, 1984.

The unit of work described is for Israeli students at a similar stage in their education. The approaches of the work are problem-centred and physics phenomenon centred (e.g. radioisotopes), or technique centred (e.g. cardiology). Hospital visits are an essential part of the course, so this paper gives details of how these were organised and briefly evaluates the benefits to students of 18 such visits.

Theme 1

Bell, J.C. *The physics of sport – a resource for teachers of physics,* Education for the industrial society project, 1985, available from Scottish Curriculum Development Service, Glasgow Centre, Lymehurst House, 74 Southbrae Drive, Glasgow G13 1SU, tel. 041 954 8287.

Short booklet of very useful data and teaching ideas, both practical and theoretical, for teaching the subject at secondary level.

Eden, John *The eye book,* Penguin 1981. (*2).

Hobson's Science Support Series *Noise,* Hobson's 1986. (*3).

Koretz, J.F. and Handelman, G.H. How the human eye focuses, *Scientific American,* p.64, 1988.

A review of the current state of knowledge of how the focusing ability of the eye declines with age. Causes include both the geometry and the biochemistry of the eye. An authoritative and well-illustrated account.

Mason, Peter *Light Fantastic,* Pelican,1982. (*2).

Nathan, Peter *The nervous system,* Oxford Paperback, 1983.

A detailed account, for a general readership, of all the senses and the brain. Also useful for Chapter 5 of this book.

Observer, *The body report,* Observer newspaper, 1988. (*1).

Copies of this seven part colour supplement may be available from The Observer, Chelsea Bridge House, Queenstown Road, London SW8 4NN, tel. 01 672 0700.

Theme 2

Absten, G.T. and Joffe, S.N. *Lasers in medicine, an introductory guide,* Chapman and Hall, 1985.

A short description of all aspects of the topic – the physics of lasers, interaction with tissue, types of lasers and delivery systems, clinical applications and safety.

Bonnett, R. A death ray for cancer? *New Scientist,* p.38. 20.2.86. (*7).

Cromwell, L. Weibell, F. and Pfeiffer, E. *Biomedical instrumentation and measurement,* Prentice Hall, 1980.

A clearly presented, technical account of the field, with details of contemporary equipment.

Devey, G.B. and Wells, P.N.T. Ultrasound in medical diagnosis, *Scientific American,* p.98, May 1978.

A detailed review of the nature of ultrasound echo techniques for exploring the human body. Well-illustrated explanation of different scanning techniques and descriptions of a wide range of applications. Useful for further reading.

Institute of Physics, *Snippets no. 15*, winter 1987/8.

This describes diagnosis of osteoporosis (brittle bones), by the measurement of attenuation through the bone.

New Scientist, 13.8.87. Ultrasound takes another leap forward.

The diagnosis of muscular dystrophy in children, by quantifying the images.

Theme 3

Anspaugh, L.R. Catlin, R.J. and Goldman, M. The global impact of the Chernobyl reactor accident, *Science (USA)*, p.1513, 16.12.88.

An impressive summary of the data collected world wide, on the effects of Chernobyl. Includes a useful summary of its medical and economic effects, as well as a discussion of the social implications.

Ball, J.L. and Moore, A.D. *Essential physics for radiographers*, Blackwell Scientific, 1980.

A brief account with chapter summaries, useful as a supplement.

Clarke, R.H. and Southwood,T.R.E. Risks from ionizing radiation, *Nature*, p.197, 16.3.89.

An authoritative summary, by the director and chairman of NRPB, of the relative importance of sources of radiation to the population of the UK and the US. In particular it reports a substantial increase in the risk from radon, following a recent progamme of monitoring and assessment. Their conclusion is quoted in a margin note in Chapter 11.

Ennis, John. Statistics, St, Petersburg and Sellafield, *New Scientist*, p.26, 2.5.85.

The mathematics of probability and some of its paradoxical results, with particular reference to the risks associated with nuclear reprocessing.

Fry, F.A. The Chernobyl reactor accident, the impact on the United Kingdom, *British Journal of Radiology*, vol. 60, p.1147, 1987.

An authoritative and clearly presented account, by a scientist at NRPB, of the nature of the accident, the actions taken in the UK, levels of irradiation and environmental effects.

Hall, E.J. *Radiation and life*, Pergamon, second edition 1984. (*9).

Hay, G. A. and Hughes, D. *First year physics for radiographers*, Balliere Tyndal, third edition 1983. (*9).

Hobson's Science Support Series *Irradiation and radioactivity*, Hobson's, 1984. (*10).

IAEA *Facts about low-level radiation*, IAEA, 1986. (*11).

Martin, A. and Harbison, S.A. *An introduction to radiation protection,* Chapman and Hall, third edition 1986.

A detailed and clearly written manual for those working in the field, in hospitals, industry, power stations and research. Useful introduction to radiation physics, and a standard reference for protection regulations and procedures.

Meredith, W.J. and Massey, J.B. *Fundamental physics of radiology,* Wright, third edition 1977.

The standard text for radiographers and radiologists. Detailed and technical, with a wealth of reference material.

Morgan, J.R. A history of pitchblende, *Atom,* vol. 329, March 1984.

A fascinating historical account of uranium ore and its decay products. The first documented reports of radiation sickness comes from the work of Agricola in Central Europe in the sixteenth century. The rapid use of radioactive substances, following the discovery of radium at the end of the nineteenth century, was not linked to the radiation sickness reports for some time. Even now large numbers of people travel to this same area of Central Europe, to take radon baths for their health.

NRPB *Living with radiation,* HMSO, third edition 1988. (*11).

Page, R.A. Environmental issues: what people think, *Chemistry in Britain,* p.559,1987. (*11).

Parker, R.P. Smith, P.H.S. and Taylor, D.M. *Basic science of nuclear medicine,* Churchill Livingstone, second edition 1984.

A clearly-written, technical treatment of the physics and chemistry of the use of radionuclides in clinical practice. Recommended for background and supplementary material for tutors.

Plant, R.D. A school investigation into Chernobyl fallout, *Physics Education,* vol. 23, p.26, 1988.

A description of the nature and extent of the contamination of a school in Kent, using only the school's Geiger tube. The contamination was found to be mainly I-131 and Cs-137, and the dose approximately 4 μSv.

Shrimpton, P.C. *et al.* Doses to patients from routine diagnostic X-ray examinations in England, *British Journal of Radiology,* vol. 59, p.749, 1986.

A major survey carried out by NRPB and the Hospital Physicists' Association to provide data on exposure levels for ten routine types of X-ray examination.

Shrimpton, P.C. and Wall, B.F. Doses to patients from medical radiological investigations in Great Britain, *Radiological Protection Bulletin,* vol.77, p.10, 1986.

A brief report of surveys of X-ray and nuclear medicine examinations of patients and of the significance of the dose in the total dose of the UK population.

Simmonds, Jane. Europe calculates the health risk, *New Scientist*, 23.4.87.

A readable summary of the contamination caused by the Chernobyl cloud. The geographical distribution and uptake in humans of iodine-131 and caesium-137, is well illustrated by the author, a scientist from NRPB.

Sutton, Christine. Nuclear medicine homes in on disease, *New Scientist*, p.48, 15.1.87. (*10).

Sutton, Christine. Subatomic surgery takes on the tumours, *New Scientist*, p.50, 25.8.88. (*10).

Sumner, David. *Radiation risks : an evaluation,* The Tarragon Press (Glasgow), 1987.

A short, well-presented review of the physics and biology of radiation effects and their incidence naturally and in medical use. The major part of the book deals with evidence of damage and estimates of risk, with particular reference to the Chernobyl accident.

Upton, A.C. The biological effects of low-level ionising radiation, *Scientific American,* p.29, Feb. 1982.

A well-illustrated explanation of the interation of radiation with human tissue. Includes striking data on the difference between risks and their public perception.

Theme 4

Goldman, Myer. *A guide to the X-ray department,* Wright (Bristol), second edition 1986.

A short, readable, and well-illustrated account of the work in an X-ray department, written for nurses, medical students and other informed visitors. Very useful as background for the case study.

Kane, J.W. and Sternheim, M.M. *Physics,* John Wiley, third edition 1988. Chapter 29 has supplementary topics on NMR.

SATIS 16-19 *X-rays and patients*, ASE 1989.

An excercise on the design of X-ray tubes, considering the differing needs of the radiographer, radiologist and patient.

Sutton, Christine. A magnetic window into bodily functions, *New Scientist*, 11.9.86.

A rather technical introduction to the development of NMR imaging. Illustrated with examples of its use in animal and human investigations.

Index

Emboldended page numbers indicate definitions, or definitive descriptions. Other entries are either more general descriptions or brief mentions.